Graduate Texts in Mathematics 94

Springer
New York
Berlin
Heidelberg
Barcelona
Hong Kong
London
Milan
Paris
Singapore
Tokyo

Graduate Texts in Mathematics

continued after index

Frank W. Warner

FOUNDATIONS of DIFFERENTIABLE MANIFOLDS and LIE GROUPS

With 57 Illustrations

Springer

Frank W. Warner
University of Pennsylvania
Department of Mathematics E1
Philadelphia, PA 19104
USA

Mathematics Subject Classification (2000): 58-01

Library of Congress Cataloging in Publication Data
Warner, Frank W. (Frank Wilson), 1938–
Foundations of differentiable manifolds and Lie groups.
(Graduate texts in mathematics; 94)
Reprint. Originally published: Glenview, Ill.: Scott, Foresman, 1971.
Bibliography: p.
Includes index.
1. Differentiable manifolds. 2. Lie groups. I. Title. II. Series.
QA614.3.W37 1983 512'.55 83-12395

Printed on acid-free paper.

Printed and bound by R. R. Donnelley & Sons, Harrisonburg, VA.
Printed in the United States of America.

9 8 7

ISBN 0-387-90894-3 Springer-Verlag New York Berlin Heidelberg
ISBN 3-540-90894-3 Springer-Verlag Berlin Heidelberg New York SPIN 10768139

Preface

This book provides the necessary foundation for students interested in any of the diverse areas of mathematics which require the notion of a differentiable manifold. It is designed as a beginning graduate-level textbook and presumes a good undergraduate training in algebra and analysis plus some knowledge of point set topology, covering spaces, and the fundamental group. It is also intended for use as a reference book since it includes a number of items which are difficult to ferret out of the literature, in particular, the complete and self-contained proofs of the fundamental theorems of Hodge and de Rham.

The core material is contained in Chapters 1, 2, and 4. This includes differentiable manifolds, tangent vectors, submanifolds, implicit function theorems, vector fields, distributions and the Frobenius theorem, differential forms, integration, Stokes' theorem, and de Rham cohomology.

Chapter 3 treats the foundations of Lie group theory, including the relationship between Lie groups and their Lie algebras, the exponential map, the adjoint representation, and the closed subgroup theorem. Many examples are given, and many properties of the classical groups are derived. The chapter concludes with a discussion of homogeneous manifolds. The standard reference for Lie group theory for over two decades has been Chevalley's *Theory of Lie Groups*, to which I am greatly indebted.

For the de Rham theorem, which is the main goal of Chapter 5, axiomatic sheaf cohomology theory is developed. In addition to a proof of the strong form of the de Rham theorem—the de Rham homomorphism given by integration is a ring isomorphism from the de Rham cohomology ring to the differentiable singular cohomology ring—it is proved that there are canonical isomorphisms of all the classical cohomology theories on manifolds. The pertinent parts of all these theories are developed in the text. The approach which I have followed for axiomatic sheaf cohomology is due to H. Cartan, who gave an exposition in his *Séminaire* 1950/1951.

For the Hodge theorem, a complete treatment of the local theory of elliptic operators is presented in Chapter 6, using Fourier series as the basic tool. Only a slight acquaintance with Hilbert spaces is presumed. I wish to thank Jerry Kazdan, who spent a large portion of the summer of 1969 educating me to the whys and wherefores of inequalities and who provided considerable assistance with the preparation of this chapter. I also benefited from notes on lectures by J. J. Kohn and Stephen Andrea, from several papers of Louis Nirenberg, and from *Partial Differential*

Equations by Bers, John, and Schechter, which the reader might wish to consult for further references to the literature.

At the end of each chapter is a set of exercises. These are an integral part of the text. Often where a claim in a chapter has been left to the reader, there is a reminder in the Exercises that the reader should provide a proof of the claim. Some exercises are routine and test general understanding of the chapter. Many present significant extensions of the text. In some cases the exercises contain major theorems. Two notable examples are properties of the eigenfunctions of the Laplacian and the Peter-Weyl theorem, which are developed in the Exercises for Chapter 6. Hints are provided for many of the difficult exercises.

There are a few notable omissions in the text. I have not treated complex manifolds, although the sheaf theory developed in Chapter 5 will provide the reader with one of the basic tools for the study of complex manifolds. Neither have I treated infinite dimensional manifolds, for which I refer the reader to Lang's *Introduction to Differentiable Manifolds*, nor Sard's theorem and imbedding theorems, which the reader can find in Sternberg's *Lectures on Differential Geometry*.

Several possible courses can be based on this text. Typical one-semester courses would cover the core material of Chapters 1, 2, and 4, and then either Chapter 3 or 5 or 6, depending on the interests of the class. The entire text can be covered in a one-year course.

Students who wish to continue with further study in differential geometry should consult such advanced texts as *Differential Geometry and Symmetric Spaces* by Helgason, *Geometry of Manifolds* by Bishop and Crittenden, and *Foundations of Differential Geometry* (2 vols.) by Kobayashi and Nomizu.

I am happy to express my gratitude to Professor I. M. Singer, from whom I learned much of the material in this book and whose courses have always generated a great excitement and enthusiasm for the subject.

Many people generously devoted considerable time and effort to reading early versions of the manuscript and making many corrections and helpful suggestions. I particularly wish to thank Manfredo do Carmo, Jerry Kazdan, Stuart Newberger, Marc Rieffel, John Thorpe, Nolan Wallach, Hung-Hsi Wu, and the students in my classes at the University of California at Berkeley and at the University of Pennsylvania. My special thanks to Jeanne Robinson, Marian Griffiths, and Mary Ann Hipple for their excellent job of typing, and to Nat Weintraub of Scott, Foresman and Company for his cooperation and excellent guidance and assistance in the final preparation of the manuscript.

Frank Warner

Preface to the Springer Edition

This Springer edition is a reproduction of the original Scott, Foresman printing with the exception that the few mathematical and typographical errors of which I am aware have been corrected. A few additional titles have been added to the bibliography.

I am especially grateful to all those colleagues who wrote concerning their experiences with the original edition. I received many fine suggestions for improvements and extensions of the text and for some time debated the possibility of writing an entirely new second edition. However, many of the extensions I contemplated are easily accessible in a number of excellent sources. Also, quite a few colleagues urged that I leave the text as it is. Thus it is reprinted here basically unchanged. In particular, all of the numbering and page references remain the same for the benefit of those who have made specific references to this text in other publications.

In the past decade there have been remarkable advances in the applications of analysis—especially the theory of elliptic partial differential equations, to geometry—and in the application of geometry, especially the theory of connections on principle fiber bundles, to physics. Some references to these exciting developments as well as several excellent treatments of topics in differential and Riemannian geometry, which students might wish to consult in conjunction with or subsequent to this text, have been included in the bibliography.

Finally, I want to thank Springer for encouraging me to republish this text in the Graduate Texts in Mathematics series. I am delighted that it has now come to pass.

Philadelphia, Pennsylvania
October, 1983

Frank Warner

Contents

1 MANIFOLDS

2 TENSORS AND DIFFERENTIAL FORMS

3 LIE GROUPS

4 INTEGRATION ON MANIFOLDS

5 SHEAVES, COHOMOLOGY, AND THE DE RHAM THEOREM

6 THE HODGE THEOREM

1
MANIFOLDS

After establishing some notational conventions which will be used throughout the book, we will begin with the notion of a differentiable manifold. These are spaces which are locally like Euclidean space and which have enough structure so that the basic concepts of calculus can be carried over. In this first chapter we shall primarily be concerned with the analogs and implications for manifolds of the fundamental theorems of differential calculus. Later, in Chapter 4, we shall consider the theory of integration on manifolds.

From the notion of directional derivative in Euclidean space we will obtain the notion of a tangent vector to a differentiable manifold. We will study mappings between manifolds and the effect that mappings have on tangent vectors. We will investigate the implications for mappings of manifolds of the classical inverse and implicit function theorems. We will see that the fundamental existence and uniqueness theorems for ordinary differential equations translate into existence and uniqueness statements for integral curves of vector fields. The chapter closes with the Frobenius theorem, which pertains to the existence and uniqueness of integral manifolds of involutive distributions on manifolds.

PRELIMINARIES

1.1 Some Basic Notation and Terminology Throughout this text we will describe sets either by listings of their elements, for example

$$\{a_1, \ldots, a_n\},$$

or by expressions of the form

$$\{x: P\},$$

which denote the set of all x satisfying property P. The expression $a \in A$ means that a is an *element* of the set A. If a set A is a *subset* of a set B (that is, $a \in B$ whenever $a \in A$), we write $A \subset B$. If $A \subset B$ and $B \subset A$, then A *equals* B, denoted $A = B$. The negations of $\in$, $\subset$ and $=$ are denoted by $\notin$, $\not\subset$, and $\neq$ respectively. A set A is a *proper subset* of B if $A \subset B$ but $A \neq B$.

We will denote the *empty set* by $\varnothing$. We will often denote a collection $\{U_\alpha : \alpha \in A\}$ of sets U_α indexed by the set A simply by $\{U_\alpha\}$ if explicit mention of the index set is not necessary. The *union* of the sets in the collection $\{U_\alpha : \alpha \in A\}$ will be denoted $\bigcup_{\alpha \in A} U_\alpha$ or simply $\bigcup U_\alpha$. Similarly, their *intersection* will be denoted $\bigcap_{\alpha \in A} U_\alpha$ or simply $\bigcap U_\alpha$.

$$\bigcup_{\alpha \in A} U_\alpha = \{a : a \text{ belongs to some } U_\alpha\}.$$

$$\bigcap_{\alpha \in A} U_\alpha = \{a : a \text{ belongs to every } U_\alpha\}.$$

The expression $f: A \to B$ means that f is a *mapping* of the set A into the set B. When describing a mapping by describing its effect on individual elements, we use the special arrow $\mapsto$; thus "the mapping $m \mapsto f(m)$ of A into B" means that f is a mapping of the set A into the set B taking the element m of A into the element $f(m)$ of B. If $U \subset A$, then $f \mid U$ denotes the *restriction of f to U*, and $f(U) = \{b \in B : f(a) = b \text{ for some } a \in U\}$. If $C \subset B$, then $f^{-1}(C) = \{a \in A : f(a) \in C\}$. A mapping f is *one-to-one* (also denoted 1 : 1), or *injective*, if whenever a and b are distinct elements of A, then $f(a) \neq f(b)$. A mapping f is *onto*, or *surjective*, if $f(A) = B$.

If $f: A \to B$ and $g: C \to D$, then the *composition* $g \circ f$ is the map

$$g \circ f : f^{-1}(B \cap C) \to D$$

defined by $g \circ f(a) = g(f(a))$ for every $a \in f^{-1}(B \cap C)$. For notational convenience, we shall not exclude the case in which $f^{-1}(B \cap C) = \varnothing$. That is, given any two mappings f and g, we shall consider their composition $g \circ f$ as being defined, with the understanding that the domain of $g \circ f$ may well be the empty set.

The *cartesian product $A \times B$ of two sets A and B* is the set of all pairs (a, b) of points $a \in A$ and $b \in B$. If $f: A \to C$ and $g: B \to D$, then the *cartesian product $f \times g$ of the maps f and g* is the map $(a, b) \mapsto (f(a), g(b))$ of $A \times B$ into $C \times D$.

We shall denote the *identity map* on any set by "id."

A diagram of maps such as

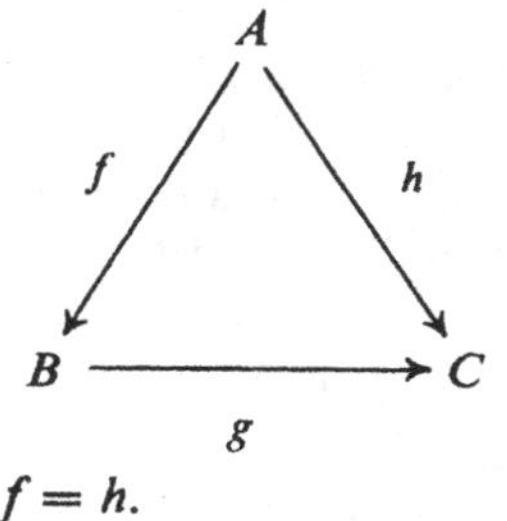

is called *commutative* if $g \circ f = h$.

We shall always use the term *function* to mean a mapping into the real numbers.

Let $d \geq 1$ be an integer, and let

$$\mathbb{R}^d = \{a\colon a = (a_1, \ldots, a_d) \text{ where the } a_i \text{ are real numbers}\}.$$

Then $\mathbb{R}^d$ is the *d-dimensional Euclidean space.* In the case $d = 1$, we denote the *real line* $\mathbb{R}^1$ simply by $\mathbb{R}$. The *origin* $(0, \ldots, 0)$ in Euclidean space of any dimension will be denoted 0. The notations $[a,b]$ and (a,b) denote as usual the intervals of the real line $a \leq t \leq b$ and $a < t < b$ respectively. The function $r_i\colon \mathbb{R}^d \to \mathbb{R}$ defined by

$$r_i(a) = a_i,$$

where $a = (a_1, \ldots, a_d) \in \mathbb{R}^d$, is called the *i*th *(canonical) coordinate function* on $\mathbb{R}^d$. The canonical coordinate function r_1 on $\mathbb{R}$ will be denoted simply by r. Thus $r(a) = a$ for each $a \in \mathbb{R}$. If $f\colon X \to \mathbb{R}^d$, then we let

$$f_i = r_i \circ f,$$

where f_i is called the *i*th *component function of f.*

If $f\colon \mathbb{R} \to \mathbb{R}$ and $t \in \mathbb{R}$, then we denote the *derivative of f at t* by

$$\left.\frac{d}{dr}\right|_t (f) = \left.\frac{df}{dr}\right|_t = \lim_{h \to 0} \frac{f(t+h) - f(t)}{h}.$$

If $f\colon \mathbb{R}^n \to \mathbb{R}$, if $1 \leq i \leq n$, and if $t = (t_1, \ldots, t_n) \in \mathbb{R}^n$, then we denote the *partial derivative of f with respect to* r_i *at t* by

$$\left.\frac{\partial}{\partial r_i}\right|_t (f) = \left.\frac{\partial f}{\partial r_i}\right|_t = \lim_{h \to 0} \frac{f(t_1, \ldots, t_{i-1}, t_i + h, t_{i+1}, \ldots, t_n) - f(t)}{h}.$$

If $p \in \mathbb{R}^d$, then $B_p(r)$ will denote the *open ball* of radius r about p. The open ball of radius r about the origin will be denoted simply by $B(r)$. $C(r)$ will denote the *open cube* with sides of length $2r$ about the origin in $\mathbb{R}^d$. That is,

$$C(r) = \{(a_1, \ldots, a_d) \in \mathbb{R}^d\colon |a_i| < r \text{ for all } i\}.$$

We shall use $\mathbb{C}$ to denote the *complex number field* and $\mathbb{C}^n$ to denote *complex n-space,*

$$\mathbb{C}^n = \{(z_1, \ldots, z_n)\colon z_i \in \mathbb{C} \text{ for } 1 \leq i \leq n\}.$$

Unless we indicate otherwise, we shall always use the term *neighborhood* in the sense of *open* neighborhood. If A is a subset of a topological space, its closure will be denoted by $\bar{A}$. If φ is a function on a topological space X, the *support* of φ is the subset of X defined by

$$\operatorname{supp} \varphi = \overline{\varphi^{-1}(\mathbb{R} - \{0\})}.$$

We use the *Kronecker index*

$$\delta_{ij} = \begin{cases} 1, & i = j \\ 0, & i \neq j. \end{cases}$$

If $\alpha = (\alpha_1, \ldots, \alpha_d)$ is a d-tuple of non-negative integers, then we set

$$[\alpha] = \sum \alpha_i,$$

$$\alpha! = \alpha_1!\,\alpha_2! \cdots \alpha_d!,$$

and

$$\frac{\partial^\alpha}{\partial r^\alpha} = \frac{\partial^{[\alpha]}}{\partial r_1^{\alpha_1} \cdots \partial r_d^{\alpha_d}}.$$

If $\alpha = (0, \ldots, 0)$, then we let

$$\frac{\partial^\alpha}{\partial r^\alpha}(f) = f.$$

DIFFERENTIABLE MANIFOLDS

1.2 Definitions Let $U \subset \mathbb{R}^d$ be open, and let $f\colon U \to \mathbb{R}$. We say that f is *differentiable of class C^k on U* (or simply that f is C^k), for k a non-negative integer, if the partial derivatives $\partial^\alpha f/\partial r^\alpha$ exist and are continuous on U for $[\alpha] \leq k$. In particular, f is C^0 iff f is continuous. If $f\colon U \to \mathbb{R}^n$, then f is *differentiable of class C^k* if each of the component functions $f_i = r_i \circ f$ is C^k. We say that f is C^∞ if it is C^k for all $k \geq 0$.

1.3 Definitions A *locally Euclidean space M of dimension d* is a Hausdorff topological space M for which each point has a neighborhood homeomorphic to an open subset of Euclidean space $\mathbb{R}^d$. If φ is a homeomorphism of a connected open set $U \subset M$ onto an open subset of $\mathbb{R}^d$, φ is called a *coordinate map*, the functions $x_i = r_i \circ \varphi$ are called the *coordinate functions*, and the pair (U,φ) (sometimes denoted by $(U, x_1, \ldots, x_d)$) is called a *coördinate system.* A coordinate system (U, φ) is called a *cubic coordinate system* if $\varphi(U)$ is an open cube about the origin in $\mathbb{R}^d$. If $m \in U$ and $\varphi(m) = 0$, then the coordinate system is said to be *centered at m*.

1.4 Definitions A *differentiable structure $\mathscr{F}$ of class C^k* $(1 \leq k \leq \infty)$ on a locally Euclidean space M is a collection of coordinate systems $\{(U_\alpha, \varphi_\alpha)\colon \alpha \in A\}$ satisfying the following three properties:

(a) $\bigcup_{\alpha \in A} U_\alpha = M$.

(b) $\varphi_\alpha \circ \varphi_\beta^{-1}$ is C^k for all $\alpha, \beta \in A$.

(c) The collection $\mathscr{F}$ is maximal with respect to (b); that is, if (U,φ) is a coordinate system such that $\varphi \circ \varphi_\alpha^{-1}$ and $\varphi_\alpha \circ \varphi^{-1}$ are C^k for all $\alpha \in A$, then $(U,\varphi) \in \mathscr{F}$.

If $\mathscr{F}_0 = \{(U_\alpha, \varphi_\alpha): \alpha \in A\}$ is any collection of coordinate systems satisfying properties (a) and (b), then there is a unique differentiable structure $\mathscr{F}$ containing $\mathscr{F}_0$. Namely, let

$$\mathscr{F} = \{(U,\varphi): \varphi \circ \varphi_\alpha^{-1} \text{ and } \varphi_\alpha \circ \varphi^{-1} \text{ are } C^k \text{ for all } \varphi_\alpha \in \mathscr{F}_0\}.$$

Then $\mathscr{F}$ contains $\mathscr{F}_0$, clearly satisfies (a), and it is easily checked that $\mathscr{F}$ satisfies (b). Now $\mathscr{F}$ is maximal by construction, and so $\mathscr{F}$ is a differentiable structure containing $\mathscr{F}_0$. Clearly $\mathscr{F}$ is the unique such structure.

We mention two other fundamental types of differentiable structures on locally Euclidean spaces, types that we shall not treat in this text, namely, the structure of class C^ω and the complex analytic structure. For a *differentiable structure of class* C^ω, one requires that the compositions in (b) are locally given by convergent power series. For a *complex analytic structure* on a $2d$-dimensional locally Euclidean space, one requires that the coordinate systems have range in complex d-space $\mathbb{C}^d$ and overlap holomorphically.

A *d-dimensional differentiable manifold of class* C^k (similarly C^ω or complex analytic) is a pair $(M,\mathscr{F})$ consisting of a d-dimensional, second countable, locally Euclidean space M together with a differentiable structure $\mathscr{F}$ of class C^k. We shall usually denote the differentiable manifold $(M,\mathscr{F})$ simply by M, with the understanding that when we speak of the "differentiable manifold M" we are considering the locally Euclidean space M with some given differentiable structure $\mathscr{F}$. Our attention will be restricted solely to the case of class C^∞, so by *differentiable* we will always mean *differentiable of class* C^∞. We also use the terminology *smooth* to indicate differentiability of class C^∞. We often refer to differentiable manifolds simply as *manifolds*, with differentiability of class C^∞ always implicitly assumed. A manifold can be viewed as a triple consisting of an underlying point set, a second countable locally Euclidean topology for this set, and a differentiable structure. If X is a set, by a *manifold structure on* X we shall mean a choice of both a second countable locally Euclidean topology for X and a differentiable structure.

Even though we shall restrict our attention to the C^∞ case, many of our theorems do, however, have C^k versions for $k < \infty$, which are essentially no more complicated than the ones we shall obtain. They simply require that one keep track of degrees of differentiability, for differentiating a C^k function may only yield a function of class C^{k-1} if $1 \leq k < \infty$.

Unless we indicate otherwise, we shall always use M and N to denote differentiable manifolds, and M^d will indicate that M is a manifold of dimension d.

1.5 Examples

(a) The standard differentiable structure on Euclidean space $\mathbb{R}^d$ is obtained by taking $\mathscr{F}$ to be the maximal collection (with respect to 1.4(b)) containing $(\mathbb{R}^d,i)$, where i: $\mathbb{R}^d \to \mathbb{R}^d$ is the identity map.

(b) Let V be a finite dimensional real vector space. Then V has a natural manifold structure. Indeed, if $\{e_i\}$ is a basis of V, then the elements of the dual basis $\{r_i\}$ are the coordinate functions of a global coordinate system on V. Such a global coordinate system uniquely determines a differentiable structure $\mathscr{F}$ on V. This differentiable structure is independent of the choice of basis, since different bases give C^∞ overlapping coordinate systems. In fact, the change of coordinates is given simply by a constant non-singular matrix.

(c) Complex n-space $\mathbb{C}^n$ is a real $2n$-dimensional vector space, and so, by Example (b), has a natural structure as a $2n$-dimensional real manifold. If $\{e_i\}$ is the canonical complex basis in which e_i is the n-tuple consisting of zeros except for a 1 in the ith spot, then

$$\{e_1, \ldots, e_n, \sqrt{-1}e_1, \ldots, \sqrt{-1}e_n\}$$

is a real basis for $\mathbb{C}^n$, and its dual basis is the canonical global coordinate system on $\mathbb{C}^n$.

(d) The *d-sphere* is the set

$$S^d = \{a \in \mathbb{R}^{d+1}: \sum_{i=1}^{d+1} a_i^2 = 1\}.$$

Let $n = (0, \ldots, 0, 1)$ and $s = (0, \ldots, 0, -1)$. Then the standard differentiable structure on S^d is obtained by taking $\mathscr{F}$ to be the maximal collection containing $(S^d - n, p_n)$ and $(S^d - s, p_s)$, where p_n and p_s are stereographic projections from n and s respectively.

(e) An open subset U of a differentiable manifold $(M, \mathscr{F}_M)$ is itself a differentiable manifold with differentiable structure

$$\mathscr{F}_U = \{(U_\alpha \cap U, \varphi_\alpha \mid U_\alpha \cap U): (U_\alpha, \varphi_\alpha) \in \mathscr{F}_M\}.$$

Unless specified otherwise, open subsets of differentiable manifolds will always be given this natural differentiable structure.

(f) The *general linear group* $Gl(n,\mathbb{R})$ is the set of all $n \times n$ non-singular real matrices. If we identify in the obvious way the points of $\mathbb{R}^{n^2}$ with $n \times n$ real matrices, then the determinant becomes a continuous function on $\mathbb{R}^{n^2}$. $Gl(n,\mathbb{R})$ receives a manifold structure as the open subset of $\mathbb{R}^{n^2}$ where the determinant function does not vanish.

(g) *Product manifolds.* Let $(M_1, \mathscr{F}_1)$ and $(M_2, \mathscr{F}_2)$ be differentiable manifolds of dimensions d_1 and d_2 respectively. Then $M_1 \times M_2$ becomes a differentiable manifold of dimension $d_1 + d_2$, with differentiable structure $\mathscr{F}$ the maximal collection containing

$$\{(U_\alpha \times V_\beta, \varphi_\alpha \times \psi_\beta: U_\alpha \times V_\beta \to \mathbb{R}^{d_1} \times \mathbb{R}^{d_2}): (U_\alpha, \varphi_\alpha) \in \mathscr{F}_1, (V_\beta, \psi_\beta) \in \mathscr{F}_2\}.$$

1.6 Definitions Let $U \subset M$ be open. We say that $f: U \to \mathbb{R}$ is a *C^∞ function on U* (denoted $f \in C^\infty(U)$) if $f \circ \varphi^{-1}$ is C^∞ for each coordinate map φ on M. A continuous map $\psi: M \to N$ is said to be *differentiable of class C^∞* (denoted $\psi \in C^\infty(M,N)$ or simply $\psi \in C^\infty$) if $g \circ \psi$ is a C^∞ function on ψ^{-1}(domain of g) for all C^∞ functions g defined on open sets in N. Equivalently, the continuous map ψ is C^∞ if and only if $\varphi \circ \psi \circ \tau^{-1}$ is C^∞ for each coordinate map τ on M and φ on N.

Clearly the composition of two differentiable maps is again differentiable. Observe that a mapping $\psi: M \to N$ is C^∞ if and only if for each $m \in M$ there exists an open neighborhood U of m such that $\psi \mid U$ is C^∞.

THE SECOND AXIOM OF COUNTABILITY

The second axiom of countability has many consequences for manifolds. Among them, manifolds are normal, metrizable, and paracompact. Paracompactness implies the existence of partitions of unity, an extremely useful tool for piecing together global functions and structures out of local ones, and conversely for representing global structures as locally finite sums of local ones. After giving the necessary definitions, we shall give a simple direct proof of paracompactness for manifolds, and shall then derive the existence of partitions of unity. It is evident that manifolds are regular topological spaces and their normality follows easily from this and the paracompactness. We shall leave the proof that manifolds are normal as an exercise. For the fact that manifolds are metrizable, see [13].

1.7 Definitions A collection $\{U_\alpha\}$ of subsets of M is a *cover* of a set $W \subset M$ if $W \subset \bigcup U_\alpha$. It is an *open cover* if each U_α is open. A subcollection of the U_α which still covers is called a *subcover*. A *refinement* $\{V_\beta\}$ of the cover $\{U_\alpha\}$ is a cover such that for each β there is an α such that $V_\beta \subset U_\alpha$. A collection $\{A_\alpha\}$ of subsets of M is *locally finite* if whenever $m \in M$ there exists a neighborhood W_m of m such that $W_m \cap A_\alpha \neq \varnothing$ for only finitely many α. A topological space is *paracompact* if every open cover has an open locally finite refinement.

1.8 Definition A *partition of unity on M* is a collection $\{\varphi_i: i \in I\}$ of C^∞ functions on M (where I is an arbitrary index set, not assumed countable) such that

(a) The collection of supports $\{\text{supp } \varphi_i: i \in I\}$ is locally finite.

(b) $\sum_{i \in I} \varphi_i(p) = 1$ for all $p \in M$, and $\varphi_i(p) \geq 0$ for all $p \in M$ and $i \in I$.

A partition of unity $\{\varphi_i: i \in I\}$ is *subordinate* to the cover $\{U_\alpha: \alpha \in A\}$ if for each i there exists an α such that $\text{supp } \varphi_i \subset U_\alpha$. We say that it is subordinate to the cover $\{U_i: i \in I\}$ *with the same index set as the partition of unity* if $\text{supp } \varphi_i \subset U_i$ for each $i \in I$.

1.9 Lemma *Let X be a topological space which is locally compact (each point has at least one compact neighborhood), Hausdorff, and second countable (manifolds, for example). Then X is paracompact. In fact, each open cover has a countable, locally finite refinement consisting of open sets with compact closures.*

PROOF We prove first that there exists a sequence $\{G_i : i = 1, 2, \ldots\}$ of open sets such that

$$\begin{aligned} &\overline{G_i} \text{ is compact,} \\ (1) \qquad &\overline{G_i} \subset G_{i+1}, \\ &X = \bigcup_{i=1}^{\infty} G_i . \end{aligned}$$

Let $\{U_i : i = 1, 2, \ldots\}$, be a countable basis of the topology of X consisting of open sets with compact closures. Such a basis can be obtained by starting with any countable basis and selecting the subcollection consisting of basic sets with compact closures. The fact that X is Hausdorff and locally compact implies that this subcollection is itself a basis. Now, let $G_1 = U_1$. Suppose that

$$G_k = U_1 \cup \cdots \cup U_{j_k} .$$

Then let j_{k+1} be the smallest positive integer greater than j_k such that

$$\overline{G_k} \subset \bigcup_{i=1}^{j_{k+1}} U_i ,$$

and define

$$G_{k+1} = \bigcup_{i=1}^{j_{k+1}} U_i .$$

This defines inductively a sequence $\{G_k\}$ satisfying (1).

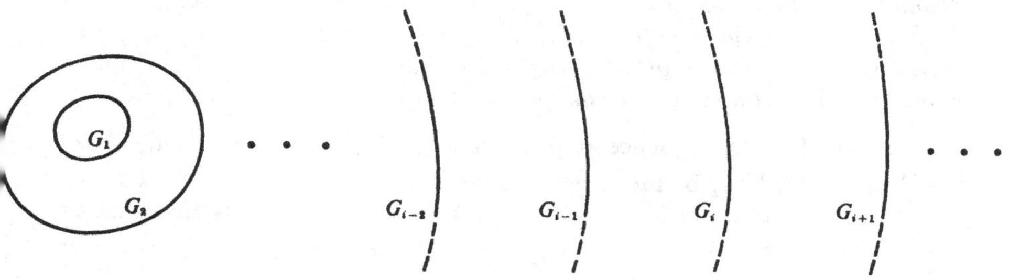

Let $\{U_\alpha: \alpha \in A\}$ be an arbitrary open cover. The set $\overline{G_i} - G_{i-1}$ is compact and contained in the open set $G_{i+1} - \overline{G_{i-2}}$. For each $i \geq 3$ choose a finite subcover of the open cover $\{U_\alpha \cap (G_{i+1} - \overline{G_{i-2}}): \alpha \in A\}$ of $\overline{G_i} - G_{i-1}$, and choose a finite subcover of the open cover $\{U_\alpha \cap G_3: \alpha \in A\}$ of the compact set $\overline{G_2}$. This collection of open sets is easily seen to be a countable, locally finite refinement of the open cover $\{U_\alpha\}$, and consists of open sets with compact closures.

1.10 Lemma *There exists a non-negative C^∞ function φ on $\mathbb{R}^d$ which equals* 1 *on the closed cube $\overline{C(1)}$ and zero on the complement of the open cube $C(2)$.*

PROOF We need only let φ be the product

(1) $$\varphi = (h \circ r_1) \cdots (h \circ r_d),$$

where h is a non-negative C^∞ function on the real line which is 1 on $[-1,1]$ and zero outside of $(-2,2)$. To construct such an h, we start with the function

(2) $$f(t) = \begin{cases} e^{-1/t} & t > 0 \\ 0 & t \leq 0 \end{cases}$$

which is non-negative, C^∞, and positive for $t > 0$. Then the function

(3) $$g(t) = \frac{f(t)}{f(t) + f(1-t)}$$

is non-negative, C^∞, and takes the value 1 for $t \geq 1$ and the value zero for $t \leq 0$. We obtain the desired function h by setting

(4) $$h(t) = g(t+2)g(2-t).$$

1.11 Theorem (Existence of Partitions of Unity) *Let M be a differentiable manifold and $\{U_\alpha: \alpha \in A\}$ an open cover of M. Then there exists a countable partition of unity $\{\varphi_i: i = 1, 2, 3, \ldots\}$ subordinate to the cover $\{U_\alpha\}$ with* supp φ_i *compact for each i. If one does not require compact supports, then there is a partition of unity $\{\varphi_\alpha\}$ subordinate to the cover $\{U_\alpha\}$ (that is,* supp $\varphi_\alpha \subset U_\alpha$*) with at most countably many of the φ_α not identically zero.*

PROOF Let the sequence $\{G_i\}$ cover M as in 1.9(1), and set $G_0 = \emptyset$. For $p \in M$, let i_p be the largest integer such that $p \in M - \overline{G_{i_p}}$. Choose an α_p such that $p \in U_{\alpha_p}$, and let (V,τ) be a coordinate system centered at p such that $V \subset U_{\alpha_p} \cap (G_{i_p+2} - \overline{G_{i_p}})$ and such that $\tau(V)$ contains the closed cube $\overline{C(2)}$. Define

(1) $$\psi_p = \begin{cases} \varphi \circ \tau & \text{on } V \\ 0 & \text{elsewhere} \end{cases}$$

where φ is the function 1.10(1). Then ψ_p is a C^∞ function on M which has the value 1 on some open neighborhood W_p of p, and has compact support lying in $V \subset U_{\alpha_p} \cap (G_{i_p+2} - \overline{G_{i_p}})$. For each $i \geq 1$, choose a finite set of points p in M whose corresponding W_p neighborhoods cover $\overline{G_i} - G_{i-1}$. Order the corresponding ψ_p functions in a sequence ψ_j, $j = 1, 2, 3, \ldots$. The supports of the ψ_j form a locally finite family of subsets of M. Thus the function

(2) $$\psi = \sum_{j=1}^{\infty} \psi_j$$

is a well-defined C^∞ function on M, and moreover $\psi(p) > 0$ for each $p \in M$. For each $i = 1, 2, 3, \ldots$ define

(3) $$\varphi_i = \frac{\psi_i}{\psi}.$$

Then the functions $\{\varphi_i : i = 1, 2, 3, \ldots\}$ form a partition of unity subordinate to the cover $\{U_\alpha\}$ with supp φ_i compact for each i. If we let φ_α be identically zero if no φ_i has support in U_α, and otherwise let φ_α be the sum of the φ_i with support in U_α, then $\{\varphi_\alpha\}$ is a partition of unity subordinate to the cover $\{U_\alpha\}$ with at most countably many of the φ_α not identically zero. To see that the support of φ_α lies in U_α, observe that if $\mathscr{A}$ is a locally finite family of closed sets, then $\overline{\bigcup_{A \in \mathscr{A}} A} = \bigcup_{A \in \mathscr{A}} A$. Observe, however, that the support of φ_α is not necessarily compact.

Corollary *Let G be open in M, and let A be closed in M, with $A \subset G$. Then there exists a C^∞ function $\varphi\colon M \to \mathbb{R}$ such that*

(a) $0 \leq \varphi(p) \leq 1$ *for all* $p \in M$.

(b) $\varphi(p) = 1$ *if* $p \in A$.

(c) supp $\varphi \subset G$.

PROOF There is a partition of unity $\{\varphi, \psi\}$ subordinate to the cover $\{G, M - A\}$ of M with supp $\varphi \subset G$ and supp $\psi \subset M - A$. Then φ is the desired function.

TANGENT VECTORS AND DIFFERENTIALS

1.12 A vector v with components $v_1, \ldots, v_d$ at a point p in Euclidean space $\mathbb{R}^d$ can be thought of as an operator on differentiable functions. Specifically, if f is differentiable on a neighborhood of p, then v assigns to f the real number $v(f)$ which is the directional derivative of f in the direction v at p. That is,

(1) $$v(f) = v_1 \left.\frac{\partial f}{\partial r_1}\right|_p + \cdots + v_d \left.\frac{\partial f}{\partial r_d}\right|_p .$$

This operation of the vector v on differentiable functions satisfies two important properties,

$$(2)\qquad \begin{aligned} v(f+\lambda g) &= v(f)+\lambda v(g),\\ v(f\cdot g) &= f(p)v(g)+g(p)v(f), \end{aligned}$$

whenever f and g are differentiable near p, and λ is a real number. The first property says that v acts linearly on functions, and the second says that v is a *derivation*. This motivates our definition of tangent vectors on manifolds. They will be directional derivatives, that is, linear derivations on functions. The operation of taking derivatives depends only on local properties of functions, properties in arbitrarily small neighborhoods of the point at which the derivative is being taken. In order to express most conveniently this dependence of the derivative on the local nature of functions, we introduce the notion of germs of functions.

1.13 Definitions Let $m \in M$. Functions f and g defined on open sets containing m are said to have the same *germ* at m if they agree on some neighborhood of m. This introduces an equivalence relation on the C^∞ functions defined on neighborhoods of m, two functions being equivalent if and only if they have the same germ. The equivalence classes are called *germs*, and we denote the set of germs at m by $\tilde{F}_m$. If f is a C^∞ function on a neighborhood of m, then $\mathbf{f}$ will denote its germ. The operations of addition, scalar multiplication, and multiplication of functions induce on $\tilde{F}_m$ the structure of an algebra over $\mathbb{R}$. A germ $\mathbf{f}$ has a well-defined value $\mathbf{f}(m)$ at m, namely, the value at m of any representative of the germ. Let $F_m \subset \tilde{F}_m$ be the set of germs which vanish at m. Then F_m is an ideal in $\tilde{F}_m$, and we let F_m^k denote its kth power. F_m^k is the ideal of $\tilde{F}_m$ consisting of all finite linear combinations of k-fold products of elements of F_m. These form a descending sequence of ideals $\tilde{F}_m \supset F_m \supset F_m^2 \supset F_m^3 \supset \cdots$.

1.14 Definition A *tangent vector v at the point $m \in M$* is a linear derivation of the algebra $\tilde{F}_m$. That is, for all $\mathbf{f}, \mathbf{g} \in \tilde{F}_m$ and $\lambda \in \mathbb{R}$,

(a) $v(\mathbf{f}+\lambda\mathbf{g}) = v(\mathbf{f})+\lambda v(\mathbf{g})$.

(b) $v(\mathbf{f}\cdot\mathbf{g}) = \mathbf{f}(m)v(\mathbf{g})+\mathbf{g}(m)v(\mathbf{f})$.

M_m denotes the set of tangent vectors to M at m and is called the *tangent space to M at m*. Observe that if we define $(v+w)(\mathbf{f})$ and $(\lambda v)(\mathbf{f})$ by

$$(1)\qquad \begin{aligned} (v+w)(\mathbf{f}) &= v(\mathbf{f})+w(\mathbf{f})\\ (\lambda v)(\mathbf{f}) &= \lambda(v(\mathbf{f})) \end{aligned}$$

whenever $v, w \in M_m$ and $\lambda \in \mathbb{R}$, then $v+w$ and λv again are tangent vectors at m. So in this way M_m becomes a real vector space. The fundamental property of the vector space M_m, which we shall establish in 1.17, is that its dimension equals the dimension of M. This definition of tangent vector is not suitable in the C^k case for $1 \leq k < \infty$. (We will discuss the C^k case further in 1.21.) We give this definition of tangent vector for several reasons. One reason is that it is intrinsic; that is, it does not depend on coordinate systems. Another reason is that it generalizes naturally to higher order tangent vectors, as we shall see in 1.26.

1.15 If $\mathbf{c}$ is the germ of a function with the constant value c on a neighborhood of m, and if v is a tangent vector at m, then $v(\mathbf{c}) = 0$, for

$$v(\mathbf{c}) = cv(\mathbf{1}),$$

and

$$v(\mathbf{1}) = v(\mathbf{1} \cdot \mathbf{1}) = 1v(\mathbf{1}) + 1v(\mathbf{1}) = 2v(\mathbf{1}).$$

1.16 Lemma *M_m is naturally isomorphic with $(F_m/F_m^2)^*$.* (The symbol * denotes dual vector space.)

PROOF If $v \in M_m$, then v is a linear function on F_m vanishing on F_m^2 because of the derivation property. Conversely, if $\ell \in (F_m/F_m^2)^*$, we define a tangent vector v_ℓ at m by setting $v_\ell(\mathbf{f}) = \ell(\{\mathbf{f} - \mathbf{f(m)}\})$ for $\mathbf{f} \in \tilde{F}_m$. (Here $\mathbf{f(m)}$ denotes the germ of the function with the constant value $f(m)$, and $\{\ \}$ is used to denote cosets in F_m/F_m^2.) Linearity of v_ℓ on $\tilde{F}_m$ is clear. It is a derivation since

$$\begin{aligned} v_\ell(\mathbf{f} \cdot \mathbf{g}) &= \ell(\{\mathbf{f} \cdot \mathbf{g} - \mathbf{f(m)g(m)}\}) \\ &= \ell(\{(\mathbf{f} - \mathbf{f(m)})(\mathbf{g} - \mathbf{g(m)}) + \mathbf{f(m)}(\mathbf{g} - \mathbf{g(m)}) \\ &\quad + (\mathbf{f} - \mathbf{f(m)})\mathbf{g(m)}\}) \\ &= \ell(\{(\mathbf{f} - \mathbf{f(m)})(\mathbf{g} - \mathbf{g(m)})\}) + f(m)\ell(\{\mathbf{g} - \mathbf{g(m)}\}) \\ &\quad + g(m)\ell(\{\mathbf{f} - \mathbf{f(m)}\}) \\ &= f(m)v_\ell(\mathbf{g}) + g(m)v_\ell(\mathbf{f}). \end{aligned}$$

Thus we obtain mappings of M_m into $(F_m/F_m^2)^*$, and vice versa. It is easily checked that these are inverses of each other and thus are isomorphisms.

1.17 Theorem $\dim(F_m/F_m^2) = \dim M$.

The proof is based on the following calculus lemma [31].

Lemma *If g is of class C^k ($k \geq 2$) on a convex open set U about p in $\mathbb{R}^d$, then for each $q \in U$,*

$$(1)\quad g(q) = g(p) + \sum_{i=1}^{d} \left.\frac{\partial g}{\partial r_i}\right|_p (r_i(q) - r_i(p))$$
$$+ \sum_{i,j} (r_i(q) - r_i(p))(r_j(q) - r_j(p)) \int_0^1 (1-t) \left.\frac{\partial^2 g}{\partial r_i\, \partial r_j}\right|_{(p+t(q-p))} dt.$$

In particular, if $g \in C^\infty$, then the second summation in (1) determines an element of F_p^2 since the integral as a function of q is of class C^∞.

PROOF OF 1.17 Let (U,φ) be a coordinate system about m with coordinate functions $x_1, \ldots, x_d$ $(d = \dim M)$. Let $\mathbf{f} \in F_m$. Apply (1) to $f \circ \varphi^{-1}$, and compose with φ to obtain

$$f = \sum_{i=1}^{d} \frac{\partial(f \circ \varphi^{-1})}{\partial r_i}\bigg|_{\varphi(m)} (x_i - x_i(m)) + \sum_{i,j} (x_i - x_i(m))(x_j - x_j(m))h$$

on a neighborhood of m, where $h \in C^\infty$. Thus

$$\mathbf{f} = \sum_{i=1}^{d} \frac{\partial(f \circ \varphi^{-1})}{\partial r_i}\bigg|_{\varphi(m)} (\mathbf{x}_i - \mathbf{x}_i(\mathbf{m})) \bmod F_m{}^2.$$

Hence $\{(\mathbf{x}_i - \mathbf{x}_i(\mathbf{m})) : i = 1, \ldots, d\}$ spans $F_m/F_m{}^2$. Consequently $\dim F_m/F_m{}^2 \leq d$. We claim that these elements are linearly independent. For suppose that

$$\sum_{i=1}^{d} a_i(\mathbf{x}_i - \mathbf{x}_i(\mathbf{m})) \in F_m{}^2.$$

Now,

$$\sum_{i=1}^{d} a_i(x_i - x_i(m)) \circ \varphi^{-1} = \sum_{i=1}^{d} a_i(r_i - r_i(\varphi(m))).$$

Thus

$$\sum_{i=1}^{d} a_i(\mathbf{r}_i - \mathbf{r}_i(\boldsymbol{\varphi}(\mathbf{m}))) \in F^2_{\varphi(m)}.$$

But this implies that

$$\frac{\partial}{\partial r_j}\bigg|_{\varphi(m)} \left(\sum a_i(r_i - r_i(\varphi(m)))\right) = 0$$

for $j = 1, \ldots, d$, which implies that the a_i must all be zero.

Corollary $\dim M_m = \dim M$.

1.18 In practice we will treat tangent vectors as operating on functions rather than on their germs. If f is a differentiable function defined on a neighborhood of m, and $v \in M_m$, we define

$$v(f) = v(\mathbf{f}). \tag{1}$$

Thus $v(f) = v(g)$ whenever f and g agree on a neighborhood of m, and clearly

$$\begin{aligned} v(f + \lambda g) &= v(f) + \lambda v(g) \qquad (\lambda \in \mathbb{R}), \\ v(f \cdot g) &= f(m)v(g) + g(m)v(f), \end{aligned} \tag{2}$$

where $f + \lambda g$ and $f \cdot g$ are defined on the intersection of the domains of definition of f and g.

1.19 Definition Let (U,φ) be a coordinate system with coordinate functions $x_1, \ldots, x_d$, and let $m \in U$. For each $i \in (1, \ldots, d)$, we define a tangent vector $(\partial/\partial x_i)|_m \in M_m$ by setting

(1) $$\left(\frac{\partial}{\partial x_i}\bigg|_m\right)(f) = \frac{\partial(f \circ \varphi^{-1})}{\partial r_i}\bigg|_{\varphi(m)}$$

for each function f which is C^∞ on a neighborhood of m. We interpret (1) as the directional derivative of f at m in the x_i coordinate direction. We also use the notation

(2) $$\frac{\partial f}{\partial x_i}\bigg|_m = \left(\frac{\partial}{\partial x_i}\bigg|_m\right)(f).$$

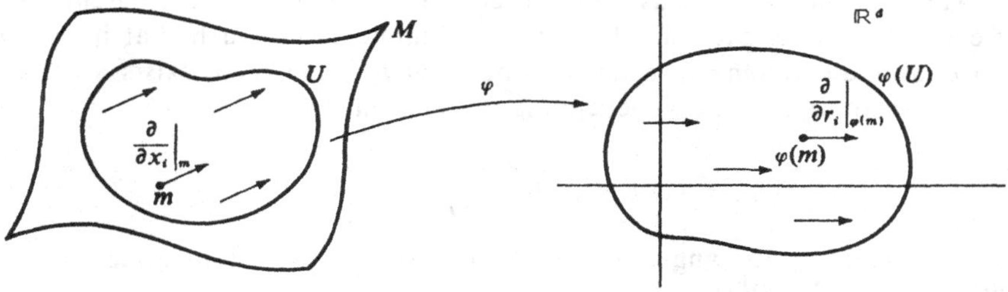

1.20 Remarks on 1.19

(a) Clearly $((\partial/\partial x_i)|_m)(f)$ depends only on the germ of f at m, and (a) and (b) of 1.14 are satisfied; so $(\partial/\partial x_i)|_m$ is a tangent vector at m. Moreover, $\{(\partial/\partial x_i)|_m : i = 1, \ldots, d\}$ is a basis of M_m. Indeed, it is the basis of M_m dual to the basis $\{\{\mathbf{x}_i - \mathbf{x}_i(\mathbf{m})\} : i = 1, \ldots, d\}$ of F_m/F_m^2 since

$$\frac{\partial}{\partial x_i}\bigg|_m (x_j - x_j(m)) = \delta_{ij}.$$

(b) If $v \in M_m$, then

$$v = \sum_{i=1}^{d} v(x_i) \frac{\partial}{\partial x_i}\bigg|_m.$$

Simply check that both sides give the same results when applied to the functions $(x_j - x_j(m))$.

(c) Suppose that (U,φ) and (V,ψ) are coordinate systems about m, with coordinate functions $x_1, \ldots, x_d$ and $y_1, \ldots, y_d$ respectively. Then it follows from remark (b) that

$$\frac{\partial}{\partial y_j}\bigg|_m = \sum_{i=1}^{d} \frac{\partial x_i}{\partial y_j}\bigg|_m \frac{\partial}{\partial x_i}\bigg|_m.$$

Observe that $(\partial/\partial x_i)$ depends on φ and not only on x_i. In particular, if x_1 were equal to y_1, it would not necessarily follow that $\partial/\partial x_1$ equals $\partial/\partial y_1$.

(d) If we apply Definition 1.19 to the canonical coordinate system $r_1, \ldots, r_d$ on $\mathbb{R}^d$, then the tangent vectors which we obtain are none other than the ordinary partial derivative operators $(\partial/\partial r_i)$.

1.21 Our proof of the finite dimensionality of F_m/F_m^2 certainly fails in the C^k case for $k < \infty$ since the remainder term in the lemma of 1.17 will not be a sum of products of C^k functions, and the lemma doesn't even make sense in the C^1 case. In fact, it turns out (see [21]) that F_m/F_m^2 is always infinite dimensional in the C^k case for $1 \leq k < \infty$. There are various ways to define tangent vectors in the C^k case in order that $\dim M_m = \dim M$ (all of which work in the C^∞ case, too). One way is to define a tangent vector v at m as a mapping which assigns to each function (defined and differentiable of class C^k on a neighborhood of m) a real number $v(f)$ such that if (U,φ) is a coordinate system on a neighborhood of m, then there exists a list of real numbers $(a_1, \ldots, a_d)$ (depending on φ) such that

$$v(f) = \sum_{i=1}^{d} a_i \left.\frac{\partial(f \circ \varphi^{-1})}{\partial r_i}\right|_{\varphi(m)}.$$

Then the space M_m of tangent vectors again turns out to be finite dimensional, with a basis $\{(\partial/\partial x_i)|_m\}$.

1.22 The Differential Let $\psi: M \to N$ be C^∞, and let $m \in M$. The *differential of* ψ *at* m is the linear map

(1) $$d\psi: M_m \to N_{\psi(m)}$$

defined as follows. If $v \in M_m$, then $d\psi(v)$ is to be a tangent vector at $\psi(m)$, so we describe how it operates on functions. Let g be a C^∞ function on a neighborhood of $\psi(m)$. Define $d\psi(v)(g)$ by setting

(2) $$d\psi(v)(g) = v(g \circ \psi).$$

It is easily checked that $d\psi$ is a linear map of M_m into $N_{\psi(m)}$. Strictly speaking, this map should be denoted $d\psi \mid M_m$, or simply $d\psi_m$. However, we omit the subscript m when there is no possibility of confusion. The map ψ is called *non-singular at m* if $d\psi_m$ is non-singular, that is, if the kernel of (1) consists of 0 alone. The *dual map*

(3) $$\delta\psi: N^*_{\psi(m)} \to M^*_m$$

is defined as usual by requiring that

(4) $$\delta\psi(\omega)(v) = \omega(d\psi(v))$$

whenever $\omega \in N^*_{\psi(m)}$ and $v \in M_m$. In the special case of a C^∞ function $f: M \to \mathbb{R}$, if $v \in M_m$ and $f(m) = r_0$, then

(5) $$df(v) = v(f) \left.\frac{d}{dr}\right|_{r_0}.$$

In this case, we usually take df to mean the element of M_m^* defined by

(6) $$df(v) = v(f).$$

That is, we identify df with $\delta f(\omega)$, where ω is the basis of the 1-dimensional space $\mathbb{R}_{r_0}^*$ dual to $(d/dr)|_{r_0}$. Particular usage will be clear from the context.

1.23 Remarks on 1.22

(a) Let $(U, x_1, \ldots, x_d)$ and $(V, y_1, \ldots, y_\ell)$ be coordinate systems about m and $\psi(m)$ respectively. Then it follows from 1.22(2) and 1.20(b) that

$$d\psi\left(\frac{\partial}{\partial x_j}\bigg|_m\right) = \sum_{i=1}^{\ell} \frac{\partial(y_i \circ \psi)}{\partial x_j}\bigg|_m \frac{\partial}{\partial y_i}\bigg|_{\psi(m)}.$$

The matrix $\{\partial(y_i \circ \psi)/\partial x_j\}$ is called the *Jacobian* of the map ψ (with respect to the given coordinate system). For maps between Euclidean spaces, the Jacobian will always be taken with respect to the canonical coordinate systems.

(b) If $(U, x_1, \ldots, x_d)$ is a coordinate system on M, and $m \in U$, then $\{dx_i|_m\}$ is the basis of M_m^* dual to $\{\partial/\partial x_i|_m\}$. If $f\colon M \to \mathbb{R}$ is a C^∞ function, then

$$df_m = \sum_{i=1}^{d} \frac{\partial f}{\partial x_i}\bigg|_m dx_i|_m.$$

(c) *Chain Rule*. Let $\psi\colon M \to N$ and $\varphi\colon N \to X$ be C^∞ maps. Then

$$d(\varphi \circ \psi)_m = d\varphi_{\psi(m)} \circ d\psi_m,$$

or simply $d(\varphi \circ \psi) = d\varphi \circ d\psi$. It is a useful exercise to check the form that this equation takes when the maps are expressed in terms of the matrices obtained by choosing coordinate systems.

(d) If $\psi\colon M \to N$ and $f\colon N \to \mathbb{R}$ are C^∞, then $\delta\psi(df_{\psi(m)}) = d(f \circ \psi)_m$, for $\delta\psi(df_{\psi(m)})(v) = df(d\psi(v)) = d(f \circ \psi)_m(v)$ whenever $v \in M_m$.

(e) A C^∞ mapping $\sigma\colon (a, b) \to M$ is called a *smooth curve* in M. Let $t \in (a, b)$. Then the *tangent vector to the curve σ at t* is the vector

$$d\sigma\left(\frac{d}{dr}\bigg|_t\right) \in M_{\sigma(t)}.$$

We shall denote the tangent vector to σ at t by $\dot\sigma(t)$.

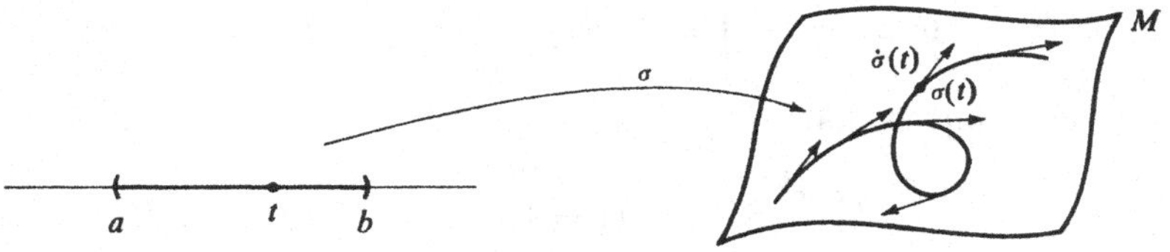

Now, if $v \neq 0$ is any element of M_m, then v is the tangent vector to a smooth curve in M. For one can simply choose a coordinate system (U,φ), centered at m, for which

$$v = d\varphi^{-1}\left(\frac{\partial}{\partial r_1}\bigg|_0\right).$$

Then v is the tangent vector at 0 to the curve $t \mapsto \varphi^{-1}(t, 0, \ldots, 0)$. One should observe that many curves can have the same tangent vector, and that two smooth curves σ and τ in M for which $\sigma(t_0) = \tau(t_0) = m$ have the same tangent vector at t_0 if and only if

$$\frac{d(f \circ \sigma)}{dr}\bigg|_{t_0} = \frac{d(f \circ \tau)}{dr}\bigg|_{t_0}$$

for all functions f which are C^∞ on a neighborhood of m.

If σ happens to be a curve in the Euclidean space $\mathbb{R}^n$, then

$$\dot{\sigma}(t) = \frac{d\sigma_1}{dr}\bigg|_t \frac{\partial}{\partial r_1}\bigg|_{\sigma(t)} + \cdots + \frac{d\sigma_n}{dr}\bigg|_t \frac{\partial}{\partial r_n}\bigg|_{\sigma(t)}.$$

If we identify this tangent vector with the element

$$\left(\frac{d\sigma_1}{dr}\bigg|_t, \ldots, \frac{d\sigma_n}{dr}\bigg|_t\right)$$

of $\mathbb{R}^n$, then we have

$$\dot{\sigma}(t) = \lim_{h \to 0} \frac{\sigma(t+h) - \sigma(t)}{h}.$$

Thus with this identification our notion of tangent vector coincides, in this special case, with the geometric notion of a tangent to a curve in Euclidean space.

1.24 Theorem *Let ψ be a C^∞ mapping of the connected manifold M into the manifold N. Suppose that for each $m \in M$, $d\psi_m \equiv 0$. Then ψ is a constant map.*

PROOF Let $n \in \psi(M)$. $\psi^{-1}(n)$ is closed. We need only show that it is open. For this, let $m \in \psi^{-1}(n)$. Choose coordinate systems $(U, x_1, \ldots, x_d)$ and $(V, y_1, \ldots, y_c)$ about m and n respectively, so that $\psi(U) \subset V$. Then on U,

$$0 = d\psi\left(\frac{\partial}{\partial x_j}\right) = \sum_{i=1}^{c} \frac{\partial(y_i \circ \psi)}{\partial x_j} \frac{\partial}{\partial y_i} \qquad (j = 1, \ldots, d),$$

which implies that

$$\frac{\partial(y_i \circ \psi)}{\partial x_j} \equiv 0 \qquad (i = 1, \ldots, c;\ j = 1, \ldots, d).$$

Thus the functions $y_i \circ \psi$ are constant on U. This implies that $\psi(U) = n$; hence $\psi^{-1}(n)$ is open and consequently $\psi^{-1}(n) = M$.

We shall now see that in a natural way the collection of all tangent vectors to a differentiable manifold itself forms a differentiable manifold called the *tangent bundle*. We have a similar dual object called the *cotangent bundle* formed from the linear functionals on the tangent spaces.

1.25 Tangent and Cotangent Bundles Let M be a C^∞ manifold with differentiable structure $\mathscr{F}$. Let

$$T(M) = \bigcup_{m \in M} M_m, \tag{1}$$
$$T^*(M) = \bigcup_{m \in M} M_m^*.$$

There are natural projections:

$$\pi\colon T(M) \to M, \qquad \pi(v) = m \quad \text{if} \quad v \in M_m, \tag{2}$$
$$\pi^*\colon T^*(M) \to M, \qquad \pi^*(\tau) = m \quad \text{if} \quad \tau \in M_m^*.$$

Let $(U,\varphi) \in \mathscr{F}$ with coordinate functions $x_1, \ldots, x_d$. Define $\tilde{\varphi}\colon \pi^{-1}(U) \to \mathbb{R}^{2d}$ and $\tilde{\varphi}^*\colon (\pi^*)^{-1}(U) \to \mathbb{R}^{2d}$ by

$$\tilde{\varphi}(v) = \big(x_1(\pi(v)), \ldots, x_d(\pi(v)), dx_1(v), \ldots, dx_d(v)\big) \tag{3}$$
$$\tilde{\varphi}^*(\tau) = \left(x_1(\pi^*(\tau)), \ldots, x_d(\pi^*(\tau)), \tau\left(\frac{\partial}{\partial x_1}\right), \ldots, \tau\left(\frac{\partial}{\partial x_d}\right)\right)$$

for all $v \in \pi^{-1}(U)$ and $\tau \in (\pi^*)^{-1}(U)$. Note that $\tilde{\varphi}$ and $\tilde{\varphi}^*$ are both one-to-one maps onto open subsets of $\mathbb{R}^{2d}$. The following steps outline the construction of a topology and a differentiable structure on $T(M)$. The construction for $T^*(M)$ goes similarly. The proofs are left as exercises.

(a) If (U,φ) and $(V,\psi) \in \mathscr{F}$, then $\tilde{\psi} \circ \tilde{\varphi}^{-1}$ is C^∞.

(b) The collection $\{\tilde{\varphi}^{-1}(W)\colon W \text{ open in } \mathbb{R}^{2d},\ (U,\varphi) \in \mathscr{F}\}$ forms a basis for a topology on $T(M)$ which makes $T(M)$ into a $2d$-dimensional, second countable, locally Euclidean space.

(c) Let $\tilde{\mathscr{F}}$ be the maximal collection, with respect to 1.4(b), containing

$$\{(\pi^{-1}(U), \tilde{\varphi})\colon (U,\varphi) \in \mathscr{F}\}.$$

Then $\tilde{\mathscr{F}}$ is a differentiable structure on $T(M)$.

$T(M)$ and $T^*(M)$ with these differentiable structures are called respectively the *tangent bundle* and the *cotangent bundle*. It will sometimes be convenient to write the points of $T(M)$ as pairs (m,v) where $m \in M$ and $v \in M_m$ (and similarly for $T^*(M)$).

If $\psi: M \to N$ is a C^∞ map, then the differential of ψ defines a mapping of the tangent bundles

(4) $$d\psi: T(M) \to T(N),$$

where $d\psi(m,v) = d\psi_m(v)$ whenever $v \in M_m$. It is easily checked that (4) is a C^∞ map.

1.26† Higher Order Tangent Vectors and Differentials It is useful to look at M_m as $(F_m/F_m^2)^*$, for this point of view allows an immediate generalization to higher order tangent vectors. We digress for a moment to give these definitions.

Recall that $\tilde{F}_m$ is the algebra of germs of functions at m. $F_m \subset \tilde{F}_m$ is the ideal of germs vanishing at m, and F_m^k (k an integer ≥ 1) is the ideal of $\tilde{F}_m$ consisting of all finite linear combinations of k-fold products of elements of F_m.

The vector space F_m/F_m^{k+1} is called the space of kth *order differentials at m*, and we denote it by ${}^kM_m$. As before, **f** denotes the germ of f at m, and { } will denote cosets in F_m/F_m^{k+1}. Let f be a differentiable function on a neighborhood of m. We define the kth *order differential* d^kf *of f at m* by

(1) $$d^kf = \{\mathbf{f} - \mathbf{f}(\mathbf{m})\}.$$

A kth *order tangent vector at m* is a real linear function on $\tilde{F}_m$ vanishing on F_m^{k+1} and vanishing also on the set of germs of functions constant on a neighborhood of m. The real linear space of kth order tangent vectors at m will be denoted by $M_m{}^k$. We have a natural identification of $M_m{}^k$ with $({}^kM_m)^*$ since any kth order tangent vector restricted to F_m yields a linear function on F_m vanishing on F_m^{k+1}, and hence yields an element of $({}^kM_m)^*$; and conversely an element of $({}^kM_m)^*$ uniquely determines a linear function on F_m vanishing on F_m^{k+1}, and this extends uniquely to a kth order tangent vector by requiring it to annihilate germs of constant functions.

We can tie up this notion of higher order tangent vector with the usual notion of higher order derivative in Euclidean space by looking at the forms that these tangent vectors and differentials take in a coordinate system. Let (U,φ) be a coordinate system about m with coordinate functions $x_1, \ldots, x_d$ such that $\varphi(U)$ is a convex open set in Euclidean space $\mathbb{R}^d$. Let $\alpha = (\alpha_1, \ldots, \alpha_d)$ be a list of non-negative integers. In addition to our conventions of 1.1, we let

$$(x - x(m))^\alpha = (x_1 - x_1(m))^{\alpha_1} \cdots (x_d - x_d(m))^{\alpha_d}.$$

Let f be a C^∞ function on U. Then it follows from the lemma of 1.17, that

(2) $$f = f(m) + \sum_{[\alpha]=1}^{k} a_\alpha (x - x(m))^\alpha + \sum_{[\alpha]=k+1} h_\alpha (x - x(m))^\alpha,$$

† The material of this section will not be used elsewhere in the book, and so it may be skipped without loss of continuity.

where the h_α are C^∞ functions on U and where

(3)
$$a_\alpha = \frac{1}{\alpha!} \frac{\partial^\alpha (f \circ \varphi^{-1})}{\partial r^\alpha}\bigg|_{\varphi(m)}.$$

Hence

(4)
$$d^k f = \sum_{1 \leq [\alpha] \leq k} a_\alpha \{(\mathbf{x} - \mathbf{x}(\mathbf{m}))^\alpha\}.$$

Thus the collection

(5)
$$[\{(\mathbf{x} - \mathbf{x}(\mathbf{m}))^\alpha\} : 1 \leq [\alpha] \leq k]$$

spans ${}^k M_m$. The proof that these elements are linearly independent in ${}^k M_m$ is the obvious generalization of the proof for the case $k = 2$ which was treated in 1.17. Thus the collection (5) forms a basis of ${}^k M_m$. Consequently ${}^k M_m$ is *finite dimensional* with dimension equal to the binomial coefficient $\sum_{j=1}^{k} \binom{d+j-1}{j}$. As the dual space of ${}^k M_m$, $M_m{}^k$ is also finite dimensional with the same dimension. Since $M_m{}^k$ is identified with $({}^k M_m)^*$, and these spaces are finite dimensional, we have a canonical isomorphism of ${}^k M_m$ with $(M_m{}^k)^*$, under which the element of $d^k f \in {}^k M_m$, considered as an element of $(M_m{}^k)^*$, satisfies

(6)
$$d^k f(v) = v(\mathbf{f}).$$

Let

(7)
$$\frac{\partial^\alpha f}{\partial x^\alpha}\bigg|_m = \frac{\partial^\alpha (f \circ \varphi^{-1})}{\partial r^\alpha}\bigg|_{\varphi(m)}.$$

Since the derivative is linear, and since the value of $\partial^\alpha f/\partial x^\alpha$ at m depends only on the germ of f at m and vanishes if f is constant on a neighborhood of m or if f is an $[\alpha] + 1$-fold product of functions which vanish at m, then $(\partial^\alpha/\partial x^\alpha)|_m$ is an $[\alpha]$th order tangent vector at m. It follows that

(8)
$$\left\{\left(\frac{1}{\alpha!}\right)\frac{\partial^\alpha}{\partial x^\alpha}\bigg|_m : 1 \leq [\alpha] \leq k\right\}$$

is the basis of $M_m{}^k$ dual to the basis (5) of ${}^k M_m$. If v is a kth order tangent vector at m, then

(9)
$$v = \sum_{[\alpha]=1}^{k} b_\alpha \frac{\partial^\alpha}{\partial x^\alpha}\bigg|_m,$$

where

(10)
$$b_\alpha = \left(\frac{1}{\alpha!}\right) v((\mathbf{x} - \mathbf{x}(\mathbf{m}))^\alpha).$$

In terms of the basis (8), equation (3) becomes

(11)
$$a_\alpha = \frac{1}{\alpha!} \frac{\partial^\alpha f}{\partial x^\alpha}\bigg|_m.$$

As in the case of first order tangent vectors, we customarily think of tangent vectors as operating on the functions themselves rather than their germs; indeed, we define

$$v(f) = v(\mathbf{f}) \tag{12}$$

whenever f is C^∞ on a neighborhood of m and v is a tangent vector of any order at m.

Finally, just as there are natural mappings of tangent vectors and differentials associated with a differentiable map $\varphi\colon M \to N$, so are there linear mappings

$$\begin{aligned} d^k\varphi &: M_m{}^k \to N^k_{\varphi(m)}, \\ \delta^k\varphi &: {}^kN_{\varphi(m)} \to {}^kM_m \end{aligned} \tag{13}$$

defined by

$$\begin{aligned} d^k\varphi(v)(g) &= v(g \circ \varphi), \\ \delta^k\varphi(d^k g) &= d^k(g \circ \varphi) \end{aligned} \tag{14}$$

whenever $v \in M_m{}^k$ and g is a C^∞ function on a neighborhood of $\varphi(m)$. It is easily checked that (14) does indeed define the mappings (13) and that the mappings $d^k\varphi$ and $\delta^k\varphi$ are dual.

Our definition of a first order tangent vector in this section agrees with Definition 1.14 in view of Lemma 1.16. Moreover, we have seen three interpretations of the first order differential df of a function f; the interpretation (13) agrees with our original definition 1.22(1), the interpretation (6) agrees with 1.22(6), and we have the additional interpretation (1).

SUBMANIFOLDS, DIFFEOMORPHISMS, AND THE INVERSE FUNCTION THEOREM

1.27 Definitions Let $\psi\colon M \to N$ be C^∞.

(a) ψ is an *immersion* if $d\psi_m$ is non-singular for each $m \in M$.

(b) The pair (M,ψ) is a *submanifold* of N if ψ is a one-to-one immersion.

(c) ψ is an *imbedding* if ψ is a one-to-one immersion which is also a homeomorphism into; that is, ψ is open as a map into $\psi(M)$ with the relative topology.

(d) ψ is a *diffeomorphism* if ψ maps M one-to-one onto N and ψ^{-1} is C^∞.

1.28 Remarks on 1.27 One can, for example, immerse the real line $\mathbb{R}$ into the plane, as illustrated in the following figure, so that the first case is an immersion which is not a submanifold, the second is a submanifold which is not an imbedding, and the third is an imbedding.

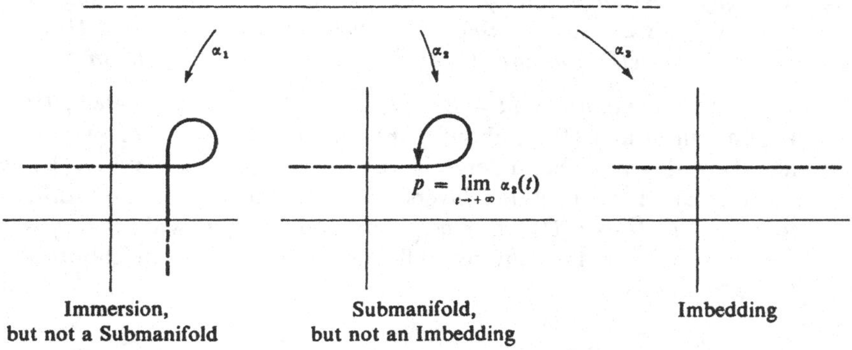

Observe that if (U,φ) is a coordinate system, then $\varphi\colon U \to \varphi(U)$ is a diffeomorphism.

The composition of diffeomorphisms is again a diffeomorphism. Thus the relation of being diffeomorphic is an equivalence relation on the collection of differentiable manifolds. It is quite possible for a locally Euclidean space to possess distinct differentiable structures which are diffeomorphic. (See Exercise 2.) In a remarkable paper, Milnor showed the existence of locally Euclidean spaces (S^7 is an example) which possess non-diffeomorphic differentiable structures [19]. There are also locally Euclidean spaces which possess no differentiable structures at all [14].

If ψ is a diffeomorphism, then $d\psi_m$ is an isomorphism since both $(d\psi \circ d\psi^{-1})|_{\psi(m)}$ and $(d\psi^{-1} \circ d\psi)|_m$ are the identity transformations. The inverse function theorem gives us a local converse of this—whenever $d\psi_m$ is an isomorphism, ψ is a diffeomorphism on a neighborhood of m. Before we recall the precise statement of the inverse function theorem, we give a definition which will be needed in the corollaries.

1.29 Definition A set $y_1, \ldots, y_j$ of C^∞ functions defined on some neighborhood of m in M is called an *independent set at m* if the differentials $dy_1, \ldots, dy_j$ form an independent set in M_m^*.

1.30 Inverse Function Theorem *Let $U \subset \mathbb{R}^d$ be open, and let $f\colon U \to \mathbb{R}^d$ be C^∞. If the Jacobian matrix*

$$\left\{\frac{\partial r_i \circ f}{\partial r_j}\right\}_{i,j=1,\ldots,d}$$

is non-singular at $r_0 \in U$, then there exists an open set V with $r_0 \in V \subset U$ such that $f|V$ maps V one-to-one onto the open set $f(V)$, and $(f|V)^{-1}$ is C^∞.

This is one of the results we shall assume from advanced calculus. For a proof, we refer the reader, for example, to [31] or [6].

Corollary (a) *Assume that $\psi: M \to N$ is C^∞, that $m \in M$, and that $d\psi: M_m \to N_{\psi(m)}$ is an isomorphism. Then there is a neighborhood U of m such that $\psi: U \to \psi(U)$ is a diffeomorphism onto the open set $\psi(U)$ in N.*

PROOF Observe that $\dim M = \dim N$, say d. Choose coordinate systems (V,φ) about m and (W,τ) about $\psi(m)$ with $\psi(V) \subset W$. Let $\varphi(m) = p$ and $\tau(\psi(m)) = q$. The differential of the map $\tau \circ \psi \circ \varphi^{-1} \mid \varphi(V)$ is non-singular at p. Thus the inverse function theorem yields a diffeomorphism $\alpha: \tilde{U} \to \alpha(\tilde{U})$ on a neighborhood $\tilde{U}$ of p with $\tilde{U} \subset \varphi(V)$. Then $\tau^{-1} \circ \alpha \circ \varphi$ is the required diffeomorphism on the neighborhood $U = \varphi^{-1}(\tilde{U})$ of m.

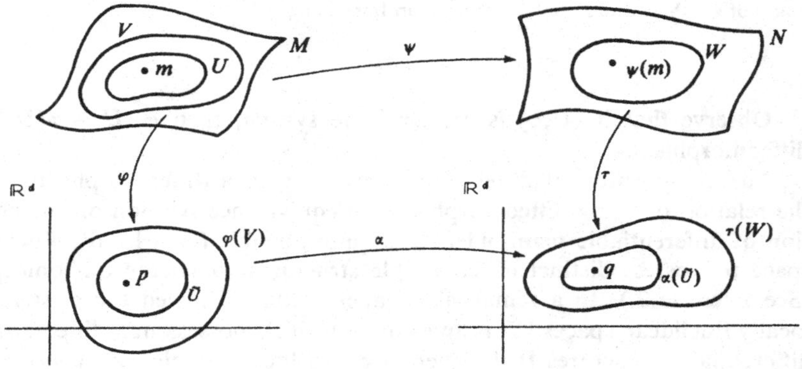

Corollary (b) *Suppose that* $\dim M = d$ *and that* $y_1, \ldots, y_d$ *is an independent set of functions at $m_0 \in M$. Then the functions $y_1, \ldots, y_d$ form a coordinate system on a neighborhood of m_0.*

PROOF Suppose that the y_i are defined on the open set U containing m_0. Define $\psi: U \to \mathbb{R}^d$ by

$$\psi(m) = (y_1(m), \ldots, y_d(m)) \qquad (m \in U).$$

Then ψ is C^∞. Now $\delta\psi$ is an isomorphism on $(\mathbb{R}^d_{\psi(m_0)})^*$ since

$$\delta\psi(dr_i) = d(r_i \circ \psi) = dy_i$$

which implies that $\delta\psi|_{\psi(m_0)}$ takes a basis to a basis. Consequently, the differential $d\psi_{m_0}$ (which is the dual of $\delta\psi|_{\psi(m_0)}$) is an isomorphism. So the inverse function theorem implies that ψ is a diffeomorphism on a neighborhood $V \subset U$ of m_0, and consequently the functions $y_1, \ldots, y_d$ yield a coordinate system when restricted to V.

Corollary (c) *Suppose that* $\dim M = d$ *and that $y_1, \ldots, y_l$, with $l < d$, is an independent set of functions at m. Then they form part of a coordinate system on a neighborhood of m.*

PROOF Let $(U, x_1, \ldots, x_d)$ be a coordinate system about m. Then $\{dy_1, \ldots, dy_l, dx_1, \ldots, dx_d\}$ spans M_m^*. Choose $d - l$ of the x_i so that $\{dy_1, \ldots, dy_l, dx_{i_1}, \ldots, dx_{i_{d-l}}\}$ is a basis of M_m^*. Then apply Corollary (b).

Corollary (d) *Let $\psi: M \to N$ be C^∞, and assume that $d\psi: M_m \to N_{\psi(m)}$ is surjective. Let $x_1, \ldots, x_l$ form a coordinate system on some neighborhood of $\psi(m)$. Then $x_1 \circ \psi, \ldots, x_l \circ \psi$ form part of a coordinate system on some neighborhood of m.*

PROOF The fact that $d\psi_m$ is surjective implies that the dual map $\delta\psi|_{\psi(m)}$ is injective. Thus the functions $\{x_i \circ \psi: i = 1, \ldots, l\}$ are independent at m since $\delta\psi(dx_i) = d(x_i \circ \psi)$. The claim now follows from Corollary (c).

Corollary (e) *Suppose that $y_1, \ldots, y_k$ is a set of C^∞ functions on a neighborhood of m such that their differentials span M_m^*. Then a subset of the y_i forms a coordinate system on a neighborhood of m.*

PROOF Simply choose a subset whose differentials form a basis of M_m^*, and apply Corollary (b).

Corollary (f) *Let $\psi: M \to N$ be C^∞, and assume that $d\psi: M_m \to N_{\psi(m)}$ is injective. Let $x_1, \ldots, x_k$ form a coordinate system on a neighborhood of $\psi(m)$. Then a subset of the functions $\{x_i \circ \psi\}$ forms a coordinate system on a neighborhood of m. In particular, ψ is one-to-one on a neighborhood of m.*

PROOF The fact that $d\psi_m$ is injective implies that $\delta\psi|_{\psi(m)}$ is surjective. This implies that $\{d(x_i \circ \psi) = \delta\psi(dx_i): i = 1, \ldots, k\}$ spans M_m^*. This corollary then follows from Corollary (e).

1.31 The situation often arises that one has a C^∞ mapping, say ψ, of a manifold N into a manifold M factoring through a submanifold (P, φ) of M. That is, $\psi(N) \subset \varphi(P)$, whence there is a uniquely defined mapping ψ_0 of N into P such that $\varphi \circ \psi_0 = \psi$. The problem is: When is ψ_0 of class C^∞? This is certainly not always the case. As an example, let N and P both be the real line, and let M be the plane. Let $(\mathbb{R}, \psi)$ and $(\mathbb{R}, \varphi)$ both be figure-8 submanifolds with precisely the same image sets, but with the difference that as $t \to \pm\infty$, $\psi(t)$ approaches the intersection along the horizontal direction, but $\varphi(t)$ approaches along the vertical. Suppose also that $\psi(0) = \varphi(0) = 0$. Then ψ_0 is not even continuous since $\psi_0^{-1}(-1,1)$ consists of the origin plus two open sets of the form $(\alpha, +\infty)$, $(-\infty, -\alpha)$ for some $\alpha > 0$.

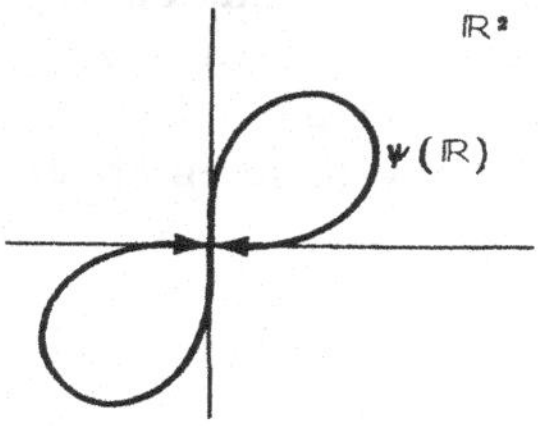

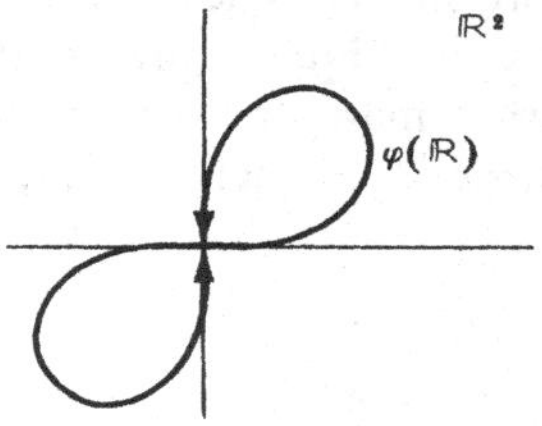

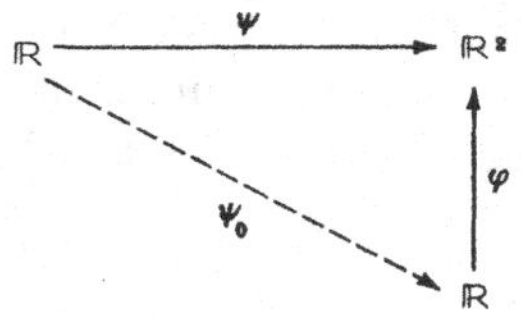

1.32 Theorem *Suppose that $\psi: N \to M$ is C^∞, that (P,φ) is a submanifold of M, and that ψ factors through (P,φ), that is, $\psi(N) \subset \varphi(P)$. Since φ is injective, there is a unique mapping ψ_0 of N into P such that $\varphi \circ \psi_0 = \psi$.*

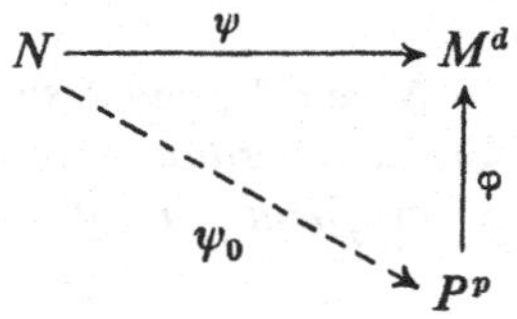

(a) *ψ_0 is C^∞ if it is continuous.*

(b) *ψ_0 is continuous if φ is an imbedding.*

Another important case in which ψ_0 is continuous occurs when (P,φ) is an integral manifold of an involutive distribution on M, as we shall see in 1.62.

PROOF Result (b) is obvious. So assume that ψ_0 is continuous. We prove that it is C^∞. It suffices to show that P can be covered by coordinate systems (U,τ) such that the map $\tau \circ \psi_0$ restricted to the *open* set $\psi_0^{-1}(U)$ is C^∞. Let $p \in P$, and let (V,γ) be a coordinate system on a neighborhood of $\varphi(p)$ in M^d. Then by Corollary (f) of 1.30 there exists a projection π of $\mathbb{R}^d$ onto a suitable subspace (obtained by setting certain of the coordinate functions equal to 0) such that the map $\tau = \pi \circ \gamma \circ \varphi$ yields a coordinate system on a neighborhood U of p. Then

$$\begin{aligned} \tau \circ \psi_0 \mid \psi_0^{-1}(U) &= \pi \circ \gamma \circ \varphi \circ \psi_0 \mid \psi_0^{-1}(U) \\ &= \pi \circ \gamma \circ \psi \mid \psi_0^{-1}(U), \end{aligned}$$

which is C^∞.

1.33 Further Remarks on Submanifolds Submanifolds (N_1,φ_1) and (N_2,φ_2) of M will be called *equivalent* if there exists a diffeomorphism $\alpha: N_1 \to N_2$ such that $\varphi_1 = \varphi_2 \circ \alpha$.

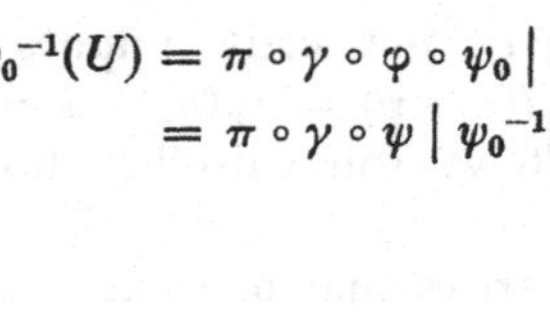

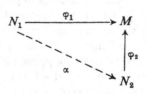

This is an equivalence relation on the collection of all submanifolds of M. Each equivalence class ξ has a unique representative of the form (A,i) where A is a subset of M with a manifold structure such that the inclusion map $i: A \to M$ is a C^∞ immersion. Namely, if (N,φ) is any representative of ξ, then the subset A of M must be $\varphi(N)$. We induce a manifold structure on A by requiring $\varphi\colon N \to A$ to be a diffeomorphism. With this manifold structure, (A,i) is a submanifold of M equivalent to (N,φ). This is the only manifold structure on A with the property that (A,i) is equivalent to (N,φ); thus this is the unique such representative of ξ.

The conclusion of some theorems in the following sections state that there exist unique submanifolds satisfying certain conditions. Uniqueness means up to equivalence as defined above. In particular, if the submanifolds of M are viewed as subsets $A \subset M$ with manifold structures for which the inclusion maps are C^∞ immersions, then uniqueness means unique subset with unique second countable locally Euclidean topology and unique differentiable structure.

In the case of a submanifold (A,i) of M where i is the inclusion map, we shall often drop the i and simply speak of the submanifold $A \subset M$.

Let A be a subset of M. Then generally there is not a unique manifold structure on A such that (A,i) is a submanifold of M, if there is one at all. For example, the diagrams in 1.31 illustrate two distinct manifold structures on the figure-8 in the plane, each of which makes the figure-8 together with the inclusion map a submanifold of $\mathbb{R}^2$. However, we have the following two uniqueness theorems which involve conditions on the topology on A.

(a) *Let M be a differentiable manifold and A a subset of M. Fix a topology on A. Then there is at most one differentiable structure on A such that (A,i) is a submanifold of M, where i is the inclusion map.*

(b) *Again let A be a subset of M. If in the relative topology, A has a differentiable structure such that (A,i) is a submanifold of M, then A has a unique manifold structure (that is, unique second countable locally Euclidean topology together with a unique differentiable structure) such that (A,i) is a submanifold of M.*

We leave these to the reader as exercises. Result (a) follows from an application of Theorem 1.32. Result (b) depends strongly on our assumption that manifolds are second countable, and for its proof you will need to use the proposition in Exercise 6 in addition to Theorem 1.32.

1.34 Slices Suppose that (U,φ) is a coordinate system on M with coordinate functions $x_1, \ldots, x_d$, and that c is an integer, $0 \le c \le d$. Let $a \in \varphi(U)$, and let

(1) $$S = \{q \in U\colon x_i(q) = r_i(a), i = c+1, \ldots, d\}.$$

The subspace S of M together with the coordinate system

(2) $$\{x_j \mid S\colon j = 1, \ldots, c\}$$

forms a manifold which is a submanifold of M called a *slice* of the coordinate system (U,φ).

1.35 Proposition *Let $\psi: M^c \to N^d$ be an immersion, and let $m \in M$. Then there exists a cubic-centered coordinate system (V,φ) about $\psi(m)$ and a neighborhood U of m such that $\psi \mid U$ is 1:1 and $\psi(U)$ is a slice of (V,φ).*

PROOF Let (W,τ) be a centered coordinate system about $\psi(m)$ with coordinate functions $y_1, \ldots, y_d$. By Corollary (f) of 1.30 we can renumber the coordinate functions so that

(1) $$\tilde{\tau} = \pi_c \circ \tau \circ \psi$$

is a coordinate map on a neighborhood V' of m where $\pi_c: \mathbb{R}^d \to \mathbb{R}^c$ is projection on the first c coordinates. Define functions $\{x_i\}$ on $(\pi_c \circ \tau)^{-1}(\tilde{\tau}(V'))$ by setting

(2) $$x_i = \begin{cases} y_i & (i = 1, \ldots, c) \\ y_i - y_i \circ \psi \circ \tilde{\tau}^{-1} \circ \pi_c \circ \tau & (i = c+1, \ldots, d). \end{cases}$$

The functions $\{x_i\}$ are independent at $\psi(m)$, since at $\psi(m)$,

(3) $$dx_i = \begin{cases} dy_i & (i = 1, \ldots, c) \\ dy_i + \sum_{j=1}^{c} a_{ij}\, dy_j & (i = c+1, \ldots, d) \end{cases}$$

for some constants a_{ij}. By Corollary (b) of 1.30 the $\{x_i\}$ form a coordinate system on a neighborhood of $\psi(m)$. Let V be a neighborhood of $\psi(m)$ on which the $x_1, \ldots, x_d$ form a cubic coordinate system. Denote the corresponding coordinate map by φ. Let $U = \psi^{-1}(V) \cap V'$. Then U and (V,φ) are the required neighborhood and coordinate system.

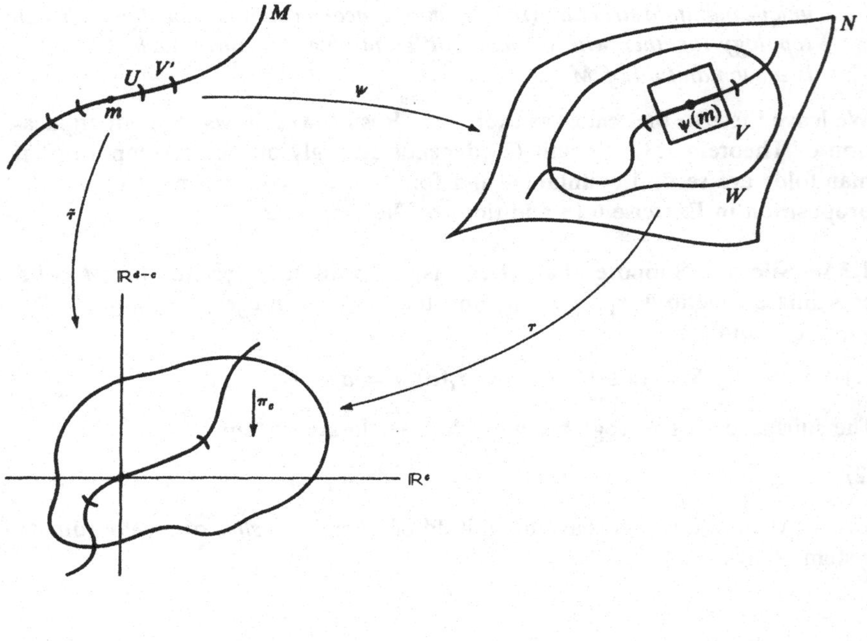

We emphasize that this proposition only says that there is a *neighborhood* U of m such that $\psi(U)$ is a slice of the coordinate system (V,φ). Even if (M,ψ) is a submanifold of N, it may well be that $\psi(M) \cap V$ is far from being a slice or even a union of slices. For an example, consider again the figure-8 submanifold of the plane:

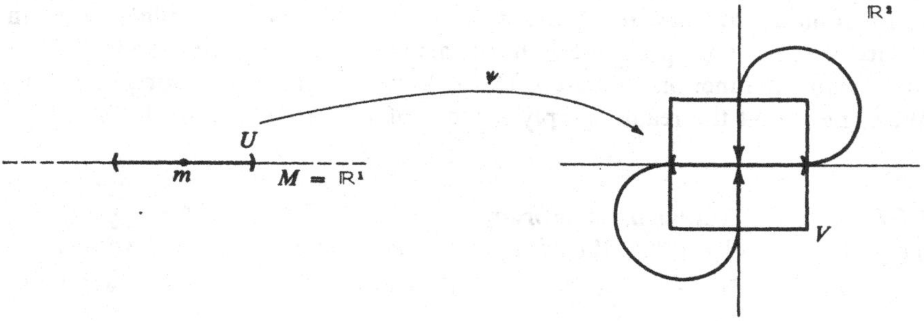

However, in the case that (M,ψ) is an imbedded submanifold, the coordinate system (V,φ) can be chosen so that all of $\psi(M) \cap V$ is a single slice of V.

Let us now consider the question of the extent to which the set of C^∞ functions on a manifold determines the set of C^∞ functions on a submanifold. Let (M,ψ) be a submanifold of N. Then, of course, if $f \in C^\infty(N)$, then $f|M$ is a C^∞ function on M. (More precisely, $f \circ \psi$ is a C^∞ function on M.) In general, however, the converse does not hold; that is, not all C^∞ functions on M arise as the restrictions to M of C^∞ functions on N. For the converse to hold, it is necessary and sufficient to assume that ψ is an imbedding and that $\psi(M)$ is closed. We prove the sufficiency in the following proposition, and leave the necessity as Exercise 11 below.

1.36 Proposition *Let $\psi: M \to N$ be an imbedding such that $\psi(M)$ is closed in N. If $g \in C^\infty(M)$, then there exists $f \in C^\infty(N)$ such that $f \circ \psi = g$.*

To simplify notation, we shall suppress the map ψ and consider $M \subset N$.

PROOF For each point $p \in M$ there exists an open set O_p in N containing p and an extension of g from $O_p \cap M$ to a C^∞ function $\tilde{g}_p$ on O_p. One simply has to take O_p to be a cubic-centered coordinate neighborhood of p for which $M \cap O_p$ is a single slice, and then define $\tilde{g}_p$ to be the composition of the natural projection of O_p onto the slice followed by g. The collection $\{O_p : p \in M\}$ together with $N - M$ forms an open cover of N. By Theorem 1.11, there exists a partition of unity $\{\varphi_j\}$, with $j = 1, 2, \ldots$, subordinate to this cover. Take the subsequence (which we shall continue to denote by $\{\varphi_j\}$) such that $\operatorname{supp} \varphi_j \cap M \neq \varnothing$. For each such j, we can choose a point p_j such that $\operatorname{supp} \varphi_j \subset O_{p_j}$. Then $f = \sum_j \varphi_j \tilde{g}_{p_j}$ is a C^∞ function on N, and $f|M = g$.

IMPLICIT FUNCTION THEOREMS

From the inverse function theorem we shall obtain two theorems which will provide us with an extremely useful way of proving that certain subsets of manifolds are submanifolds. Under suitable conditions on a differentiable map, the inverse image of a submanifold of its range will be a submanifold of its domain. We first recall the statement of the classical implicit function theorem. This is simply a local (but somewhat more explicit) version of the first "implicit function" theorem (1.38) that we shall prove for manifolds. We suggest that the reader supply a proof of 1.37 after reading 1.38.

1.37 Implicit Function Theorem *Let $U \subset \mathbb{R}^{c-d} \times \mathbb{R}^d$ be open, and let $f\colon U \to \mathbb{R}^d$ be C^∞. We denote the canonical coordinate system on $\mathbb{R}^{c-d} \times \mathbb{R}^d$ by $(r_1, \ldots, r_{c-d}, s_1, \ldots, s_d)$. Suppose that at the point $(r_0, s_0) \in U$*

$$f(r_0, s_0) = 0,$$

and that the matrix

$$\left\{\frac{\partial f_i}{\partial s_j}\bigg|_{(r_0,s_0)}\right\}_{i,j=1,\ldots,d}$$

is non-singular. Then there exists an open neighborhood V of r_0 in $\mathbb{R}^{c-d}$ and an open neighborhood W of s_0 in $\mathbb{R}^d$ such that $V \times W \subset U$, and there exists a C^∞ map $g\colon V \to W$ such that for each $(p,q) \in V \times W$

(1) $$f(p,q) = 0 \iff q = g(p).$$

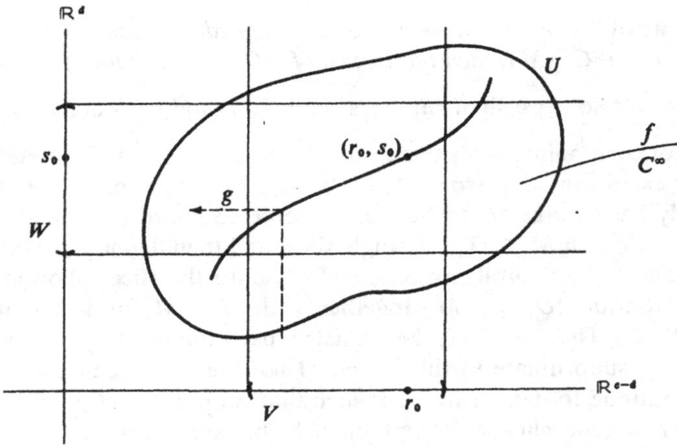

1.38 Theorem *Assume that $\psi: M^c \to N^d$ is C^∞, that n is a point of N, that $P = \psi^{-1}(n)$ is non-empty, and that $d\psi: M_m \to N_{\psi(m)}$ is surjective for all $m \in P$. Then P has a unique manifold structure such that (P,i) is a submanifold of M, where i is the inclusion map. Moreover, $i: P \to M$ is actually an imbedding, and the dimension of P is $c - d$.*

PROOF According to result (b) of 1.33, it is sufficient to prove that in the relative topology, P has a differentiable structure such that (P,i) is a submanifold of M of dimension $c - d$. For this it is sufficient to prove that if $m \in P$, then there exists a coordinate system on a neighborhood U of m in M for which $P \cap U$ is a single slice of the correct dimension. Let $x_1, \ldots, x_d$ be a coordinate system centered at n in N. Then since $d\psi: M_m \to N_n$ is surjective, it follows from Corollary (d) of 1.30 that the collection of functions

$$\{y_i = x_i \circ \psi: i = 1, \ldots, d\},$$

forms part of a coordinate system about $m \in M$. Complete to a coordinate system $y_1, \ldots, y_d, y_{d+1}, \ldots, y_c$ on a neighborhood U of m. Then $P \cap U$ is precisely the slice of this coordinate system given by

$$y_1 = y_2 = \cdots = y_d = 0.$$

In this theorem the inverse image of a point is shown to be a submanifold as long as the differential is surjective at each point of the inverse image. A point can be thought of as a 0-dimensional submanifold. We now generalize this theorem by proving that under suitable conditions the inverse images of higher dimensional submanifolds are themselves submanifolds.

1.39 Theorem *Assume that $\psi: M \to N^d$ is C^∞ and that (O^c,φ) is a submanifold of N. Suppose that whenever $m \in \psi^{-1}(\varphi(O))$, then*

$$N_{\psi(m)} = d\psi(M_m) + d\varphi(O_{\varphi^{-1}(\psi(m))}) \tag{1}$$

(not necessarily a direct sum). Then if $P = \psi^{-1}(\varphi(O))$ and is non-empty, P can be given a manifold structure so that (P,i) is a submanifold of M, where i is the inclusion map, with

$$\dim M - \dim P = \dim N - \dim O. \tag{2}$$

Moreover, if (O,φ) is an imbedded submanifold, then so is (P,i), and in this case there is a unique manifold structure on P such that (P,i) is a submanifold of M.

In general, if (O,φ) is not an imbedding, P need not have the relative topology, and there is no unique manifold structure on P such that (P,i) is a submanifold. We leave it to the reader to supply examples.

PROOF The proof will consist of locally reducing this case to the case of Theorem 1.38. Let $p \in O$. By Proposition 1.35, we can pick a neighborhood W of p and a centered coordinate system (V,τ) with coordinate functions $x_1, \ldots, x_d$ about $\varphi(p)$ such that $\varphi(W)$ is the slice

(3) $$x_{c+j} = 0 \qquad (j = 1, \ldots, d - c).$$

Let

(4) $$\pi\colon \mathbb{R}^d \to \mathbb{R}^{d-c}, \quad \pi(a) = (a_{c+1}, \ldots, a_d).$$

So

(5) $$\pi \circ \tau(\varphi(W)) = \{0\}.$$

Let

(6) $$U = \psi^{-1}(V),$$

and let

(7) $$\psi_1 = \pi \circ \tau \circ \psi \mid U\colon U \to \mathbb{R}^{d-c}.$$

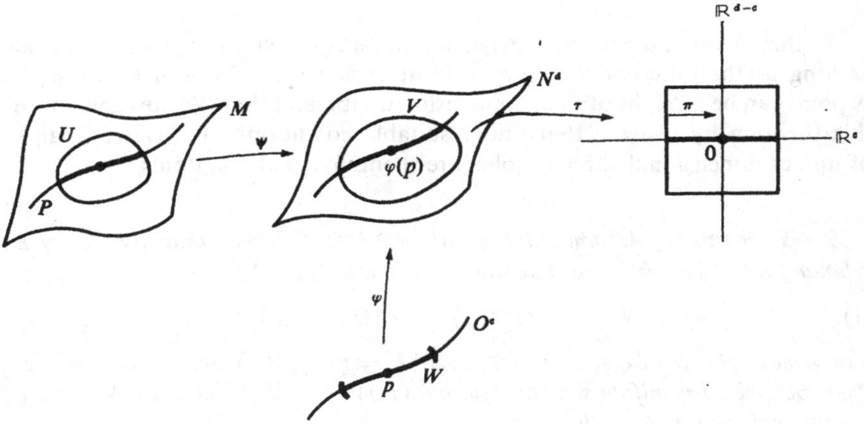

Now, (3) through (7) imply that

(8) $$\psi_1^{-1}(0) = \psi^{-1}(\varphi(W)).$$

For each point $m \in \psi_1^{-1}(0)$, $d\psi_1 \mid M_m$ is surjective since $d(\pi \circ \tau) \mid N_{\psi(m)}$ is surjective and since by (1), (5), and (7)

$$d(\pi \circ \tau)(N_{\psi(m)}) = d\psi_1(M_m) + \{0\}.$$

Thus by 1.38, $\psi^{-1}(\varphi(W))$ has a unique manifold structure such that $(\psi^{-1}(\varphi(W)),i)$ is a submanifold of M; and in this manifold structure,

$\psi^{-1}(\varphi(W))$ has the relative topology and has dimension equal to $\dim M - \dim N + \dim O$. Cover O by a countable collection $\{W_i\colon i = 1, 2, 3, \ldots\}$ of such sets. Then

$$P = \bigcup_{i=1}^{\infty} \psi^{-1}(\varphi(W_i)). \tag{9}$$

If $i \neq j$, the submanifolds $\psi^{-1}(\varphi(W_i))$ and $\psi^{-1}(\varphi(W_j))$ intersect in open subsets of each other. It follows that the union of the topologies on the various $\psi^{-1}(\varphi(W_i))$ forms a basis for a topology on P. With this topology, P is a locally Euclidean space with dimension equal to $\dim M - \dim N + \dim O$; and, moreover, P is second countable since (9) expresses P as a countable union of second countable open subsets. The manifold structures on the various $\psi^{-1}(\varphi(W_i))$ are compatible on overlaps because of the uniqueness of the manifold structure on $\psi^{-1}(\varphi(W_i))$ (or any open subset of $\psi^{-1}(\varphi(W_i))$) such that $(\psi^{-1}(\varphi(W_i)),i)$ is a submanifold of M (by result (b) of 1.33). Thus the collection of coordinate systems on P containing the coordinate systems on the various $\psi^{-1}(\varphi(W_i))$ and maximal with respect to 1.4(b) forms a differentiable structure on P. That (P,i) is a submanifold of M now follows immediately from the fact that the $(\psi^{-1}(\varphi(W_i)),i)$ are submanifolds. If (O,φ) is an imbedding, the coordinate neighborhood V can be chosen small enough so that $\varphi(O) \cap V$ consists only of the single slice $\varphi(W)$, and thus $U \cap P = \psi^{-1}(\varphi(W))$. It follows in this case that P has the relative topology; hence (P,i) is an imbedding. The uniqueness of the manifold structure in this case is guaranteed by result (b) of 1.33.

1.40 Examples

(a) The differential of the function $f(p) = \sum_{i=1}^{d} r_i(p)^2$ on $\mathbb{R}^d$ is surjective except at the origin. Thus it follows from 1.38 that the sphere $f^{-1}(r^2)$, for a constant $r > 0$, has a unique manifold structure for which it is a submanifold of $\mathbb{R}^d$ under the inclusion map. In particular, this is the same manifold structure as the one defined in Example 1.5(d).

(b) We define a map ψ from the general linear group $Gl(d,\mathbb{R})$ (Example 1.5(f)) to the vector space of all real symmetric $d \times d$ matrices by

$$\psi(A) = AA^t,$$

where A^t is the transpose of the matrix A. Let

$$O(d) = \psi^{-1}(I)$$

where I is the $d \times d$ identity matrix. $O(d)$ is a subgroup of $Gl(d,\mathbb{R})$ under matrix multiplication called the *orthogonal group*. To apply 1.38 to conclude that $O(d)$ has a unique manifold structure such that $(O(d),i)$ is a submanifold of $Gl(d,\mathbb{R})$, and that in this manifold structure i is an imbedding and $O(d)$ has dimension $\frac{1}{2}(d(d-1))$, one need only check that $d\psi_\sigma$ is surjective at each $\sigma \in O(d)$. For this, it suffices to check that $d\psi_I$ is surjective, since whenever $\sigma \in O(d)$,

$$\psi = \psi \circ r_\sigma$$

where r_σ (*right translation* by σ) is the diffeomorphism of $Gl(d,\mathbb{R})$ defined by $r_\sigma(\tau) = \tau\sigma$. We leave the details to the reader as an exercise.

VECTOR FIELDS

1.41 Definitions Smooth curves $\sigma\colon (a,b) \to M$ and their tangent vectors $\dot\sigma(t)$ were defined in 1.23(e). We say that a mapping $\sigma\colon [a,b] \to M$ is a *smooth curve in M* if σ extends to be a C^∞ mapping of $(a - \varepsilon, b + \varepsilon)$ into M for some $\varepsilon > 0$. The curve $\sigma\colon [a,b] \to M$ is said to be *piecewise smooth* if there exists a partition $a = \alpha_0 < \alpha_1 < \cdots < \alpha_n = b$ such that $\sigma \mid [\alpha_i, \alpha_{i+1}]$ is smooth for each $i = 0, \ldots, n-1$. Observe that piecewise smooth curves are necessarily continuous. If $\sigma\colon [a,b] \to M$ is a smooth curve in M, then its tangent vector

$$\dot\sigma(t) = d\sigma\left(\frac{d}{dr}\bigg|_t\right) \in M_{\sigma(t)}$$

is well-defined for each $t \in [a,b]$.

1.42 Definitions A *vector field X along a curve* $\sigma\colon [a,b] \to M$ is a mapping $X\colon [a,b] \to T(M)$ which *lifts* σ; that is, $\pi \circ X = \sigma$. A vector field

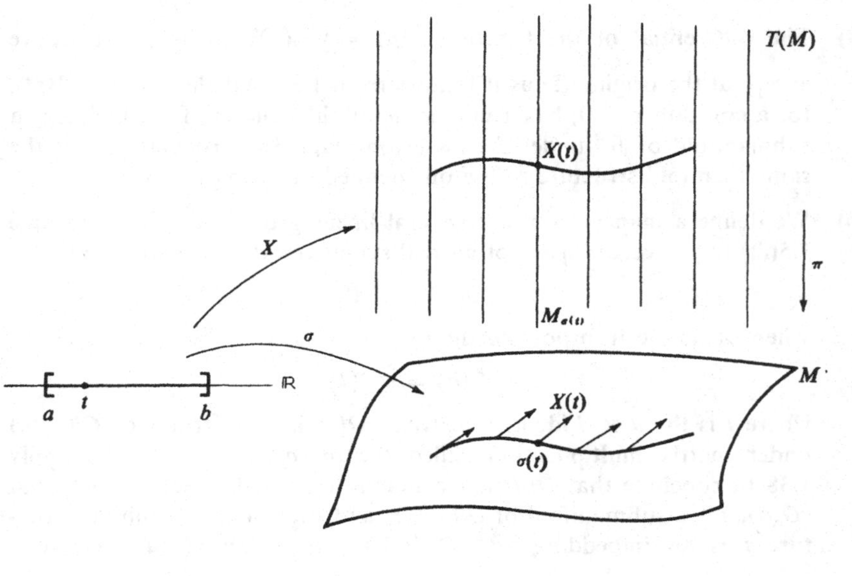

X is called a smooth (C^∞) vector field along σ if the mapping $X\colon [a,b] \to T(M)$ is C^∞. A *vector field X on an open set U in M* is a *lifting* of U into $T(M)$, that is, a map $X\colon U \to T(M)$ such that

$$\pi \circ X = \text{identity map on } U.$$

Again, for the vector field X to be smooth (C^∞) means that $X \in C^\infty(U,T(M))$. The set of smooth vector fields over U forms in the obvious way a vector space over $\mathbb{R}$ and a module over the ring $C^\infty(U)$ of C^∞ functions on U. If X is a vector field on U and $m \in U$, then $X(m)$ (often denoted X_m) is an element of M_m. If f is a C^∞ function on U, then $X(f)$ is the function on U whose value at m is $X_m(f)$.

1.43 Proposition *Let X be a vector field on M. Then the following are equivalent:*

(a) *X is C^∞.*

(b) *If $(U, x_1, \ldots, x_d)$ is a coordinate system on M, and if $\{a_i\}$ is the collection of functions on U defined by*

$$X \mid U = \sum_{i=1}^{d} a_i \frac{\partial}{\partial x_i},$$

then $a_i \in C^\infty(U)$.

(c) *Whenever V is open in M and $f \in C^\infty(V)$, then $X(f) \in C^\infty(V)$.*

PROOF (a) $\Rightarrow$ (b) The fact that X is smooth implies that $X \mid U$ is smooth; and since the composition of differentiable maps is again differentiable, it follows that $dx_i \circ X \mid U$ is smooth. (Recall that the dx_i are coordinate functions on $\pi^{-1}(U) \subset T(M)$, 1.25(3).) But

$$dx_i \circ X \mid U = a_i.$$

Hence the a_i are C^∞ functions on U.

(b) $\Rightarrow$ (c) It suffices to prove that $X(f) \mid U \in C^\infty(U)$ where $(U, x_1, \ldots, x_d)$ is an arbitrary coordinate system on M for which $U \subset V$. But, by (b),

$$X(f) \mid U = \sum_{i=1}^{d} a_i \frac{\partial f}{\partial x_i},$$

and the right-hand side is a C^∞ function on U.

(c) $\Rightarrow$ (a) To prove that X is C^∞, it suffices to prove that $X \mid U$ is C^∞ where $(U, x_1, \ldots, x_d)$ is an arbitrary coordinate system on M. To prove that $X \mid U$ is C^∞, we need only check that $X \mid U$ composed with the canonical coordinate functions 1.25(3) on $\pi^{-1}(U)$ are C^∞ functions. Now, $x_i \circ \pi \circ X \mid U = x_i$ and $dx_i \circ X \mid U = X(x_i)$, all of which are C^∞ functions on U.

1.44 Lie Bracket If X and Y are smooth vector fields on M, we define a vector field $[X,Y]$ called the *Lie bracket* of X and Y by setting

(1) $$[X,Y]_m(f) = X_m(Yf) - Y_m(Xf).$$

1.45 Proposition

(a) *$[X,Y]$ is indeed a smooth vector field on M.*

(b) *If $f, g \in C^\infty(M)$, then $[fX,gY] = fg[X,Y] + f(Xg)Y - g(Yf)X$.*

(c) $[X,Y] = -[Y,X]$.

(d) $[[X,Y],Z] + [[Y,Z],X] + [[Z,X],Y] = 0$ *for all smooth vector fields X, Y, and Z on M.*

We leave the proof as an exercise. Part (d) is known as the *Jacobi identity.* A vector space with a bilinear operation satisfying (c) and (d) is called a *Lie algebra.*

1.46 Definition Let X be a smooth vector field on M. A smooth curve σ in M is an *integral curve of X* if

(1) $$\dot\sigma(t) = X(\sigma(t))$$

for each t in the domain of σ.

1.47 Let X be a C^∞ vector field on M, and let $m \in M$. Let us now consider the question: Does there exist an integral curve of X through m, and if so, is there a unique one?

A curve $\gamma\colon (a,b) \to M$ is an integral curve of X if and only if

(1) $$d\gamma\left(\frac{d}{dr}\bigg|_t\right) = X(\gamma(t)) \qquad (t \in (a,b)).$$

Let us interpret this in coordinates. Suppose that $0 \in (a,b)$ and $\gamma(0) = m$. Choose a coordinate system (U,φ) with coordinate functions $x_1, \ldots, x_d$ about m. By 1.43(b),

(2) $$X \mid U = \sum_{i=1}^{d} f_i \frac{\partial}{\partial x_i}$$

where the f_i are C^∞ functions on U. Moreover, for each t such that $\gamma(t) \in U$,

(3) $$d\gamma\left(\frac{d}{dr}\bigg|_t\right) = \sum_{i=1}^{d} \frac{d(x_i \circ \gamma)}{dr}\bigg|_t \frac{\partial}{\partial x_i}\bigg|_{\gamma(t)}.$$

Thus, in view of (2) and (3), equation (1) becomes

(4) $$\sum_{i=1}^{d} \frac{d(x_i \circ \gamma)}{dr}\bigg|_t \frac{\partial}{\partial x_i}\bigg|_{\gamma(t)} = \sum_{i=1}^{d} f_i(\gamma(t)) \frac{\partial}{\partial x_i}\bigg|_{\gamma(t)}.$$

Thus γ is an integral curve of X on $\gamma^{-1}(U)$ if and only if

$$(5)\quad \left.\frac{d\gamma_i}{dr}\right|_t = f_i \circ \varphi^{-1}(\gamma_1(t), \ldots, \gamma_d(t)) \qquad (i = 1, \ldots, d \quad \text{and} \quad t \in \gamma^{-1}(U)),$$

where $\gamma_i = x_i \circ \gamma$. Equation (5) is a system of first order ordinary differential equations for which there exist fundamental existence and uniqueness theorems [11]. These theorems, when translated into manifold terminology, give the following.

1.48 Theorem *Let X be a C^∞ vector field on a differentiable manifold M. For each $m \in M$ there exists $a(m)$ and $b(m)$ in $\mathbb{R} \cup \{\pm\infty\}$, and a smooth curve*

$$(1)\qquad \gamma_m\colon (a(m),b(m)) \to M$$

such that

(a) $0 \in (a(m),b(m))$ *and* $\gamma_m(0) = m$.

(b) γ_m *is an integral curve of* X.

(c) *If* $\mu\colon (c,d) \to M$ *is a smooth curve satisfying conditions* (a) *and* (b), *then* $(c,d) \subset (a(m),b(m))$ *and* $\mu = \gamma_m \mid (c,d)$.

We continue with the statement of the theorem after the following.

Definition For each $t \in \mathbb{R}$, we define a transformation X_t with domain

$$(2)\qquad \mathscr{D}_t = \{m \in M\colon t \in (a(m),b(m))\}$$

by setting

$$(3)\qquad X_t(m) = \gamma_m(t).$$

(d) *For each $m \in M$, there exists an open neighborhood V of m and an $\varepsilon > 0$ such that the map*

$$(4)\qquad (t,p) \mapsto X_t(p)$$

is defined and is C^∞ from $(-\varepsilon,\varepsilon) \times V$ into M.

(e) $\mathscr{D}_t$ *is open for each* t.

(f) $\bigcup_{t>0} \mathscr{D}_t = M$.

(g) $X_t\colon \mathscr{D}_t \to \mathscr{D}_{-t}$ *is a diffeomorphism with inverse* X_{-t}.

(h) *Let s and t be real numbers. Then the domain of $X_s \circ X_t$ is contained in but generally not equal to $\mathscr{D}_{s+t}$. However, the domain of $X_s \circ X_t$ is $\mathscr{D}_{s+t}$ in the case in which s and t both have the same sign. Moreover, on the domain of $X_s \circ X_t$ we have*

$$(5)\qquad X_s \circ X_t = X_{s+t}.$$

PROOF We let $(a(m),b(m))$ be the union of all the open intervals which contain the origin and which are domains of integral curves of X satisfying the initial condition that the origin maps to m. That $(a(m),b(m)) \neq \varnothing$ (and hence part (f) holds) follows from an application of the fundamental existence theorem [11, THEOREM 4, p. 28] to the system 1.47(5). Now if α and β are integral curves of X with domains the open intervals A and B (with $A \cap B \neq \varnothing$), and if α and β have the same initial conditions $\alpha(t_0) = \beta(t_0)$ at some point $t_0 \in A \cap B$, then the subset of $A \cap B$ on which α and β agree is nonempty, open by the basic uniqueness theorem [11, THEOREM 3, p. 28], and closed by continuity; and hence this subset is equal to $A \cap B$ by the connectedness of $A \cap B$. It follows that there exists a curve γ_m defined on $(a(m),b(m))$ and satisfying parts (a), (b), and (c).

The existence of an $\varepsilon > 0$ and a neighborhood V of m such that the map (4) is defined on $(-\varepsilon,\varepsilon) \times V$ is the content of THEOREM 7 on p. 29 of [11]. That the map (4) is smooth (and hence part (d) holds) follows from THEOREM 9 on p. 29 of [11] on the differentiability of the solutions of 1.47(5) with respect to their initial values.

Next we prove part (h). Let $t \in (a(m),b(m))$. Then $s \mapsto \gamma_m(t + s)$ is an integral curve of X with the initial condition $0 \mapsto \gamma_m(t)$ and with *maximal* domain $(a(m) - t, b(m) - t)$. It follows from part (c) that

$$(6) \qquad (a(m) - t, b(m) - t) = (a(\gamma_m(t)),b(\gamma_m(t))),$$

and for s in the interval (6),

$$(7) \qquad \gamma_{\gamma_m(t)}(s) = \gamma_m(t + s).$$

Now let m belong to the domain of $X_s \circ X_t$. Then $t \in (a(m),b(m))$ and $s \in (a(\gamma_m(t)),b(\gamma_m(t)))$, so by (6), $s + t \in (a(m),b(m))$. Thus $m \in \mathscr{D}_{s+t}$, and (5) follows from (7). It is easy to construct examples to show that the domain of $X_s \circ X_t$ is generally not equal to $\mathscr{D}_{s+t}$. (Consider, for example, the vector field $\partial/\partial r_1$ on $\mathbb{R}^2 - \{0\}$ with $s = -1$ and $t = 1$.) If, however, s and t both have the same sign, and if $m \in \mathscr{D}_{s+t}$, that is, if $s + t \in (a(m),b(m))$, then it follows that $t \in (a(m),b(m))$ and, by (6), $s \in (a(\gamma_m(t)),b(\gamma_m(t)))$; hence m is in the domain of $X_s \circ X_t$.

Parts (e) and (g) are trivial if $t = 0$, so assume that $t > 0$ and that $m \in \mathscr{D}_t$. (A similar argument will prove (e) and (g) if $t < 0$.) It follows from part (d) and the compactness of $[0,t]$ that there exists a neighborhood W of $\gamma_m([0,t])$ and an $\varepsilon > 0$ such that the map (4) is defined and C^∞ on $(-\varepsilon,\varepsilon) \times W$. Choose a positive integer n large enough so that $t/n \in (-\varepsilon,\varepsilon)$. Let $\alpha_1 = X_{t/n} \mid W$, and let $W_1 = \alpha_1^{-1}(W)$. Then for $i = 2, \ldots, n$ we inductively define

$$\alpha_i = X_{t/n} \mid W_{i-1}$$

and

$$W_i = \alpha_i^{-1}(W_{i-1}).$$

α_i is a C^∞ map on the open set $W_{i-1} \subset W$. It follows that W_n is an open subset of W, that W_n contains m (since if $X_{t/n}$ composed with itself n times is applied to m, we obtain $\gamma_m(t)$, which lies in W), and that by part (h),

$$\alpha_1 \circ \alpha_2 \circ \cdots \circ \alpha_n \mid W_n = X_t \mid W_n. \tag{8}$$

Consequently, $W_n \subset \mathscr{D}_t$; hence $\mathscr{D}_t$ is open, which proves part (e).

Finally, X_t is a 1:1 map of $\mathscr{D}_t$ onto $\mathscr{D}_{-t}$ with inverse X_{-t}. That X_t is C^∞ (similarly for X_{-t}) follows from (8), which locally expresses X_t as a composition of C^∞ maps. Hence X_t is a diffeomorphism from $\mathscr{D}_t$ to $\mathscr{D}_{-t}$, which proves part (g) and finishes Theorem 1.48.

1.49 Definitions A smooth vector field X on M is *complete* if $\mathscr{D}_t = M$ for all t (that is, the domain of γ_m is $(-\infty,\infty)$ for each $m \in M$). In this case, the transformations X_t form a group of transformations of M parametrized by the real numbers called the *1-parameter group of* X. If X is not complete, the transformations X_t do not form a group since their domains depend on t. In this case, we shall refer to the collection of transformations X_t as the *local* 1-parameter group of X.

1.50 Remarks A simple example of a non-complete vector field is obtained by considering the vector field $\partial/\partial r_1$ on the plane with the origin removed. If $a > 0$, the domain of the maximal integral curve through $(a,0)$ is $(-a,+\infty)$. In the case in which the manifold M is compact, any C^∞ vector field on M is complete. We leave the proof as an exercise.

1.51 Definition Let $\psi\colon M \to N$ be C^∞. A smooth vector field X along ψ (that is, $X \in C^\infty(M,T(N))$ and $\pi \circ X = \psi$) has *local C^∞ extensions in* N if given $m \in M$ there exist a neighborhood U of m and a neighborhood V of $\psi(m)$ such that $\psi(U) \subset V$, and there also exists a C^∞ vector field $\tilde{X}$ on V such that

$$\tilde{X} \circ \psi \mid U = X \mid U. \tag{1}$$

1.52 Remark It is easy to prove that a C^∞ vector field X along an immersion $\psi\colon M \to N$ always has local C^∞ extensions in N. However, if ψ is not an immersion, such extensions generally do not exist. Consider the following example. We shall first define a smooth vector field X along a smooth curve $\alpha\colon \mathbb{R} \to \mathbb{R}$ in the real line. Let

$$\alpha(t) = t^3, \tag{1}$$

(that is, $\alpha = r^3$ where r is the canonical coordinate function on $\mathbb{R}$), and let

$$X(t) = \dot{\alpha}(t) = d\alpha\left(\frac{d}{dr}\bigg|_t\right). \tag{2}$$

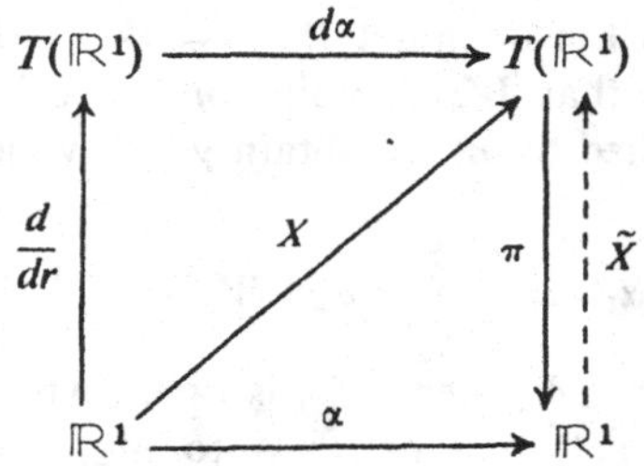

Since α is a homeomorphism, there is induced a vector field $\tilde{X}$ on $\mathbb{R}^1$ so that the above diagram commutes. Now, X is a smooth vector field along α, but $\tilde{X}$ *is not a smooth vector field on* $\mathbb{R}^1$. To show this, let $u = t^3$. Then

$$\tilde{X}_u = \tilde{X}_{\alpha(t)} = X(t) = d\alpha\left(\frac{d}{dr}\bigg|_t\right) \tag{3}$$

$$= \frac{d\alpha}{dr}\bigg|_t \frac{d}{dr}\bigg|_{\alpha(t)} = 3t^2 \frac{d}{dr}\bigg|_{\alpha(t)} = 3u^{2/3} \frac{d}{dr}\bigg|_u .$$

Thus

$$\tilde{X} = 3r^{2/3} \frac{d}{dr}, \tag{4}$$

and the function $r^{2/3}$ is not differentiable at the origin. If we extend α to be a mapping into the plane by setting

$$\alpha(t) = (t^3,0), \tag{5}$$

and again we let $X(t) = \dot{\alpha}(t)$, then X is a smooth vector field along the one-to-one C^∞ curve α in the plane, which admits no local C^∞ extension to a neighborhood of (0,0).

1.53 Proposition *Let $m \in M^d$, and let X be a smooth vector field on M such that $X(m) \neq 0$. Then there exists a coordinate system (U,φ) with coordinate functions $x_1, \ldots, x_d$ on a neighborhood of m such that*

$$X \mid U = \frac{\partial}{\partial x_1}\bigg| U. \tag{1}$$

PROOF Choose a coordinate system (V,τ) centered at m with coordinate functions $y_1, \ldots, y_d$, such that

$$X_m = \frac{\partial}{\partial y_1}\bigg|_m . \tag{2}$$

It follows from 1.48(d) that there exists an $\varepsilon > 0$ and a neighborhood W of the origin in $\mathbb{R}^{d-1}$ such that the map

$$\sigma(t, a_2, \ldots, a_d) = X_t(\tau^{-1}(0, a_2, \ldots, a_d))$$

is well-defined and smooth for $(t, a_2, \ldots, a_d) \in (-\varepsilon,\varepsilon) \times W \subset \mathbb{R}^d$.

Now, σ is non-singular at the origin since

$$d\sigma\left(\frac{\partial}{\partial r_1}\bigg|_0\right) = X_m = \frac{\partial}{\partial y_1}\bigg|_m \quad \text{and} \quad d\sigma\left(\frac{\partial}{\partial r_i}\bigg|_0\right) = \frac{\partial}{\partial y_i}\bigg|_m \qquad (i \geq 2).$$

Thus by Corollary (a) of 1.30, $\varphi = \sigma^{-1}$ is a coordinate map on some neighborhood U of m. Let $x_1, \ldots, x_d$ denote the coordinate functions of the coordinate system (U,φ). Then since

$$d\sigma\left(\frac{\partial}{\partial r_1}\bigg|_{(t,a_2,\ldots,a_n)}\right) = X_{\sigma(t,a_2,\ldots,a_n)},$$

we have

$$X \mid U = \frac{\partial}{\partial x_1}\bigg| U.$$

1.54 Definition Let $\varphi: M \to N$ be C^∞. Smooth vector fields X on M and Y on N are called *φ-related* if $d\varphi \circ X = Y \circ \varphi$.

1.55 Proposition *Let $\varphi: M \to N$ be C^∞. Let X and X_1 be smooth vector fields on M, and let Y and Y_1 be smooth vector fields on N. If X is φ-related to Y, and if X_1 is φ-related to Y_1, then $[X,X_1]$ is φ-related to $[Y,Y_1]$.*

PROOF We must show that $d\varphi \circ [X,X_1] = [Y,Y_1] \circ \varphi$. For this, let $m \in M$ and $f \in C^\infty(N)$. Then we must show that

$$(1) \qquad d\varphi([X,X_1]_m)(f) = [Y,Y_1]_{\varphi(m)}(f).$$

We simply unwind the definitions:

$$\begin{aligned}(2) \quad d\varphi([X,X_1]_m)(f) &= [X,X_1]_m(f \circ \varphi) \\ &= X_m(X_1(f \circ \varphi)) - X_1|_m(X(f \circ \varphi)) \\ &= X_m((d\varphi \circ X_1)(f)) - X_1|_m((d\varphi \circ X)(f)) \\ &= X_m(Y_1(f) \circ \varphi) - X_1|_m(Y(f) \circ \varphi) \\ &= d\varphi(X_m)(Y_1(f)) - d\varphi(X_1|_m)(Y(f)) \\ &= Y_{\varphi(m)}(Y_1(f)) - Y_1|_{\varphi(m)}(Y(f)) \\ &= [Y,Y_1]_{\varphi(m)}(f).\end{aligned}$$

DISTRIBUTIONS AND THE FROBENIUS THEOREM

1.56 Definitions Let c be an integer, $1 \leq c \leq d$. A *c-dimensional distribution* $\mathscr{D}$ on a d-dimensional manifold M is a choice of a c-dimensional subspace $\mathscr{D}(m)$ of M_m for each m in M. $\mathscr{D}$ is *smooth* if for each m in M there is a neighborhood U of m and there are c vector fields $X_1, \ldots, X_c$ of class C^∞ on U which span $\mathscr{D}$ at each point of U. A vector field X on M is said to *belong to* (or *lie in*) the distribution $\mathscr{D}$ ($X \in \mathscr{D}$) if $X_m \in \mathscr{D}(m)$ for each

$m \in M$. A smooth distribution $\mathscr{D}$ is called *involutive* (or *completely integrable*) if $[X,Y] \in \mathscr{D}$ whenever X and Y are smooth vector fields lying in $\mathscr{D}$.

1.57 Definition A submanifold (N,ψ) of M is an *integral manifold* of a distribution $\mathscr{D}$ on M if

(1) $$d\psi(N_n) = \mathscr{D}(\psi(n)) \quad \text{for each } n \in N.$$

1.58 Remarks Our object in this section is to prove that a necessary and sufficient condition for there to exist integral manifolds of $\mathscr{D}$ through each point of M is that $\mathscr{D}$ be involutive. Perhaps a word of explanation is in order about the expression "*completely integrable*" sometimes used in place of "involutive." We have required integral manifolds to be submanifolds whose tangent spaces coincide with the subspaces determined by the distribution. One could define a weaker notion of integral manifold by requiring only that the tangent spaces of the submanifold be contained in but not necessarily equal to the distribution at each point. It is possible for a distribution $\mathscr{D}$ to be "integrable" in the sense that it has low-dimensional "integral manifolds," but not completely integrable in the sense that $\mathscr{D}$ does not have integral manifolds of the maximal dimension. For us, unless specified otherwise, integral manifolds of distributions will always be taken to mean integral manifolds of maximal dimension, that is, as defined in 1.57.

1.59 Proposition *Let $\mathscr{D}$ be a smooth distribution on M such that through each point of M there passes an integral manifold of $\mathscr{D}$. Then $\mathscr{D}$ is involutive.*

PROOF Let X and Y be smooth vector fields lying in $\mathscr{D}$, and let $m \in M$. We must prove that $[X,Y]_m \in \mathscr{D}(m)$. Let (N,ψ) be an integral manifold of $\mathscr{D}$ through m, and suppose that $\psi(n_0) = m$. Since $d\psi \colon N_n \to \mathscr{D}(\psi(n))$ is an isomorphism at each n in N, there exist vector fields $\tilde{X}$, $\tilde{Y}$ on N such that

(1) $$\begin{aligned} d\psi \circ \tilde{X} &= X \circ \psi, \\ d\psi \circ \tilde{Y} &= Y \circ \psi. \end{aligned}$$

It is easily checked that $\tilde{X}$ and $\tilde{Y}$ are smooth. By 1.55, $[\tilde{X},\tilde{Y}]$ and $[X,Y]$ are ψ-related. Hence $[X,Y]_m = d\psi([\tilde{X},\tilde{Y}]_{n_0}) \in \mathscr{D}(m)$.

1.60 Theorem (Frobenius) *Let $\mathscr{D}$ be a c-dimensional, involutive, C^∞ distribution on M^d. Let $m \in M$. Then there exists an integral manifold of $\mathscr{D}$ passing through m. Indeed, there exists a cubic coordinate system (U,φ) which is centered at m, with coordinate functions $x_1, \ldots, x_d$ such that the slices*

(1) $$x_i = \text{constant} \qquad \text{for all } i \in \{c+1, \ldots, d\}$$

are integral manifolds of $\mathscr{D}$; and if (N,ψ) is a connected integral manifold of $\mathscr{D}$ such that $\psi(N) \subset U$, then $\psi(N)$ lies in one of these slices.

PROOF We shall prove the existence part of the theorem by induction on c. For the case $c = 1$, choose a vector field X lying in $\mathscr{D}$, defined on an open neighborhood of m, such that $X(m) \neq 0$. Then Proposition 1.53 yields a coordinate system (U,φ) about m, which can be taken to be cubic centered, for which $X \mid U = \partial/\partial x_1$. Hence the theorem holds for $c = 1$.

Now assume that the theorem holds for $c - 1$; we prove it for a distribution $\mathscr{D}$ of dimension c. Since $\mathscr{D}$ is smooth, there exist smooth vector fields $X_1, \ldots, X_c$ spanning $\mathscr{D}$ on a neighborhood $\tilde{V}$ of m. By 1.53, there exists a coordinate system $(V, y_1, \ldots, y_d)$ centered at m, with $V \subset \tilde{V}$, such that

(2) $$X_1 \mid V = \frac{\partial}{\partial y_1}.$$

On V, let

(3) $$\begin{aligned} Y_1 &= X_1, \\ Y_i &= X_i - X_i(y_1)X_1 \qquad (i = 2, \ldots, c). \end{aligned}$$

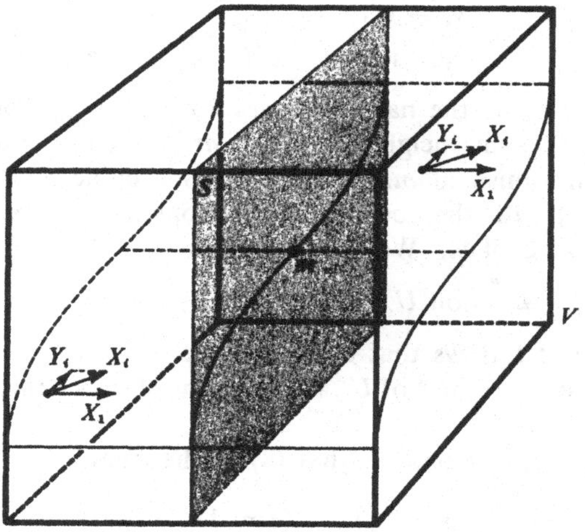

Then the vector fields $Y_1, \ldots, Y_c$ are independent C^∞ vector fields spanning $\mathscr{D}$ in V. Let S be the slice $y_1 = 0$, and let

(4) $$Z_i = Y_i \mid S \qquad (i = 2, \ldots, c).$$

Then since (2) and (3) imply that

(5) $$Y_i(y_1) = 0 \qquad (i = 2, \ldots, c),$$

the Z_i are actually vector fields on S; that is, $Z_i(q) \in S_q$ whenever $q \in S$. The Z_i span a smooth $c - 1$ dimensional distribution on S.

We claim this is involutive. Indeed, the Z_i are i-related (inclusion map of S in M) to the Y_i, and therefore, by 1.55, their Lie brackets are also i-related to the corresponding brackets of the Y_i. But $[Y_i, Y_j]$, $(i, j \geq 2)$, has no component in the Y_1 direction (apply to y_1 and get 0). Therefore, there exist C^∞ functions c_{ijk} such that

$$[Y_i, Y_j] = \sum_{k=2}^{c} c_{ijk} Y_k \tag{6}$$

on V, and thus

$$[Z_i, Z_j] = \sum_{k=2}^{c} c_{ijk}\Big|_S Z_k. \tag{7}$$

This proves that the distribution on S is involutive. By the induction hypothesis, there exists a centered coordinate system $w_2, \ldots, w_d$ on some neighborhood of m in S such that the slices defined by $w_i =$ constant for all $i \in \{c+1, \ldots, d\}$ are precisely the integral manifolds of the distribution spanned by $Z_2, \ldots, Z_c$ on this neighborhood.

The functions

$$\begin{aligned} x_1 &= y_1, \\ x_j &= w_j \circ \pi \qquad (j = 2, \ldots, d), \end{aligned} \tag{8}$$

where $\pi\colon V \to S$ is the natural projection in the y coordinate system, are defined on some neighborhood of m in M, are independent at m, and they all vanish at m. Thus there is a cubic-centered coordinate system (U, φ) with the coordinate functions $x_1, \ldots, x_d$ on a suitable neighborhood U of m. We now prove that

$$Y_i(x_{c+r}) \equiv 0 \quad \text{on } U \qquad (i = 1, \ldots, c;\ r = 1, \ldots, d-c). \tag{9}$$

From this it follows that the vector fields $\partial/\partial x_1, \ldots, \partial/\partial x_c$ form a basis for $\mathscr{D}$ at each point of U, and thus the slices (1) are integral manifolds of $\mathscr{D}$.

To prove (9), first observe that (8) implies that

$$\frac{\partial x_j}{\partial y_1} = \begin{cases} 1 & (j = 1) \\ 0 & (j = 2, \ldots, d) \end{cases} \tag{10}$$

on U; and thus (2), (3), and (10) imply that

$$Y_1 = \frac{\partial}{\partial x_1} \quad \text{on } U, \tag{11}$$

so certainly (9) holds for $i = 1$. Now let $i \in \{2, \ldots, c\}$ and $r \in \{1, \ldots, d-c\}$. By (11),

$$\frac{\partial}{\partial x_1}(Y_i(x_{c+r})) = Y_1(Y_i(x_{c+r})) = [Y_1, Y_i](x_{c+r}). \tag{12}$$

The involutivity of $\mathscr{D}$ implies that there are C^∞ functions c_{ik} such that

$$(13) \qquad [Y_1, Y_i] = \sum_{k=1}^{c} c_{ik} Y_k .$$

Using (13), (12) becomes

$$(14) \qquad \frac{\partial}{\partial x_1}(Y_i(x_{c+r})) = \sum_{k=2}^{c} c_{ik} Y_k(x_{c+r}) \quad (i = 2, \ldots, c;\ r = 1, \ldots, d - c).$$

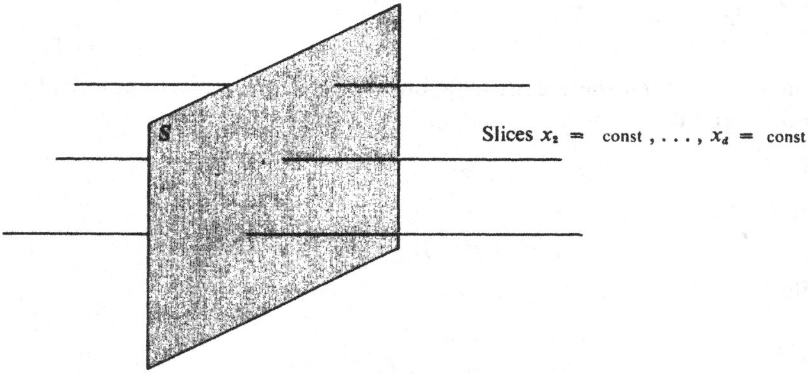

Fix a slice of U of the form $x_2 =$ constant, . . . , $x_d =$ constant. On such a slice, $Y_i(x_{c+r})$ is a function of x_1 alone, and (14) becomes a system of $c - 1$ homogeneous linear differential equations with respect to x_1. Such a system has a unique solution with given initial values [11]. Since the system is homogeneous, the 0 functions give a solution. But each such slice has a unique point in $S \cap U$, and on $S \cap U$,

$$(15) \qquad Y_i(x_{c+r}) = Z_i(w_{c+r}) = 0 \qquad (i = 2, \ldots, c).$$

The first equality follows from (4) and (8), and the second from the fact that the integral manifolds of the distribution on S determined by the Z_i are given by suitable slices in the w coordinate system. It follows from (14) and (15) that the functions $Y_i(x_{c+r})$ must be identically zero on U. Thus (9) holds, and the induction step is completed.

Finally, suppose that (N,ψ) is a connected integral manifold of $\mathscr{D}$ such that $\psi(N) \subset U$. Let π be the projection of $\mathbb{R}^d$ onto the last $d - c$ coordinates. Then vectors in $\mathscr{D}$ are annihilated by $d(\pi \circ \varphi)$. Thus

$$d(\pi \circ \varphi \circ \psi)\big|_n \equiv 0$$

for each $n \in N$. By 1.24, $\pi \circ \varphi \circ \psi$ is a constant map since N is connected. Thus $\psi(N)$ is contained in one of the slices (1).

1.61 Remarks The classical version of the Frobenius theorem appears quite different from our version in 1.60. The classical Frobenius theorem can be formulated as follows.

Let U and V be open sets in $\mathbb{R}^m$ and $\mathbb{R}^n$ respectively. We use coordinates $r_1, \ldots, r_m$ on $\mathbb{R}^m$ and $s_1, \ldots, s_n$ on $\mathbb{R}^n$. Let

$$b\colon U \times V \to M(n,m) \tag{1}$$

be a C^∞ map of $U \times V$ into the set of all $n \times m$ real matrices, and let $(r_0,s_0) \in U \times V$. If

$$\frac{\partial b_{i\beta}}{\partial r_\gamma} - \frac{\partial b_{i\gamma}}{\partial r_\beta} + \sum_{j=1}^{n}\left(\frac{\partial b_{i\beta}}{\partial s_j} b_{j\gamma} - \frac{\partial b_{i\gamma}}{\partial s_j} b_{j\beta}\right) = 0 \qquad (i = 1, \ldots, n;\ \gamma, \beta = 1, \ldots, m) \tag{2}$$

on $U \times V$, then there exist neighborhoods U_0 of r_0 in U and V_0 of s_0 in V and a unique C^∞ map

$$\alpha\colon U_0 \times V_0 \to V \tag{3}$$

such that if

$$\alpha_s(r) = \alpha(r,s) \qquad (s \in V_0,\ \ r \in U_0),$$

then

$$\begin{aligned} \alpha_s(r_0) &= s, \\ d\alpha_s\big|_r &= b(r,\alpha(r,s)) \end{aligned} \tag{4}$$

for all $(r,s) \in U_0 \times V_0$.

Equation (4) is a so-called *total differential equation*. We specify in (1) what the differential of a map should be as a function of the graph, and in (2) we have a necessary and sufficient condition for the existence of such a map with the specified initial conditions. It can be shown that this version is equivalent to 1.60. For example, if we start with a c-dimensional, involutive, C^∞ distribution $\mathscr{D}$ on M^d and a point $m \in M$, then we can obtain 1.60 from the classical version as follows. We can first choose a coordinate system (W,τ) about m with coordinate functions $y_1, \ldots, y_d$ and with $\tau(W) = U \times V \subset \mathbb{R}^c \times \mathbb{R}^{d-c}$, for which there exist C^∞ functions f_{ji} on $U \times V$ $(i = 1, \ldots, c;\ j = 1, \ldots, d - c)$ such that the vector fields

$$Y_i = \frac{\partial}{\partial y_i} + \sum_{j=1}^{d-c} f_{ji} \circ \tau \frac{\partial}{\partial y_{c+j}} \qquad (i = 1, \ldots, c) \tag{5}$$

span $\mathscr{D}$ on W. Then we define a map b as in (1) by setting

$$b(r,s) = \{f_{ji}(r,s)\}. \tag{6}$$

It turns out that the involutivity of $\mathscr{D}$ implies that (2) is satisfied, and from the map α one can obtain the desired coordinate system 1.60(1). Conversely, one can obtain the classical version from 1.60 in a similar way.

We shall give in Chapter 2 yet another version of the Frobenius theorem in terms of differential forms and differential ideals.

In 1.32 we considered the situation in which a C^∞ map $\psi: N \to M$ factors through a submanifold (P,φ) of M so that $\psi = \varphi \circ \psi_0$ where $\psi_0: N \to P$, and we sought sufficient conditions for ψ_0 to be C^∞. An important case occurs when (P,φ) is an integral manifold of an involutive distribution on M.

1.62 Theorem *Suppose that $\psi: N \to M^d$ is C^∞, that (P^c,φ) is an integral manifold of an involutive distribution $\mathscr{D}$ on M, and that ψ factors through (P,φ), that is, $\psi(N) \subset \varphi(P)$. Let $\psi_0: N \to P$ be the (unique) mapping such that $\varphi \circ \psi_0 = \psi$.*

(1)

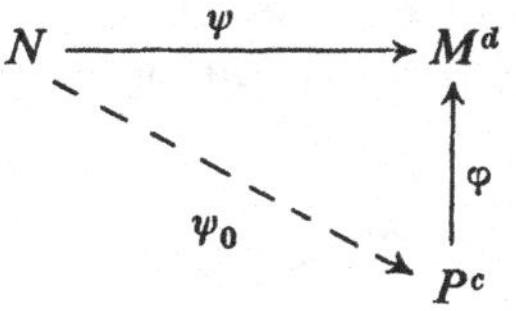

Then ψ_0 is continuous (and hence C^∞ by 1.32(a)).

PROOF Let p belong to an open set U in P, and let $n \in \psi_0^{-1}(p)$. Using 1.60, we can obtain an open set $\tilde{U}$ with $p \in \tilde{U} \subset U$ and a cubic coordinate system (V,τ) centered at $\varphi(p)$ with coordinate functions $x_1, \ldots, x_d$ such that the slices

(2) $$x_i = \text{constant} \qquad \text{for all } i \in \{c+1, \ldots, d\}$$

are the integral manifolds of $\mathscr{D}$ in V, and such that $\varphi(\tilde{U})$ is the slice

(3) $$x_{c+1} = \cdots = x_d = 0.$$

$\psi^{-1}(V)$ is open in N. Let W be the component of $\psi^{-1}(V)$ containing n. W is open. To prove that ψ_0 is continuous, we need only prove that $\psi_0(W) \subset \tilde{U} \subset U$. By the commutativity of (1) and the injectivity of φ, it suffices to prove that $\psi(W)$ lies in the slice (3) of V. Now, ψ is continuous and W connected; hence $\psi(W)$ is connected. Moreover, $\psi(W)$ has at least the point $\psi(n)$ in common with the slice (3). So since $\psi(W)$ lies in a component of $\varphi(P) \cap V$, it is sufficient to prove that components of $\varphi(P) \cap V$ are contained in slices of the form (2).

Let C be a component of $\varphi(P) \cap V$, and let $\pi: V \to \mathbb{R}^{d-c}$ be defined by

(4) $$\pi(m) = (x_{c+1}(m), \ldots, x_d(m)).$$

Then since P is second countable, and since $\varphi(P) \cap V$ is a union of the slices (2) due to the fact that (P,φ) is an integral manifold of $\mathscr{D}$, it follows that $\pi(\varphi(P) \cap V)$ consists of a countable number of points in $\mathbb{R}^{d-c}$. Thus $\pi(C)$ is a connected countable subset of $\mathbb{R}^{d-c}$; hence $\pi(C)$ is a single point, and C lies in a single slice.

1.63 Definition A *maximal integral manifold* (N,ψ) of a distribution $\mathscr{D}$ on a manifold M is a connected integral manifold of $\mathscr{D}$ whose image in M is not a proper subset of any other connected integral manifold of $\mathscr{D}$. That is, there does not exist a connected integral manifold (N_1,ψ_1) of $\mathscr{D}$ such that $\psi(N)$ is a proper subset of $\psi_1(N_1)$.

1.64 Theorem *Let $\mathscr{D}$ be a c-dimensional, involutive, C^∞ distribution on M^d. Let $m \in M$. Then through m there passes a unique maximal connected integral manifold of $\mathscr{D}$, and every connected integral manifold of $\mathscr{D}$ through m is contained in the maximal one.*

PROOF *Existence* Let K be the set of all those points p in M for which there is a piecewise smooth curve joining m to p whose smooth portions

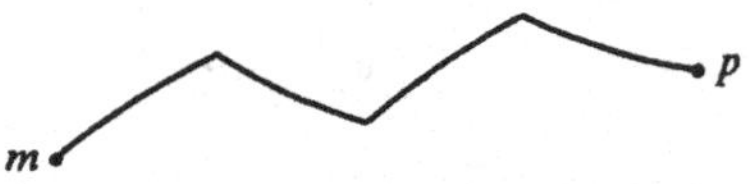

Integral curve of $\mathscr{D}$

are 1-dimensional integral curves of $\mathscr{D}$, that is, their tangent vectors belong to $\mathscr{D}$. By 1.60 and the second countability of M, there is a countable covering of M by cubic coordinate systems $\{(U_i, x_1{}^i, \ldots, x_d{}^i)\colon i = 0, 1, 2, \ldots\}$ such that the integral manifolds of $\mathscr{D}$ in U_i are the slices

(1) $\qquad x^i_{c+j} = \text{constant} \qquad \text{for all } j \in \{1, \ldots, d-c\}.$

We shall assume that $m \in U_0$.

Now let $p \in K$. Then there exists an index i_p such that $p \in U_{i_p}$, and there is a slice S_{i_p} of U_{i_p} of the form (1) containing p. Observe that $S_{i_p} \subset K$. It follows from 1.60 that the collection of all open subsets of all such S_{i_p} as p runs over K forms a basis for a locally Euclidean topology on K, and that we obtain a differentiable structure on K if we take the maximal family of coordinate systems (with respect to 1.4(b)) containing the collection

(2) $\qquad \{(S_{i_p}, x_1{}^{i_p} \mid S_{i_p}, \ldots, x_c{}^{i_p} \mid S_{i_p})\colon p \in K\}.$

We claim that K with this topology and differentiable structure is a connected differentiable manifold of dimension c. K is clearly connected since it is pathwise connected by construction. We have only to prove that the topology on K is second countable. For this, fix an $i \in (0, 1, 2, \ldots)$. We need only show that there are at most countably many slices of U_i in K. Each point of U_i which lies in K is joinable to m by a piecewise smooth curve whose range also lies in K. To each such curve from m to points in U_i there corresponds (although not uniquely) a *finite* sequence

(3) $\qquad U_0, U_{i_1}, \ldots, U_{i_n}, U_i$

of the coordinate neighborhoods through which the curve passes in order. The curve thus begins in the slice of U_0 containing m, passes through some slice of U_{i_1}, then through some slice of U_{i_2}, and so on, until in a finite number of steps it reaches a slice in U_i. Since there are at most countably many such sequences (3) from U_0 to U_i, we need only show that for each such sequence there are at most countably many slices of U_i reachable in the above manner. For this, we need only observe that for any $j, k \in (0, 1, \ldots)$ a single slice of U_j can intersect at most countably many slices of U_k; for if S is a slice of U_j, then $S \cap U_k$ is an open submanifold of S and therefore consists of at most countably many components, each such component being a connected integral manifold of $\mathscr{D}$ in U_k, and hence lying in a slice of U_k. This proves the second countability of K.

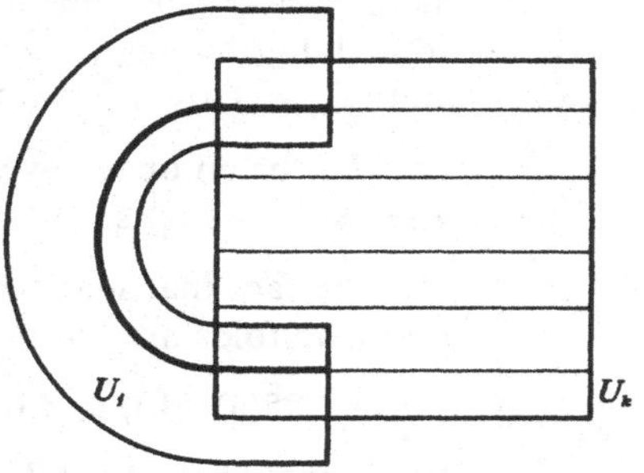

(K,i), where $i\colon K \to M$ is the inclusion map, is now a submanifold of M and is a connected integral manifold of $\mathscr{D}$ passing through m. Moreover, (K,i) is a maximal connected integral manifold of $\mathscr{D}$. For let (N,ψ) be any connected integral manifold of $\mathscr{D}$ passing through m, and let $p \in \psi(N)$. There is a piecewise smooth curve $c\colon [0,1] \to N$ joining $\psi^{-1}(m)$ to $\psi^{-1}(p)$. (Connected manifolds are pathwise connected.) Then $\psi \circ c$ is a piecewise smooth 1-dimensional integral curve of $\mathscr{D}$ connecting m to p. Thus $p \in K$, and so $\psi(N) \subset K$, which proves that K is maximal. Thus we have proved the existence of a maximal connected integral manifold (K,i) of $\mathscr{D}$ passing through m, and have proved that every connected integral manifold of $\mathscr{D}$ through m has its image in K.

Uniqueness (*cf.* 1.33) Let (N,ψ) be any other maximal connected integral manifold of $\mathscr{D}$ passing through m. As we have observed above, $\psi(N) \subset K$; thus ψ factors as follows:

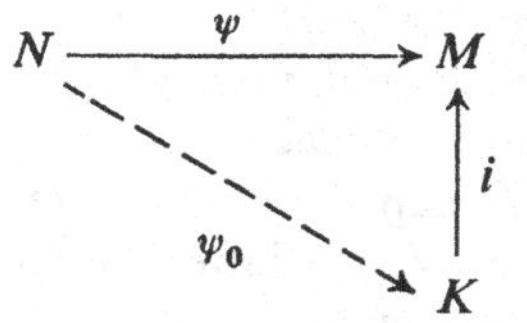

ψ_0 is C^∞ by 1.62, and is 1:1 and non-singular since ψ is 1:1 and non-singular. ψ_0 is onto since the fact that (N,ψ) is a maximal integral manifold of $\mathscr{D}$ through m implies that $\psi(N)$ cannot be a proper subset of K. It follows from Corollary (a) of 1.30 that ψ_0 is a diffeomorphism. Thus (N,ψ) and (K,i) are equivalent, and the maximal connected integral manifold of $\mathscr{D}$ through m is unique.

EXERCISES

1 Prove that in Example 1.5(d) one does indeed obtain a differentiable structure on S^d.

2 The usual differentiable structure on the real line $\mathbb{R}$ was obtained by taking $\mathscr{F}$ to be the maximal collection containing the identity map. Let $\mathscr{F}_1$ be the maximal collection (with respect to 1.4(b)) containing the map $t \mapsto t^3$. Prove that $\mathscr{F} \neq \mathscr{F}_1$, but that $(\mathbb{R},\mathscr{F})$ and $(\mathbb{R},\mathscr{F}_1)$ are diffeomorphic.

3 Let $\{U_\alpha\}$ be an open cover of a manifold M. Prove that there exists a refinement $\{V_\alpha\}$ such that $\overline{V}_\alpha \subset U_\alpha$ for each α.

4 Use the fact that manifolds are regular and paracompact to prove that manifolds are normal topological spaces.

5 Prove 1.25(a), (b), and (c).

6 Prove that if $\psi: M \to N$ is C^∞, one-to-one, onto, and everywhere non-singular, then ψ is a diffeomorphism. (This proposition depends strongly on the second countability of M. Here is an outline for a proof. This map ψ is a diffeomorphism if and only if $d\psi$ is surjective everywhere. If $d\psi$ is not surjective at some point, then the dimension of M is less than the dimension of N. Let dim $M = p$ and dim $N = d$. Assuming that $p < d$, one arrives at a contradiction as follows. Let (U,φ) be a coordinate system on N such that $\varphi(U) = \mathbb{R}^d$. Since ψ maps M onto N, the range of $\varphi \circ \psi$ is all of $\mathbb{R}^d$. Now we show that this yields a contradiction. One way is to use 1.35 to observe that the range of $\varphi \circ \psi$ is a countable union of nowhere dense sets in $\mathbb{R}^d$, and therefore, by the Baire category theorem, could not possibly be all of $\mathbb{R}^d$. Another way is to use the fact that $p < d$ to prove that the range of $\varphi \circ \psi$ has measure 0 in $\mathbb{R}^d$, which also is a contradiction (where a set in $\mathbb{R}^d$ has measure 0 if it can be covered by a sequence of balls, the union of whose volumes is arbitrarily small). That the range of $\varphi \circ \psi$ has measure 0 in $\mathbb{R}^d$ follows from the second countability of M and the fact that a C^1 map $\mathbb{R}^p \to \mathbb{R}^d$ has image of measure 0, and this, in turn, follows from the fact that a C^1 map $\mathbb{R}^d \to \mathbb{R}^d$ takes sets of measure 0 to sets of measure 0. To prove the latter, observe that if $f: \mathbb{R}^d \to \mathbb{R}^d$ is C^1, and if A is a compact set in $\mathbb{R}^d$, then f has a Lipschitz constant K on A,

$$\|f(x) - f(y)\| \leq K \|x - y\| \qquad (x, y \in A),$$

so that f magnifies the volume of balls in A by at most K^d, and hence takes sets of measure 0 in A into sets of measure 0.)

7 Prove 1.33(a) and (b).

8 Obtain the classical implicit function theorem 1.37 from 1.38.

9 Let $f: \mathbb{R}^2 \to \mathbb{R}$ be defined by

$$f(x,y) = x^3 + xy + y^3 + 1.$$

For which points $p = (0,0)$, $p = (\frac{1}{3},\frac{1}{3})$, $p = (-\frac{1}{3},-\frac{1}{3})$ is $f^{-1}(f(p))$ an imbedded submanifold in $\mathbb{R}^2$?

10 Let M be a compact manifold of dimension n, and let $f: M \to \mathbb{R}^n$ be C^∞. Prove that f cannot everywhere be non-singular.

11 Find counterexamples to show that Proposition 1.36 would fail if either of the hypotheses *closed* or *imbedded* were deleted. In fact, one can prove more; namely, if (M,ψ) is a submanifold of N such that whenever $g \in C^\infty(M)$ there is a C^∞ function f on N such that $f \circ \psi = g$, then ψ is an imbedding and $\psi(M)$ is closed in N.

12 Supply the details for 1.40(a) and (b).

13 Prove Proposition 1.45.

14 Is every vector field on the real line complete?

15 Prove that if $(U, x_1, \ldots, x_d)$ is a coordinate system on M, then $[\partial/\partial x_i, \partial/\partial x_j] = 0$ on U.

16 Let $N \subset M$ be a submanifold. Let $\gamma: (a,b) \to M$ be a C^∞ curve such that $\gamma(a,b) \subset N$. Show that it is not necessarily true that $\dot{\gamma}(t) \in N_{\gamma(t)}$ for each $t \in (a,b)$.

17 Prove that any C^∞ vector field on a compact manifold is complete.

18 Prove that a C^∞ map $f: \mathbb{R}^2 \to \mathbb{R}^1$ cannot be one-to-one.

19 Supply the details of the equivalence of the two versions 1.60 and 1.61 of the Frobenius theorem.

20 Let $\varphi: N \to M$ be C^∞, and let X be a C^∞ vector field on N. Suppose that $d\varphi(X(p)) = d\varphi(X(q))$ whenever $\varphi(p) = \varphi(q)$. Is there a smooth vector field Y on M which is φ-related to X?

21 The torus is the manifold $S^1 \times S^1$. Consider S^1 as the unit circle in the complex plane. We define a mapping $\varphi: \mathbb{R} \to S^1 \times S^1$ by setting $\varphi(t) = (e^{2\pi it}, e^{2\pi i\alpha t})$ where α is an irrational number. Prove that $(\mathbb{R}, \varphi)$ is a dense submanifold of $S^1 \times S^1$. This submanifold is known as the *skew line on the torus*.

22 Let $\gamma(t)$ be an integral curve of a vector field X on M. Suppose that $\dot{\gamma}(t) = 0$ for some t. Prove that γ is a constant map, that is, its range consists of one point.

23 A Riemannian structure on a differentiable manifold M is a smooth choice of a positive definite inner product $\langle\ ,\ \rangle_m$ on each tangent space M_m, smooth in the sense that whenever X and Y are C^∞ vector fields on M, then $\langle X, Y\rangle$ is a C^∞ function on M. Prove that there exists a Riemannian structure on every differentiable manifold. You will need to use a partition of unity argument. A *Riemannian manifold* is a differentiable manifold together with a Riemannian structure.

24 Consider the product manifold $M \times N$ with the canonical projections $\pi_1\colon M \times N \to M$ and $\pi_2\colon M \times N \to N$.

(a) Prove that $\alpha\colon \tilde{M} \to M \times N$ is C^∞ if and only if $\pi_1 \circ \alpha$ and $\pi_2 \circ \alpha$ are C^∞.

(b) Prove that the map $v \mapsto (d\pi_1(v), d\pi_2(v))$ is an isomorphism of $(M \times N)_{(m,n)}$ with $M_m \oplus N_n$.

(c) Let X and Y be C^∞ vector fields on M and N respectively. Then, by (b), X and Y canonically determine vector fields $\tilde{X} = (X,0)$ and $\tilde{Y} = (0,Y)$ on $M \times N$. Prove that $[\tilde{X}, \tilde{Y}] = 0$.

(d) Let $(m_0,n_0) \in M \times N$, and define injections $i_{n_0}\colon M \to M \times N$ and $i_{m_0}\colon N \to M \times N$ by setting

$$i_{n_0}(m) = (m,n_0),$$
$$i_{m_0}(n) = (m_0,n).$$

Let $v \in (M \times N)_{(m_0,n_0)}$, and let $v_1 = d\pi_1(v) \in M_{m_0}$ and $v_2 = d\pi_2(v) \in N_{n_0}$. Let $f \in C^\infty(M \times N)$. Prove that

$$v(f) = v_1(f \circ i_{n_0}) + v_2(f \circ i_{m_0}).$$

2

TENSORS and DIFFERENTIAL FORMS

There are a number of vector spaces and algebras naturally associated with the tangent space M_m. Suitably smooth assignments of elements of these spaces to the points in M yield tensor fields and differential forms of various types. We shall first develop some of the pertinent facts from multilinear algebra, and then beginning with 2.14 we shall apply these concepts to manifolds.

TENSOR AND EXTERIOR ALGEBRAS

Throughout 2.1–2.13, V, W, and U will denote finite dimensional real vector spaces. As usual, V^* will denote the dual space of V consisting of all real-valued linear functions on V.

2.1 Definitions Let $F(V,W)$ be the free vector space over $\mathbb{R}$ whose generators are the points of $V \times W$. Thus $F(V,W)$ consists of all finite linear combinations of pairs (v,w) with $v \in V$ and $w \in W$. Let $R(V,W)$ be the subspace of $F(V,W)$ generated by the set of all elements of $F(V,W)$ of the following forms:

$$(1) \qquad \begin{array}{l} (v_1 + v_2, w) - (v_1,w) - (v_2,w) \\ (v, w_1 + w_2) - (v,w_1) - (v,w_2) \\ (av,w) - a(v,w) \\ (v,aw) - a(v,w) \end{array} \qquad \left(\begin{array}{r} a \in \mathbb{R} \\ v, v_1, v_2 \in V \\ w, w_1, w_2 \in W \end{array} \right).$$

The quotient space $F(V,W)/R(V,W)$ is called the *tensor product* of V and W and is denoted by $V \otimes W$. The coset of $V \otimes W$ containing the element (v,w) of $F(V,W)$ is denoted by $v \otimes w$. It follows from (1) that we have the following identities in $V \otimes W$:

$$\begin{aligned} (v_1 + v_2) \otimes w &= v_1 \otimes w + v_2 \otimes w \\ v \otimes (w_1 + w_2) &= v \otimes w_1 + v \otimes w_2 \\ a(v \otimes w) &= av \otimes w = v \otimes aw. \end{aligned}$$

2.2 The following properties of the tensor product are easily established, and are left to the reader as exercises.

(a) *Universal Mapping Property.* Let φ denote the bilinear map $(v,w) \mapsto v \otimes w$ of $V \times W$ into $V \otimes W$. Then whenever U is a vector space and $l: V \times W \to U$ is a bilinear map, there exists a unique linear map $\tilde{l}: V \otimes W \to U$ such that the following diagram commutes:

(1)

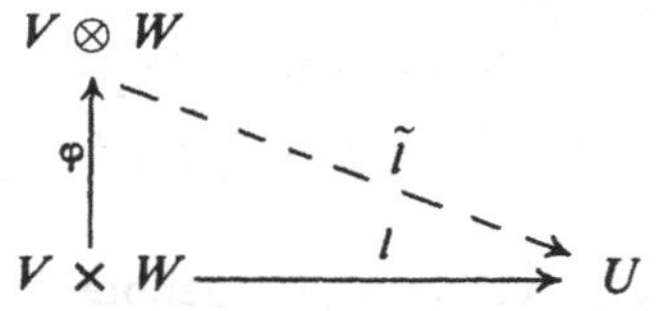

The pair consisting of $V \otimes W$ and φ is said to solve the *universal mapping problem* for bilinear maps with domain $V \times W$. Moreover, $V \otimes W$ and φ are unique with this property in the sense that if X is a vector space and $\tilde{\varphi}: V \times W \to X$ a bilinear map with the above universal mapping property, then there exists an isomorphism $\alpha: V \otimes W \to X$ such that $\alpha \circ \varphi = \tilde{\varphi}$.

(b) $V \otimes W$ is canonically isomorphic with $W \otimes V$.

(c) $V \otimes (W \otimes U)$ is canonically isomorphic with $(V \otimes W) \otimes U$.

(d) By property (a), the bilinear map of $V^* \times W$ into the vector space $\mathrm{Hom}(V,W)$ of linear transformations from V to W defined by $(f,w)(v) = f(v) \cdot w$ for $f \in V^*$, $v \in V$, and $w \in W$ determines uniquely a linear map $\alpha: V^* \otimes W \to \mathrm{Hom}(V,W)$. α is an isomorphism. As a consequence,

$$\dim V \otimes W = (\dim V)(\dim W). \tag{2}$$

(e) Let $\{e_i: i = 1, \ldots, c\}$ and $\{f_j: j = 1, \ldots, d\}$ be bases for V and W respectively. Then $\{e_i \otimes f_j: i = 1, \ldots, c \text{ and } j = 1, \ldots, d\}$ is a basis of $V \otimes W$.

2.3 Definitions The *tensor space $V_{r,s}$ of type* (r,s) associated with V is the vector space

$$\underbrace{V \otimes \cdots \otimes V}_{r \text{ copies}} \otimes \underbrace{V^* \otimes \cdots \otimes V^*}_{s \text{ copies}}. \tag{1}$$

The direct sum

$$T(V) = \sum V_{r,s} \qquad (r, s \geq 0), \tag{2}$$

where $V_{0,0} = \mathbb{R}$, is called the *tensor algebra of V*. Elements of $T(V)$ are finite linear combinations over $\mathbb{R}$ of elements of the various $V_{r,s}$ and are called *tensors*. $T(V)$ is a non-commutative, associative, graded algebra under $\otimes$ multiplication, where if $u = u_1 \otimes \cdots \otimes u_{r_1} \otimes u_1^* \otimes \cdots \otimes u_{s_1}^*$ belongs to V_{r_1,s_1} and $v = v_1 \otimes \cdots \otimes v_{r_2} \otimes v_1^* \otimes \cdots \otimes v_{s_2}^*$ belongs to V_{r_2,s_2}, then their product $u \otimes v$ is defined by

$$u \otimes v = u_1 \otimes \cdots \otimes u_{r_1} \otimes v_1 \otimes \cdots \otimes v_{r_2} \otimes u_1^* \otimes \cdots \otimes u_{s_1}^* \otimes v_1^* \otimes \cdots \otimes v_{s_2}^*$$

and belongs to $V_{r_1+r_2,s_1+s_2}$. Tensors in a particular tensor space $V_{r,s}$ are called *homogeneous of degree* (r,s). A homogeneous tensor (of degree (r,s), say) is called *decomposable* if it can be written in the form

$$v_1 \otimes \cdots \otimes v_r \otimes v_1^* \otimes \cdots \otimes v_s^*$$

where $v_i \in V \quad (i = 1, \ldots, r)$ and $v_j^* \in V^* \quad (j = 1, \ldots, s)$.

2.4 Definitions We let $C(V)$ denote the subalgebra $\sum_{k=0}^{\infty} V_{k,0}$ of $T(V)$. Let $I(V)$ be the two-sided ideal in $C(V)$ generated by the set of elements of the form $v \otimes v$ for $v \in V$, and set

(1) $$I_k(V) = I(V) \cap V_{k,0}.$$

It follows that

(2) $$I(V) = \sum_{k=0}^{\infty} I_k(V),$$

and is a graded ideal in $C(V)$. The *exterior algebra* $\Lambda(V)$ of V is the graded algebra $C(V)/I(V)$. If we set

(3) $\Lambda_k(V) = V_{k,0}/I_k(V) \quad (k \geq 2), \qquad \Lambda_0(V) = \mathbb{R}, \qquad \Lambda_1(V) = V,$

then

(4) $$\Lambda(V) = \sum_{k=0}^{\infty} \Lambda_k(V).$$

We shall denote multiplication in the algebra $\Lambda(V)$ by $\wedge$. This is called the *wedge* or *exterior product*. In particular, the residue class containing $v_1 \otimes \cdots \otimes v_k$ is $v_1 \wedge \cdots \wedge v_k$.

2.5 Definition A multilinear map

(1) $$h\colon \underbrace{V \times \cdots \times V}_{r \text{ copies}} \to W$$

is called *alternating if*

(2) $h(v_{\pi(1)}, \ldots, v_{\pi(r)}) = (\operatorname{sgn} \pi) h(v_1, \ldots, v_r) \qquad (v_1, \ldots, v_r \in V),$

for all permutations π in the permutation group S_r on r letters. Sgn π is the sign of the permutation π ($+1$ if π is even, -1 if π is odd). The vector space of all alternating multilinear functions

$$\underbrace{V \times \cdots \times V}_{r \text{ copies}} \to \mathbb{R}$$

will be denoted by $A_r(V)$, and for convenience we set $A_0(V) = \mathbb{R}$.

2.6 The following properties of the exterior algebra are left to the reader as exercises:

(a) If $u \in \Lambda_k(V)$ and $v \in \Lambda_l(V)$, then $u \wedge v \in \Lambda_{k+l}(V)$ and $u \wedge v = (-1)^{kl} v \wedge u$.

(b) If $e_1, \ldots, e_d$ is a basis of V, then

$$\{e_\Phi\} \tag{1}$$

is a basis of $\Lambda(V)$, where Φ runs over all subsets of $\{1, \ldots, d\}$, including the empty set; where $e_\Phi = e_{i_1} \wedge \cdots \wedge e_{i_r}$ with $i_1 < \cdots < i_r$ when Φ is the subset $\{i_1, \ldots, i_r\}$ of $\{1, \ldots, d\}$; and where $e_\Phi = 1$ when $\Phi = \varnothing$. In particular,

$$\begin{aligned} \Lambda_d(V) &\cong \mathbb{R}, \\ \Lambda_{d+j}(V) &= \{0\} \qquad (j > 0). \end{aligned} \tag{2}$$

Moreover, it follows that

$$\begin{aligned} &\dim \Lambda(V) = 2^d, \\ &\dim \Lambda_k(V) = \binom{d}{k} = \frac{d!}{k!(d-k)!} \qquad (0 \le k \le d). \end{aligned} \tag{3}$$

(*Hint:* Observe that the elements $\{e_\Phi\}$ span $\Lambda(V)$. To prove that they also are linearly independent, first prove that $e_1 \wedge \cdots \wedge e_d$ is not zero in $\Lambda_d(V)$. For this, one must show that $e_1 \otimes \cdots \otimes e_d$ does not belong to $I(V)$. Express an *arbitrary* element of $I(V)$ in terms of the basis vectors $e_1, \ldots, e_d$, and show that it could not equal $e_1 \otimes \cdots \otimes e_d$. Then for linear independence of the entire set $\{e_\Phi\}$, multiply the equation $\sum a_\Phi e_\Phi = 0$ by suitable products of the e_i to land in $\Lambda_d(V)$, and conclude that the various a_Φ are all zero.)

(c) *Universal Mapping Property.* Let φ denote the mapping $(v_1, \ldots, v_k) \mapsto v_1 \wedge \cdots \wedge v_k$ of $V \times \cdots \times V$ (k copies) into $\Lambda_k(V)$. Then φ is an alternating multilinear map. Now to each alternating multilinear map h of $V \times \cdots \times V$ (k copies) into a vector space W, there corresponds uniquely a linear map $\tilde{h}\colon \Lambda_k(V) \to W$ such that $\tilde{h} \circ \varphi = h$.

(4)

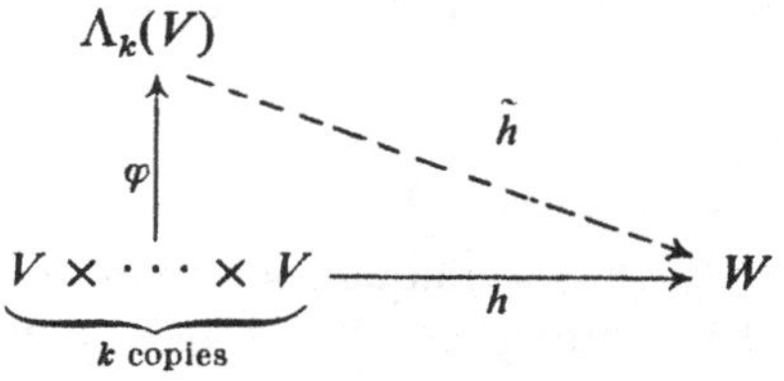

The pair consisting of $\Lambda_k(V)$ and φ is said to solve the *universal mapping problem* for alternating multilinear maps with domain $V \times \cdots \times V$; and this is the unique solution in the sense that if X is a vector space and $\tilde{\varphi}\colon V \times \cdots \times V \to X$ an alternating multilinear map also possessing the universal mapping property for alternating multilinear maps with domain $V \times \cdots \times V$, then there is an isomorphism $\alpha: \Lambda_k(V) \to X$ such that $\alpha \circ \varphi = \tilde{\varphi}$.

In the special case in which $W = \mathbb{R}$, the diagram (4) establishes a natural isomorphism

$$\Lambda_k(V)^* \cong A_k(V) \tag{5}$$

of $\Lambda(V_k)^*$ with the vector space $A_k(V)$ of all alternating multilinear functions on $V \times \cdots \times V$ (k copies). It follows from property (b) that $A_k(V) = \{0\}$ for $k > \dim V$.

We shall now consider various dualities between the spaces $V_{r,s}$, $\Lambda_k(V)$, $\Lambda(V)$ and the corresponding spaces $(V^*)_{r,s}$, $\Lambda_k(V^*)$, $\Lambda(V^*)$ built on the dual space V^* of V.

2.7 Definition Let V and W be real finite dimensional vector spaces. A *pairing* of V and W is a bilinear map $(\ ,\ ): V \times W \to \mathbb{R}$. A pairing is called *non-singular* if whenever $w \neq 0$ in W, there exists an element $v \in V$ such that $(v,w) \neq 0$, and whenever $v \neq 0$ in V, there exists an element $w \in W$ such that $(v,w) \neq 0$.

Let V and W be non-singularly paired by $(\ ,\ )$, and define

$$\varphi: V \to W^* \quad \text{by} \quad \varphi(v)(w) = (v,w) \qquad (v \in V;\ w \in W). \tag{1}$$

It follows that φ is 1:1. Similarly, there is a 1:1 map $W \to V^*$. Therefore V and W have the same dimension, and hence φ is an isomorphism of V with W^*. Thus a non-singular pairing of V and W in a canonical way yields an isomorphism $\varphi: V \to W^*$ and similarly an isomorphism $W \to V^*$.

2.8 Definition *A non-singular pairing of* $(V^*)_{r,s}$ *with* $V_{r,s}$. This pairing is to be the bilinear map $(V^*)_{r,s} \times V_{r,s} \to \mathbb{R}$ which on decomposable elements

$$v^* = v_1^* \otimes \cdots \otimes v_r^* \otimes u_{r+1} \otimes \cdots \otimes u_{r+s} \in (V^*)_{r,s}$$

and

$$u = u_1 \otimes \cdots \otimes u_r \otimes v_{r+1}^* \otimes \cdots \otimes v_{r+s}^* \in V_{r,s}$$

yields the following:

$$(v^*,u) = v_1^*(u_1) \cdots v_{r+s}^*(u_{r+s}). \tag{1}$$

It is easily checked that there is a unique such bilinear map and that it is a non-singular pairing. This pairing establishes an isomorphism

$$(V^*)_{r,s} \cong (V_{r,s})^*. \tag{2}$$

On the other hand, the obvious extension of the universal mapping property 2.2(a) shows that there is a natural isomorphism

$$(V_{r,s})^* \cong M_{r,s}(V) \tag{3}$$

where $M_{r,s}(V)$ is the vector space of all multilinear functions

$$\underbrace{V \times \cdots \times V}_{r \text{ copies}} \times \underbrace{V^* \times \cdots \times V^*}_{s \text{ copies}} \to \mathbb{R}.$$

Under the isomorphism (3), if $\tilde{h} \in (V_{r,s})^*$, then the corresponding multilinear function h in $M_{r,s}(V)$ satisfies

(4) $h(v_1, \ldots, v_r, v_1^*, \ldots, v_s^*) = \tilde{h}(v_1 \otimes \cdots \otimes v_r \otimes v_1^* \otimes \cdots \otimes v_s^*).$

Finally, from (2) and (3) we obtain an isomorphism

(5) $$(V^*)_{r,s} \cong M_{r,s}(V).$$

2.9 Definition *A non-singular pairing of* $\Lambda_k(V^*)$ *with* $\Lambda_k(V)$. This pairing is to be the bilinear map of $\Lambda_k(V^*) \times \Lambda_k(V) \to \mathbb{R}$ which on decomposable elements $v^* = v_1^* \wedge \cdots \wedge v_k^*$ in $\Lambda_k(V^*)$ and $u = u_1 \wedge \cdots \wedge u_k$ in $\Lambda_k(V)$ yields

(1) $$(v^*,u) = \det\big(v_i^*(u_j)\big).$$

Again it is easily checked that there is a unique such bilinear map and that it is a non-singular pairing. For $k = 0$, the pairing is simply multiplication of real numbers. This pairing establishes an isomorphism

(2) $$\Lambda_k(V^*) \cong \Lambda_k(V)^*.$$

Composing (2) with the natural isomorphism

(3) $$\Lambda_k(V)^* \cong A_k(V)$$

of 2.6(5) yields an isomorphism

(4) $$\Lambda_k(V^*) \cong A_k(V).$$

Using the isomorphism (2), and observing that the dual space of a finite direct sum is canonically isomorphic to the direct sum of the dual spaces, we obtain isomorphisms

(5) $$\Lambda(V^*) = \sum_{k=0}^{\infty} \Lambda_k(V^*) \cong \sum_{k=0}^{\infty} \Lambda_k(V)^* \cong \Lambda(V)^*,$$

and from (3) we obtain an isomorphism

(6) $$\Lambda(V)^* \cong A(V) = \sum_{k=0}^{\infty} A_k(V).$$

Henceforth we shall make use of the identification (5) of $\Lambda(V^*)$ with $\Lambda(V)^*$ via the pairing (1) without further comment.

2.10 Remarks on 2.9

(a) If $\{e_1, \ldots, e_d\}$ is a basis of V with dual basis $\{\gamma_1, \ldots, \gamma_d\}$ in V^*, then the bases $\{e_\Phi\}$ and $\{\gamma_\Phi\}$ defined in 2.6(b) are dual bases of $\Lambda(V)$ and $\Lambda(V^*)$ under the isomorphism 2.9(5).

(b) In 2.9(5) and (6), the following isomorphisms were established:

$$\Lambda(V^*) \cong \Lambda(V)^* \cong A(V).$$

The second of these is the natural isomorphism arising from the universal mapping property 2.6(c). The first, call it α for the moment, arose from our choice of the pairing 2.9(1). There is another pairing in common use, this one obtained by replacing 2.9(1) by

$$(1) \qquad (v^*,u) = \frac{1}{k!}\det\big(v_i^*(u_j)\big).$$

This pairing gives a different isomorphism of $\Lambda(V^*)$ with $\Lambda(V)^*$, call it β. Now $\Lambda(V^*)$ is an algebra under wedge multiplication. Via the above isomorphisms we obtain two algebra structures $\wedge_\alpha$ and $\wedge_\beta$ on $A(V)$. It is not difficult to see that if $f \in A_p(V)$ and $g \in A_q(V)$, then these induced algebra structures on $A(V)$ take the form

$$(2) \quad f \wedge_\alpha g(v_1, \ldots, v_{p+q}) = \sum_{p,q \text{ shuffles}} (\operatorname{sgn} \pi) f(v_{\pi(1)}, \ldots, v_{\pi(p)}) g(v_{\pi(p+1)}, \ldots, v_{\pi(p+q)})$$

and

$$(3) \quad f \wedge_\beta g(v_1, \ldots, v_{p+q}) = \frac{1}{(p+q)!} \sum_{\pi \in S_{p+q}} (\operatorname{sgn} \pi) f(v_{\pi(1)}, \ldots, v_{\pi(p)}) g(v_{\pi(p+1)}, \ldots, v_{\pi(p+q)}).$$

Here a permutation $\pi \in S_{p+q}$ is called a "p, q shuffle" if $\pi(1) < \pi(2) < \cdots < \pi(p)$ and $\pi(p+1) < \cdots < \pi(p+q)$. It follows from (2) and (3) that

$$(4) \qquad f \wedge_\alpha g = \frac{(p+q)!}{p!q!} f \wedge_\beta g.$$

For example, consider the case in which $p = q = 1$. Let γ and δ belong to $V^* = A_1(V)$, and let $v, w \in V$. Then $\gamma \wedge_\alpha \delta$ and $\gamma \wedge_\beta \delta \in A_2(V)$ and

$$(5) \qquad \gamma \wedge_\alpha \delta(v,w) = \gamma(v)\,\delta(w) - \gamma(w)\,\delta(v),$$

whereas

$$(6) \qquad \gamma \wedge_\beta \delta(v,w) = \tfrac{1}{2}\big(\gamma(v)\,\delta(w) - \gamma(w)\,\delta(v)\big).$$

We shall use only the pairing 2.9(1). It has the advantage of avoiding factors such as the $\frac{1}{2}$ in (6).

2.11 Linear Transformations of $\Lambda(V)$ End$(\Lambda(V))$ will denote the vector space of all endomorphisms of $\Lambda(V)$ (i.e. linear transformations from $\Lambda(V)$ into $\Lambda(V)$). Let $u \in \Lambda(V)$. *Left multiplication by u* is the endomorphism $\varepsilon(u)$ of $\Lambda(V)$ defined by

$$\varepsilon(u)v = u \wedge v \qquad (v \in \Lambda(V)). \tag{1}$$

The transpose $i(u)$ of $\varepsilon(u)$ is an endomorphism of $\Lambda(V)^*$. Under our identification of $\Lambda(V)^*$ with $\Lambda(V^*)$, the transpose $i(u)$ can be considered as an endomorphism of $\Lambda(V^*)$, and this endomorphism is called *interior multiplication by u*. In terms of our pairing of $\Lambda(V)$ with $\Lambda(V^*)$, $i(u)$ is defined by

$$(i(u)v^*, w) = (v^*, \varepsilon(u)w) = (v^*, u \wedge w) \qquad (v^* \in \Lambda(V^*);\ w \in \Lambda(V)). \tag{2}$$

If $u \in V$, then $i(u)$ maps $\Lambda_k(V^*)$ into $\Lambda_{k-1}(V^*)$ for each k. In particular, if $v^* \in V^*$, then $i(u)v^* \in \mathbb{R}$ and

$$i(u)v^* = (i(u)v^*, 1) = (v^*, \varepsilon(u)\cdot 1) = (v^*, u) = v^*(u).$$

An endomorphism l of $\Lambda(V)$ (or in general of any graded algebra) is

(a) a *derivation* if $l(u \wedge v) = l(u) \wedge v + u \wedge l(v) \qquad (u, v \in \Lambda(V))$,

(b) an *anti-derivation* if $l(u \wedge v) = l(u) \wedge v + (-1)^p u \wedge l(v)$
$(u \in \Lambda_p(V);\ v \in \Lambda(V))$,

(c) *of degree k* if $l\colon \Lambda_j(V) \to \Lambda_{j+k}(V)$ for all j.
(We assume that $\Lambda_i(V) = \{0\}$ if $i < 0$.)

It is easy to see that $l \in \operatorname{End}\Lambda(V)$ is an anti-derivation if and only if on decomposable elements we have

$$l(v_1 \wedge \cdots \wedge v_j) = \sum_{i=1}^{j} (-1)^{i+1} v_1 \wedge \cdots \wedge l(v_i) \wedge \cdots \wedge v_j. \tag{3}$$

2.12 Proposition *If $u \in V$, then $i(u)$ is an anti-derivation of degree -1.*

PROOF That the degree of $i(u)$ is -1 if $u \in V$ is clear from the definition 2.11(2). To prove that $i(u)$ is an anti-derivation, we check that 2.11(3) holds. For this, it is sufficient to see that both sides of 2.11(3) give the same result when paired with an element of $\Lambda(V)$ of the form $w_2 \wedge \cdots \wedge w_j$. That is, we must show that

$$\begin{aligned}(1)\quad &(i(u)(v_1^* \wedge \cdots \wedge v_j^*), w_2 \wedge \cdots \wedge w_j)\\ &= \left(\sum_{i=1}^{j} (-1)^{i+1} v_1^* \wedge \cdots \wedge i(u)v_i^* \wedge \cdots \wedge v_j^*, w_2 \wedge \cdots \wedge w_j\right).\end{aligned}$$

Now, the left-hand side of (1) equals

$$(v_1^* \wedge \cdots \wedge v_j^*, u \wedge w_2 \wedge \cdots \wedge w_j) = \det(v_i^*(w_l)),$$

where we have put $w_1 = u$. The right-hand side of (1) is

$$\sum_{i=1}^{j}(-1)^{i+1}v_i^*(u)(v_1^* \wedge \cdots \wedge \widehat{v_i^*} \wedge \cdots v_j^*, w_2 \wedge \cdots \wedge w_j)$$

$$= \sum_{i=1}^{j}(-1)^{i+1}v_i^*(u)\det(v_k^*(w_l)) \quad \begin{pmatrix} k = 1, \ldots, \widehat{i}, \ldots, j \\ l = 2, \ldots, j \end{pmatrix}$$

$$= \det(v_i^*(w_l)) \quad \begin{pmatrix} i = 1, \ldots, j \\ l = 1, \ldots, j \end{pmatrix}.$$

(Here the circumflex over a term means that it is to be omitted.)

2.13 Effect of Linear Transformations Let $l: V \to W$ be a linear transformation. Then l extends to an algebra homomorphism

(1) $$l: \Lambda(V) \to \Lambda(W),$$

where $$l(v_1 \wedge \cdots \wedge v_k) = l(v_1) \wedge \cdots \wedge l(v_k) \quad \text{and} \quad l(1) = 1.$$

The transpose $\delta l: W^* \to V^*$ also extends to an algebra homomorphism

(2) $$\delta l: \Lambda(W^*) \to \Lambda(V^*).$$

Our isomorphisms 2.9(5) of $\Lambda(V^*)$ with $\Lambda(V)^*$ and of $\Lambda(W^*)$ with $\Lambda(W)^*$ are natural in the sense that (2) is the transpose of (1); that is,

(3) $$(\delta l(w^*), v) = (w^*, l(v)) \qquad (w^* \in \Lambda(W^*); v \in \Lambda(v)).$$

TENSOR FIELDS AND DIFFERENTIAL FORMS

2.14 Definitions Let M be a differentiable manifold. We define

(1) $T_{r,s}(M) = \bigcup_{m \in M} (M_m)_{r,s}$ *tensor bundle of type (r,s) over M*

(2) $\Lambda_k^*(M) = \bigcup_{m \in M} \Lambda_k(M_m^*)$ *exterior k bundle over M*

(3) $\Lambda^*(M) = \bigcup_{m \in M} \Lambda(M_m^*)$ *exterior algebra bundle over M.*

In the cases in which $k = 0$ and $(r,s) = (0,0)$, the unions in (1) and (2) are disjoint unions of copies of the real line—one copy for each point in M. $T_{r,s}(M)$, $\Lambda_k^*(M)$, and $\Lambda^*(M)$ have natural manifold structures such that the canonical projection maps to M are C^∞. If (U,φ) is a coordinate system on M with coordinate functions $y_1, \ldots, y_d$, then the bases $\{\partial/\partial y_i\}$ of M_m and $\{dy_i\}$ of M_m^*, for $m \in U$, yield bases of $(M_m)_{r,s}$, $\Lambda_k(M_m^*)$, and $\Lambda(M_m^*)$. For example, the basis of $\Lambda_k(M_m^*)$ is $\{dy_{i_1} \wedge \cdots \wedge dy_{i_k}: i_1 < \cdots < i_k\}$.

Using these bases, one can define maps of the inverse images of U in $T_{r,s}(M)$, $\Lambda_k^*(M)$, and $\Lambda^*(M)$ under the respective projection maps to $\varphi(U) \times$ Euclidean spaces of the proper dimensions. By requiring these maps to be coordinate systems, one obtains the natural manifold structures on $T_{r,s}(M)$, $\Lambda_k^*(M)$, and $\Lambda^*(M)$ just as we did previously in 1.25 for $T(M)$ and $T^*(M)$ (which incidentally is simply $\Lambda_1^*(M)$).

2.15 Definition A C^∞ mapping of M into $T_{r,s}(M)$, $\Lambda_k^*(M)$, or $\Lambda^*(M)$ whose composition with the canonical projection is the identity map is called a *(smooth) tensor field of type* (r,s) *on* M, a *(differential) k-form on* M, or a *(differential) form on* M respectively. Since our tensor fields and differential forms will always be smooth in this sense, we shall drop the adjectives "smooth" and "differential" unless needed for emphasis.

2.16 Remarks A lifting $\alpha: M \to T_{r,s}(M)$ is a smooth tensor field of type (r,s) if and only if for each coordinate system $(U, y_1, \ldots, y_d)$ on M,

(1) $$\alpha \mid U = \sum a_{i_1,\ldots,i_r;j_1,\ldots,j_s} \frac{\partial}{\partial y_{i_1}} \otimes \cdots \otimes \frac{\partial}{\partial y_{i_r}} \otimes dy_{j_1} \otimes \cdots \otimes dy_{j_s},$$

where the $a_{i_1,\ldots,i_r;j_1,\ldots,j_s} \in C^\infty(U)$. (In the classical tensor notation one uses lower indices on tangent vectors, upper indices on functions and differentials, and the reverse on coefficients; thus the individual terms of (1) would appear as

$$a^{i_1,\ldots,i_r}_{j_1,\ldots,j_s} \frac{\partial}{\partial y^{i_1}} \otimes \cdots \otimes \frac{\partial}{\partial y^{i_r}} \otimes dy^{j_1} \otimes \cdots \otimes dy^{j_s}.)$$

A lifting $\beta: M \to \Lambda_k^*(M)$ is a differential k-form if and only if for each coordinate system $(U, y_1, \ldots, y_d)$ on M,

(2) $$\beta \mid U = \sum_{i_1 < \cdots < i_k} b_{i_1,\ldots,i_k}\, dy_{i_1} \wedge \cdots \wedge dy_{i_k},$$

where the $b_{i_1,\ldots,i_k}$ are C^∞ functions on U.

2.17 Definitions $E^k(M)$ shall denote the set of all smooth k-forms on M, and $E^*(M)$ the set of all differential forms. $E^0(M)$ can be identified with $C^\infty(M)$; indeed, the differentiable manifold $\Lambda_0^*(M)$ is simply $M \times \mathbb{R}$, and smooth liftings of M into $M \times \mathbb{R}$ are simply graphs of C^∞ functions on M. Forms can be added, multiplied by scalars, and given a product ($\wedge$). If $\omega, \varphi \in E^*(M)$ and $c \in \mathbb{R}$, then $\omega + \varphi$, $c\omega$, and $\omega \wedge \varphi$ are the forms which at m have the values $\omega_m + \varphi_m$, $c\omega_m$, and $\omega_m \wedge \varphi_m$ respectively. In the case in which f is a 0-form and $\omega \in E^*(M)$, we write $f \wedge \omega$ simply as $f\omega$. $E^*(M)$ thus has the structure both of a module over the ring $C^\infty(M)$ and of a graded algebra over $\mathbb{R}$ with wedge multiplication.

2.18 Let $\omega \in E^k(M)$. Then $\omega_m \in \Lambda_k(M_m^*)$, and can be considered (via the duality 2.9(4)) as an alternating multilinear function on M_m. So if $X_1, \ldots, X_k$ are vector fields on M, $\omega(X_1, \ldots, X_k)$ makes sense—it is the function whose value at m is

$$\omega(X_1, \ldots, X_k)(m) = \omega_m(X_1(m), \ldots, X_k(m)). \tag{1}$$

Thus, if we let $\mathfrak{X}(M)$ denote the $C^\infty(M)$ module of smooth vector fields on M, then

$$\omega\colon \underbrace{\mathfrak{X}(M) \times \cdots \times \mathfrak{X}(M)}_{k \text{ copies}} \to C^\infty(M) \tag{2}$$

and is an alternating multilinear map of the module $\mathfrak{X}(M)$ into $C^\infty(M)$. We stress that ω is multilinear over the $C^\infty(M)$ *module* $\mathfrak{X}(M)$; that is

$$\begin{aligned} \omega(X_1, \ldots, X_{i-1}, fX + gY, X_{i+1}, \ldots, X_k) &\\ = f\omega(X_1, \ldots, X_{i-1}, X, X_{i+1}, \ldots, X_k)&\\ + g\omega(X_1, \ldots, X_{i-1}, Y, X_{i+1}, \ldots, X_k)& \end{aligned} \tag{3}$$

whenever $f, g \in C^\infty(M)$ and $X_1, \ldots, X_{i-1}, X, Y, X_{i+1}, \ldots, X_k \in \mathfrak{X}(M)$.

Conversely, it is useful to observe that *any alternating* $C^\infty(M)$ *multilinear map* (2) *of the module* $\mathfrak{X}(M)$ *into* $C^\infty(M)$ *defines a form;* for we claim that if ω is such a map, then $\omega(X_1, \ldots, X_k)(m)$ depends only on the values of the vector fields X_i at m. Assuming this for the moment, it follows that ω defines an alternating multilinear function ω_m on M_m, and hence defines an element of $\Lambda_k(M_m^*)$; namely, given $(v_1, \ldots, v_k) \in M_m \times \cdots \times M_m$, choose $V_1, \ldots, V_k \in \mathfrak{X}(M)$ such that $V_i(m) = v_i$ $(i = 1, \ldots, k)$, and define

$$\omega_m(v_1, \ldots, v_k) = \omega(V_1, \ldots, V_k)(m). \tag{4}$$

By our claim, $\omega_m(v_1, \ldots, v_k)$ is well-defined, independent of the choice of the extensions V_i. Thus ω gives a lifting $m \mapsto \omega_m$ of M into $\Lambda^*(M)$, and this is easily seen to be smooth; hence ω is a form.

For simplicity of notation, we illustrate the claim with the case in which ω is a linear map of the module $\mathfrak{X}(M)$ into $C^\infty(M)$. Let $X \in \mathfrak{X}(M)$. We wish to show that $\omega(X)(m)$ depends only on $X(m)$. It suffices to show that $\omega(X)(m) = 0$ if $X(m) = 0$. Let $(U, x_1, \ldots, x_d)$ be a coordinate system about m. Then on U, we have $X = \sum a_i(\partial/\partial x_i)$ where the $a_i(m) = 0$. Now, let φ be a C^∞ function which takes the value 1 on a neighborhood $V \subset U$ of m and is zero on a neighborhood of $M - U$ (see 1.10). Then the vector field X_i which is $\varphi(\partial/\partial x_i)$ on U and 0 elsewhere is a C^∞ vector field on M; the function $\tilde{a}_i$ which is φa_i on U and 0 elsewhere belongs to $C^\infty(M)$; and

$$X = \sum \tilde{a}_i X_i + (1 - \varphi^2)X. \tag{5}$$

Thus

(6) $\omega(X)(m) = \sum \tilde{a}_i(m)\omega(X_i)(m) + ((1 - \varphi^2)(m))(\omega(X)(m)) = 0.$

So $\omega(X)(m) = 0$ if $X(m) = 0$, as was to be shown.

Finally, we remark that via the duality 2.8(5), tensor fields can be given a similar interpretation. If T is a tensor field of type (r,s), then we can consider T as a map

(7) $$T\colon \underbrace{E^1(M) \times \cdots \times E^1(M)}_{r \text{ copies}} \times \underbrace{\mathfrak{X}(M) \times \cdots \times \mathfrak{X}(M)}_{s \text{ copies}} \to C^\infty(M),$$

which is $C^\infty(M)$ multilinear with respect to the $C^\infty(M)$ modules $E^1(M)$ and $\mathfrak{X}(M)$.

Observe the simple form which formula (2) of 2.10(b) takes when $\omega, \varphi \in E^1(M)$ and $X, Y \in \mathfrak{X}(M)$. In this case, we have

(8) $$\omega \wedge \varphi(X, Y) = \omega(X)\varphi(Y) - \omega(Y)\varphi(X).$$

2.19 Definition If $f \in C^\infty(M)$, the differential df is a smooth mapping of $T(M)$ into $\mathbb{R}$ which is linear on each tangent space. Thus df can be considered as a 1-form, $df\colon M \to \Lambda_1^*(M)$. The 1-form df is called the *exterior derivative* of the 0-form f, and this *exterior differentiation operator d* has an important extension to $E^*(M)$ given by the following.

2.20 Theorem (Exterior Differentiation) *There exists a unique anti-derivation $d\colon E^*(M) \to E^*(M)$ of degree $+1$ such that*

(1) $d^2 = 0.$

(2) *Whenever $f \in C^\infty(M) = E^0(M)$, df is the differential of f.*

PROOF *Existence.* Let $p \in M$. Let $E^*(p)$ be the set of all smooth forms defined on open subsets of M containing p, with $E^k(p)$ the corresponding set of k-forms. We fix a coordinate system $(U, x_1, \ldots, x_d)$ about p. If $\omega \in E^*(p)$, then

(3) $$\omega|_{(\text{domain } \omega) \cap U} = \sum a_\Phi \, dx_\Phi$$

where the $a_\Phi \in C^\infty((\text{domain } \omega) \cap U)$, where Φ runs over all subsets of $\{1, \ldots, d\}$, and where the dx_Φ are either $dx_{i_1} \wedge \cdots \wedge dx_{i_r}$ when $\Phi = \{i_1 < \cdots < i_r\}$ or the constant function 1 when $\Phi = \varnothing$. We define $d\omega$ at p by setting

(4) $$d\omega_p = \sum da_\Phi|_p \wedge dx_\Phi|_p \in \Lambda(M_p^*).$$

We will have to show that the definition of $d\omega_p$ is independent of the choice of coordinates. But first, we give the following properties:

(a) $\omega \in E^r(p) \Rightarrow d\omega_p \in \Lambda_{r+1}(M_p^*)$.

(b) $d\omega_p$ depends only on the germ of ω at p.

(c) $d(a_1\omega_1 + a_2\omega_2)|_p = a_1(d\omega_1)|_p + a_2(d\omega_2)|_p$, $(a_i \in \mathbb{R};\ \omega_i \in E^*(p))$, where the domain of $a_1\omega_1 + a_2\omega_2$ is (domain $\omega_1 \cap$ domain ω_2).

(d) $d(\omega_1 \wedge \omega_2)|_p = d\omega_1|_p \wedge \omega_2|_p + (-1)^r\omega_1|_p \wedge d\omega_2|_p$
$$(\omega_1 \in E^r(p);\ \omega_2 \in E^*(p)).$$

In view of properties (b) and (c), it suffices to check (d) for $\omega_1 = f\,dx_{i_1} \wedge \cdots \wedge dx_{i_r}$ and $\omega_2 = g\,dx_{j_1} \wedge \cdots \wedge dx_{j_s}$ on some neighborhood of p. For the case $r = s = 0$, property (d) is simply $d(f \cdot g)|_p = df_p \cdot g(p) + f(p) \cdot dg_p$; and the case in which only one of r or s is 0 is similar. Now suppose that $r > 0$ and $s > 0$. If $\{i_1, \ldots, i_r\} \cap \{j_1, \ldots, j_s\} \neq \varnothing$, both sides are 0. So assume that this intersection is empty. Then

$$(f\,dx_{i_1} \wedge \cdots \wedge dx_{i_r}) \wedge (g\,dx_{j_1} \wedge \cdots \wedge dx_{j_s}) = \varepsilon f \cdot g\,dx_{l_1} \wedge \cdots \wedge dx_{l_{r+s}}$$

where $l_1 < \cdots < l_{r+s}$ and ε is the sign of the permutation that has been carried out. So we have that

$$\begin{aligned}
d(\omega_1 \wedge \omega_2)|_p &= d(\varepsilon f \cdot g\,dx_{l_1} \wedge \cdots \wedge dx_{l_{r+s}})|_p \\
&= \varepsilon(df_p \cdot g(p) + f(p)\,dg_p) \wedge dx_{l_1}|_p \wedge \cdots \wedge dx_{l_{r+s}}|_p \\
&= (df_p \wedge dx_{i_1}|_p \wedge \cdots \wedge dx_{i_r}|_p) \wedge (g(p)\,dx_{j_1}|_p \wedge \cdots \wedge dx_{j_s}|_p) \\
&\quad + (-1)^r(f(p)\,dx_{i_1}|_p \wedge \cdots \wedge dx_{i_r}|_p) \\
&\qquad\qquad \wedge (dg_p \wedge dx_{j_1}|_p \wedge \cdots \wedge dx_{j_s}|_p) \\
&= d\omega_1|_p \wedge \omega_2|_p + (-1)^r\omega_1|_p \wedge d\omega_2|_p.
\end{aligned}$$

(e) If f is a C^∞ function on a neighborhood of p, then $d(df)|_p = 0$. For on (domain f) $\cap U$, $df = \sum (\partial f/\partial x_i)\,dx_i$, so that

$$d(df)|_p = \sum d\left(\frac{\partial f}{\partial x_i}\right)\bigg|_p \wedge dx_i|_p = \sum_{i,j} \frac{\partial^2 f}{\partial x_j\,\partial x_i}\bigg|_p dx_j|_p \wedge dx_i|_p.$$

But $(\partial^2 f/\partial x_j\,\partial x_i)(p) = (\partial^2 f/\partial x_i\,\partial x_j)(p)$, whereas $dx_j|_p \wedge dx_i|_p = -dx_i|_p \wedge dx_j|_p$. So $d(df)|_p = 0$.

Now we claim that the definition of d at p is independent of the coordinates chosen. For let d' be defined on $E^*(p)$ relative to another coordinate system, and let $\omega \in E^*(p)$. Then ω on (domain ω) $\cap U$ is given by (3), and since d' must also satisfy properties (a)–(e), it follows that

$$\begin{aligned}(5)\quad d'(\omega)\big|_p &= d'(\textstyle\sum a_\Phi\, dx_{i_1} \wedge \cdots \wedge dx_{i_r})\big|_p && \text{(by (b))}\\ &= \textstyle\sum d'(a_\Phi\, dx_{i_1} \wedge \cdots \wedge dx_{i_r})\big|_p && \text{(by (c))}\\ &= \textstyle\sum d'(a_\Phi)\big|_p \wedge dx_{i_1}\big|_p \wedge \cdots \wedge dx_{i_r}\big|_p \\ &\quad + \textstyle\sum (-1)^{k-1} a_\Phi\big|_p\, dx_{i_1}\big|_p \wedge \cdots \wedge d'(dx_{i_k})\big|_p \wedge \cdots \wedge dx_{i_r}\big|_p \\ &&& \text{(by (d))}\\ &= \textstyle\sum d(a_\Phi)\big|_p \wedge dx_{i_1}\big|_p \wedge \cdots \wedge dx_{i_r}\big|_p && \text{(by (e))}\\ &= d\omega_p .\end{aligned}$$

Now, if $\omega \in E^*(M)$, we define $d\omega$ to be the form which as a lifting of M into $\Lambda^*(M)$ sends p to $d\omega_p$. It follows that $d^2 = 0$, since if $\omega \in E^*(M)$ and $p \in M$, $d\omega$ has the form $\sum da_\Phi \wedge dx_\Phi$ on a coordinate neighborhood of p, and thus

$$d(d\omega)\big|_p = \sum d(da_\Phi \wedge dx_\Phi)\big|_p = 0$$

by properties (e) and (d). It follows that d is an anti-derivation of $E^*(M)$ of degree $+1$ satisfying (1) and (2).

Uniqueness Let d' also be an anti-derivation of $E^*(M)$ of degree $+1$ satisfying (1) and (2). We first show that if $\omega \in E^*(M)$ and ω vanishes on a neighborhood W of p, then $d'\omega\big|_p = 0$. Choose a C^∞ function φ which is 0 on a neighborhood of p and 1 on a neighborhood of $M - W$. Then $\varphi\omega = \omega$ and

$$d'(\omega)\big|_p = d'(\varphi\omega)\big|_p = d'(\varphi)\big|_p \wedge \omega_p + \varphi(p)\, d'\omega\big|_p = 0.$$

Now d' is defined only on elements of $E^*(M)$, that is, on globally defined forms on M. We wish to define d' on $E^*(p)$ for each $p \in M$. If $\omega \in E^*(p)$, we can extend ω to a form on M having the same germ at p as does ω. Simply let φ be a C^∞ function which is 1 on a neighborhood of p and has support in the domain of ω. Then $\varphi\omega \in E^*(M)$ ($\varphi\omega$ is defined to be 0 outside of the domain of ω), and $\varphi\omega$ agrees with ω on a neighborhood of p. Thus we can define

$$d'(\omega)\big|_p = d'(\varphi\omega)\big|_p,$$

and by the above remarks this definition is independent of the extension chosen. Thus $d'(\omega)\big|_p$ is defined for all ω in $E^*(p)$ and clearly satisfies properties (a)–(e). (In (d), extend $\omega_1 \wedge \omega_2$ to a form on M by using $\varphi\omega_1 \wedge \varphi\omega_2$ for a suitable φ; and in (e), observe that $d'(df)\big|_p = d'(\varphi\, df)\big|_p = d'(d(\varphi f))\big|_p = 0$, since $d\varphi(p) = 0$ and since d' satisfies (1) and (2).) The equations (5) now imply that whenever $\omega \in E^*(p)$, in particular whenever $\omega \in E^*(M)$, $d'(\omega)\big|_p = d(\omega)\big|_p$. This proves uniqueness.

Observe that it is clear from the above proof that $d\omega \mid U = d(\omega \mid U)$ whenever U is an open set in M.

2.21 Interior Multiplication by Vector Fields Let X be a smooth vector field on M, and let $\omega \in E^*(M)$. Interior multiplication of ω by X is the form $i(X)\omega$ whose value at m is the interior multiple of ω_m by X_m (see 2.11(2)):

$$(1) \qquad (i(X)\omega)|_m = i(X_m)(\omega_m).$$

That $i(X)\omega$ is smooth follows easily from 2.16(2); and from 2.12 it follows that $i(X)$: $E^*(M) \to E^*(M)$ is an anti-derivation of degree -1.

2.22 Effect of Mappings Let $\psi: M \to N$ be a smooth map, and let $m \in M$. Then we have the differential $d\psi: M_m \to N_{\psi(m)}$, its transpose $\delta\psi: N^*_{\psi(m)} \to M^*_m$, and the induced algebra homomorphism $\delta\psi: \Lambda(N^*_{\psi(m)}) \to \Lambda(M^*_m)$. If ω is a form on N, then we can pull ω back to a form on M by setting

$$(1) \qquad \delta\psi(\omega)\big|_m = \delta\psi(\omega|_{\psi(m)}).$$

This is one of the particularly nice features of differential forms. Under a smooth mapping, they can be pulled back from the range to the domain of the map. Vector fields, on the other hand, do not display such pleasing behavior under mappings.

2.23 Proposition *Let $\psi: M \to N$ be a smooth map. Then*

(a) *$\delta\psi$: $E^*(N) \to E^*(M)$ and is an algebra homomorphism.*

(b) *$\delta\psi$ commutes with d; that is, $d(\delta\psi(\omega)) = \delta\psi(d\omega) \quad (\omega \in E^*(N))$.*

(c) *$\delta\psi(\omega)(X_1, \ldots, X_k)(m) = \omega_{\psi(m)}(d\psi(X_1(m)), \ldots, d\psi(X_k(m)))$ for $\omega \in E^k(N)$ and for vector fields $X_1, \ldots, X_k$ on M.*

PROOF Result (c) is clear from 2.13(3); and 2.13(2) and the definition 2.22(1) imply that $\delta\psi$ is an algebra homomorphism. Before we check that $\delta\psi(\omega)$ is actually a smooth form for $\omega \in E^*(N)$, observe the following special case of (b): If $f \in C^\infty(N)$, then $\delta\psi(f) = f \circ \psi \in C^\infty(M)$, and 1.23(d) implies that

$$(1) \qquad \delta\psi(df) = d(f \circ \psi) = d(\delta\psi(f)).$$

Now let $\omega \in E^*(N)$, and let $m \in M$. Choose a coordinate system $(U, x_1, \ldots, x_d)$ about $\psi(m)$ and a neighborhood V of m such that $\psi(V) \subset U$. Then there are C^∞ functions a_Φ on U such that

$$(2) \qquad \omega \mid U = \sum a_\Phi \, dx_{i_1} \wedge \cdots \wedge dx_{i_r}.$$

It follows that

$$(3) \qquad \delta\psi(\omega) \mid V = \sum a_\Phi \circ \psi \, d(x_{i_1} \circ \psi) \wedge \cdots \wedge d(x_{i_r} \circ \psi),$$

which is a smooth form on V. Hence $\delta\psi(\omega) \in E^*(M)$, and thus (a) is proved. To complete (b), we use (3):

$$\begin{aligned}
(4)\qquad d(\delta\psi(\omega))\big|_m &= d\left(\sum a_\Phi \circ \psi\, d(x_{i_1} \circ \psi) \wedge \cdots \wedge d(x_{i_r} \circ \psi)\right)\big|_m \\
&= \sum \left(d(a_\Phi \circ \psi) \wedge d(x_{i_1} \circ \psi) \wedge \cdots \wedge d(x_{i_r} \circ \psi)\right)\big|_m \\
&= \delta\psi\left(\sum da_\Phi \wedge dx_{i_1} \wedge \cdots \wedge dx_{i_r}\right)\big|_m \\
&= \delta\psi(d\omega)\big|_m .
\end{aligned}$$

Thus (b) is proved, and the proof is complete.

THE LIE DERIVATIVE

2.24 Definition Tensor fields and differential forms can be differentiated with respect to a vector field. The resulting derivative is known as the *Lie derivative* and is defined as follows. Fix a smooth vector field X on a manifold M. Recall (see 1.48) that we use X_t to denote the local 1-parameter group of transformations associated with X. Let Y be another smooth vector field on M. We shall define the derivative of Y with respect to X at the point $m \in M$. First we follow the integral curve of X through m out to the point $X_t(m)$ and evaluate Y there. Then we transfer $Y_{X_t(m)}$ back to M_m via the differential dX_{-t} of the diffeomorphism X_{-t}. In M_m we take the difference of the vectors $dX_{-t}(Y_{X_t(m)})$ and Y_m, divide the difference by t, and then take the limit as $t \to 0$. In other words, we consider the smooth M_m-valued function $t \mapsto dX_{-t}(Y_{X_t(m)})$, and we take its derivative at $t = 0$. The result is a vector in M_m which is called the *Lie derivative of Y with respect to X at m* and which is denoted by $(L_X Y)_m$. Thus we define

$$(1)\qquad (L_X Y)_m = \lim_{t \to 0} \frac{dX_{-t}(Y_{X_t(m)}) - Y_m}{t} = \frac{d}{dt}\bigg|_{t=0} \left(dX_{-t}(Y_{X_t(m)})\right).$$

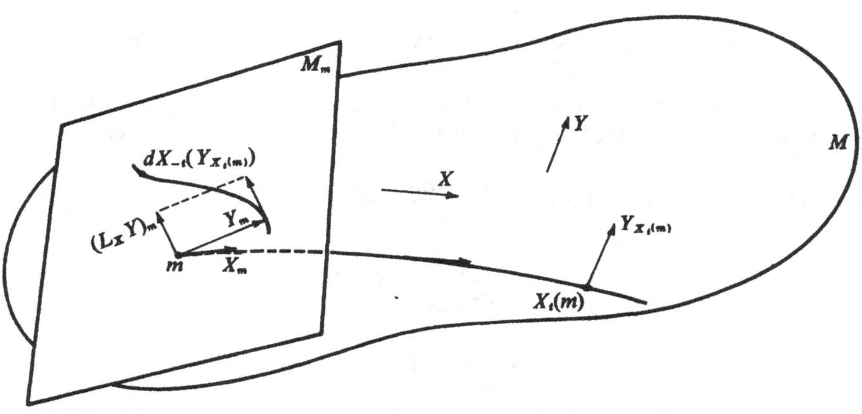

In a similar way we define the *Lie derivative of a differential form* ω *with respect to the vector field* X, except that in this case we evaluate ω at $X_t(m)$ and then pull back via δX_t to $\Lambda(M_m^*)$ where we take the difference with $\omega(m)$, divide by t, and take the limit as $t \to 0$. Thus we define

$$(2) \qquad (L_X\omega)_m = \lim_{t\to 0} \frac{\delta X_t(\omega_{X_t(m)}) - \omega_m}{t} = \frac{d}{dt}\bigg|_{t=0} (\delta X_t(\omega_{X_t(m)})).$$

Smoothness of $L_X\omega$ and of $L_X Y$ is asserted in Proposition 2.25. The Lie derivative L_X can be extended to arbitrary tensor fields in the obvious way. If T is a tensor field of type (r,s), then $(L_X T)_m$ is the derivative at $t = 0$ of the $(M_m)_{(r,s)}$-valued function whose value at t is

$$dX_{-t}(v_1 \otimes \cdots \otimes v_r) \otimes \delta X_t(v_1^* \otimes \cdots \otimes v_s^*)$$

if

$$T|_{X_t(m)} = v_1 \otimes \cdots \otimes v_r \otimes v_1^* \otimes \cdots \otimes v_s^*.$$

2.25 Proposition *Let X be a C^∞ vector field on M. Then*

(a) $L_X f = X(f)$ *whenever* $f \in C^\infty(M)$.

(b) $L_X Y = [X, Y]$ *for each C^∞ vector field Y on M.*

(c) L_X: $E^*(M) \to E^*(M)$, *and is a derivation which commutes with d.*

(d) *On* $E^*(M)$, $L_X = i(X) \circ d + d \circ i(X)$.

(e) *Let* $\omega \in E^p(M)$, *and let* $Y_0, \ldots, Y_p$ *be C^∞ vector fields on M. Then*

$$L_{Y_0}(\omega(Y_1, \ldots, Y_p)) = (L_{Y_0}\omega)(Y_1, \ldots, Y_p) + \sum_{i=1}^{p} \omega(Y_1, \ldots, Y_{i-1}, L_{Y_0}Y_i, Y_{i+1}, \ldots, Y_p).$$

(f) Assumption as in (e). *Then*

$$d\omega(Y_0, \ldots, Y_p) = \sum_{i=0}^{p} (-1)^i Y_i \omega(Y_0, \ldots, \widehat{Y_i}, \ldots, Y_p) + \sum_{i<j} (-1)^{i+j} \omega([Y_i, Y_j], Y_0, \ldots, \widehat{Y_i}, \ldots, \widehat{Y_j}, \ldots, Y_p).$$

PROOF We leave result (a) as an exercise. For (b), we need only show that $L_X Y(f) = [X, Y]f$ for each $f \in C^\infty(M)$. Let $m \in M$. Then

$$(1) \qquad (L_X Y)(f) = \left(\lim_{t\to 0} \frac{dX_{-t} Y_{X_t(m)} - Y_m}{t}\right)(f)$$

$$= \frac{d}{dt}\bigg|_{t=0} [(dX_{-t}(Y_{X_t(m)}))(f)]$$

$$= \frac{d}{dt}\bigg|_{t=0} [Y_{X_t(m)}(f \circ X_{-t})].$$

Define a real-valued function H on a neighborhood of (0,0) in $\mathbb{R}^2$ by setting

(2) $$H(t,u) = f(X_{-t}(Y_u(X_t(m)))).$$

Then by (a),

(3) $$Y_{X_t(m)}(f \circ X_{-t}) = \frac{\partial}{\partial r_2}\bigg|_{(t,0)} H(t,u),$$

and (1) becomes

(4) $$(L_X Y)_m(f) = \frac{\partial^2 H}{\partial r_1\, \partial r_2}\bigg|_{(0,0)}.$$

To evaluate this derivative, we set

(5) $$K(t,u,s) = f(X_s(Y_u(X_t(m)))).$$

Then $H(t,u) = K(t,u,-t)$. It follows from the chain rule that

(6) $$\frac{\partial^2 H}{\partial r_1\, \partial r_2}\bigg|_{(0,0)} = \frac{\partial^2 K}{\partial r_1\, \partial r_2}\bigg|_{(0,0,0)} - \frac{\partial^2 K}{\partial r_3\, \partial r_2}\bigg|_{(0,0,0)}.$$

Now, $K(t,u,0) = f(Y_u(X_t(m)))$. Hence

(7) $$\frac{\partial K}{\partial r_2}\bigg|_{(t,0,0)} = Y_{X_t(m)}(f), \qquad \frac{\partial^2 K}{\partial r_1\, \partial r_2}\bigg|_{(0,0,0)} = X_m(Yf).$$

Also $K(0,u,s) = f(X_s(Y_u(m)))$. Hence

(8) $$\frac{\partial K}{\partial r_3}\bigg|_{(0,u,0)} = Xf(Y_u(m)), \qquad \frac{\partial^2 K}{\partial r_2\, \partial r_3}\bigg|_{(0,0,0)} = Y_m(Xf).$$

Part (b) now follows from (4), (6), (7), and (8). One immediate consequence of (b) is that $L_X Y$ is a *smooth* vector field.

For result (c), first consider L_X as a mapping of $E^*(M)$ into forms which are not apriori smooth, and observe that the derivation property follows immediately from adding and subtracting suitable terms before taking the limit in the definition 2.24(2). Next we check that L_X commutes with d when applied to functions; that is,

$$\text{(9)} \qquad (L_X(df))_m = d(L_X f)_m \qquad (f \in C^\infty(M);\ m \in M).$$

Since both sides of (9) are elements of M_m^*, we need only prove that they have the same effect when applied to an arbitrary vector Y_m in M_m. The right-hand side gives

$$\text{(10)} \qquad d(L_X f)_m(Y_m) = Y_m(L_X f) = Y_m\left(\frac{d}{dt}\bigg|_{t=0} (f \circ X_t)\right),$$

where $f \circ X_t$ can be considered as a C^∞ function on $(-\varepsilon,\varepsilon) \times W$ for some $\varepsilon > 0$ and some neighborhood W of m in M (1.48(d)). The left-hand side of (9) gives

$$\begin{aligned}\text{(11)} \quad (L_X(df))_m(Y_m) &= \left\{\frac{d}{dt}\bigg|_{t=0} (\delta X_t(df_{X_t(m)}))\right\}(Y_m) \\ &= \frac{d}{dt}\bigg|_{t=0} (\delta X_t(df_{X_t(m)})(Y_m)) \\ &= \frac{d}{dt}\bigg|_{t=0} (df(dX_t(Y_m))) = \frac{d}{dt}\bigg|_{t=0} (Y_m(f \circ X_t)).\end{aligned}$$

Now let Y be an extension of Y_m to a vector field on W. Then, according to Exercise 24(c) of Chapter 1, d/dt and Y have canonical extensions to vector fields on $(-\varepsilon,\varepsilon) \times W$ where they satisfy $[d/dt, Y] \equiv 0$. This fact together with (10) and (11) implies (9). Finally, to see that the form $L_X\omega$ is actually smooth and to check that L_X commutes with d on all of $E^*(M)$, simply express an arbitrary form ω in local coordinates as in 2.16(2), and compute using equation (9), result (a), and the fact that L_X is a derivation.

For result (d), observe that both L_X and $i(X) \circ d + d \circ i(X)$ are derivations on $E^*(M)$ which commute with d, and both have the same effect on functions. Then (d) follows from a simple computation in local coordinates.

We shall leave (e) as an exercise. The proof is not difficult—one simply has to be persistent enough in unwinding the definitions. Again, one checks the identity by restricting to a local coordinate system. Try it first for the case $\omega = f\, dx_1 \wedge dx_2$ in a coordinate neighborhood.

Finally, result (f) follows from (e) by using (d) and induction on p. For the case $p = 1$, we have from (e) that

$$\text{(12)} \qquad L_{Y_0}(\omega(Y_1)) = (L_{Y_0}\omega)(Y_1) + \omega[Y_0, Y_1].$$

Applying (a) and (d) to (12), one obtains

$$\begin{aligned} Y_0\omega(Y_1) &= ((i(Y_0) \circ d + d \circ i(Y_0))\omega)(Y_1) + \omega[Y_0, Y_1] \\ &= d\omega(Y_0, Y_1) + Y_1(\omega(Y_0)) + \omega[Y_0, Y_1],\end{aligned}$$

which is result (f) in the case $p = 1$. Now assume that (f) holds for $p - 1$. Then (f) is obtained for p by again starting with (e) and applying (d) and the induction hypothesis.

DIFFERENTIAL IDEALS

Our objective here is to give a version of the Frobenius theorem in terms of differential forms, and then to describe E. Cartan's useful method of obtaining maps by looking for their graphs.

2.26 Definitions Let $\mathscr{D}$ be a p-dimensional C^∞ distribution on M. A q-form ω is said to *annihilate* $\mathscr{D}$ if for each $m \in M$

(1) $$\omega_m(v_1, \ldots, v_q) = 0 \quad \text{whenever } v_1, \ldots, v_q \in \mathscr{D}(m).$$

A form $\omega \in E^*(M)$ is said to annihilate $\mathscr{D}$ if each of the homogeneous parts of ω annihilates $\mathscr{D}$. We let

(2) $$\mathscr{I}(\mathscr{D}) = \{\omega \in E^*(M)\colon \omega \text{ annihilates } \mathscr{D}\}.$$

2.27 Definition A collection $\omega_1, \ldots, \omega_n$ of 1-forms on M is called *independent* if they form an independent set in M_m^* for each $m \in M$.

2.28 Proposition *Let $\mathscr{D}$ be a smooth p-dimensional distribution on M. Then*

(a) *$\mathscr{I}(\mathscr{D})$ is an ideal in $E^*(M)$.*

(b) *$\mathscr{I}(\mathscr{D})$ is locally generated by $d - p$ independent 1-forms. (That is, to each $m \in M$ there corresponds a neighborhood U of m and a set of independent 1-forms $\omega_1, \ldots, \omega_{d-p}$ on U such that:*
   (i) *If $\omega \in \mathscr{I}(\mathscr{D})$, then $\omega \mid U$ belongs to the ideal in $E^*(U)$ generated by $\omega_1, \ldots, \omega_{d-p}$.*
   (ii) *If $\omega \in E^*(M)$, and if there is a cover of M by sets U (as above) such that for each U in the cover, $\omega \mid U$ belongs to the ideal generated by $\omega_1, \ldots, \omega_{d-p}$, then $\omega \in \mathscr{I}(\mathscr{D})$.)*

(c) *If $\mathscr{I} \subset E^*(M)$ is an ideal locally generated by $d - p$ independent 1-forms, then there exists a unique C^∞ distribution $\mathscr{D}$ of dimension p on M for which $\mathscr{I} = \mathscr{I}(\mathscr{D})$.*

PROOF Part (a) follows from the definition of $\mathscr{I}(\mathscr{D})$ and the definition of multiplication in $E^*(M)$.

Let $m \in M$. Since $\mathscr{D}$ is smooth and p-dimensional, there exist C^∞ vector fields $X_{d-p+1}, \ldots, X_d$ defined and spanning $\mathscr{D}$ at each point of a neighborhood of m. This collection can be completed to a collection $X_1, \ldots, X_d$ of smooth vector fields forming a basis of M_n for each n in a neighborhood U of m. Let $\omega_1, \ldots, \omega_d$ be the dual 1-forms; that is,

$$\omega_i(X_j)(n) = \delta_{ij} \qquad \text{(Kronecker index)}$$

for each n in U. Then $\omega_1, \ldots, \omega_{d-p}$ are the desired 1-forms on U.

They are independent smooth 1-forms on U. If $\omega \in \mathscr{I}(\mathscr{D})$, $\omega \mid U = \sum a_\Phi \omega_{i_1} \wedge \cdots \wedge \omega_{i_r}$, where Φ runs over nonempty subsets $\{i_1, \ldots, i_r\} \subset \{1, \ldots, d\}$, and where the a_Φ must be identically zero unless $\{i_1, \ldots, i_r\} \cap \{1, \ldots, d-p\} \neq \varnothing$. Thus $\omega \mid U$ belongs to the ideal in $E^*(U)$ generated by $\omega_1, \ldots, \omega_{d-p}$. Conversely, if ω is a form such that for each such U in some covering of M, $\omega \mid U$ belongs to the ideal in $E^*(U)$ generated by $\omega_1, \ldots, \omega_{d-p}$, then clearly $\omega \in \mathscr{I}(\mathscr{D})$. This proves (b).

For part (c), let $m \in M$, and let the independent 1-forms $\omega_1, \ldots, \omega_{d-p}$ generate $\mathscr{I}$ on a neighborhood U of m. Define $\mathscr{D}(m)$ to be the subspace of M_m whose annihilator is the subspace of M_m^* spanned by the collection $\{\omega_i(m)\colon i = 1, \ldots, d-p\}$. It follows that $\mathscr{D}$ is a smooth p-dimensional distribution on M and that $\mathscr{I} = \mathscr{I}(\mathscr{D})$. Uniqueness of $\mathscr{D}$ follows from the fact that $\mathscr{D} \neq \mathscr{D}_1$ implies $\mathscr{I}(\mathscr{D}) \neq \mathscr{I}(\mathscr{D}_1)$.

2.29 Definition An ideal $\mathscr{I} \subset E^*(M)$ is called a *differential ideal* if it is closed under exterior differentiation d; that is,

(1) $$d(\mathscr{I}) \subset \mathscr{I}.$$

2.30 Proposition *A C^∞ distribution $\mathscr{D}$ on M is involutive if and only if the ideal $\mathscr{I}(\mathscr{D})$ is a differential ideal.*

PROOF Let ω be a q-form in $\mathscr{I}(\mathscr{D})$, and let $X_0, \ldots, X_q$ be smooth vector fields lying in $\mathscr{D}$. Then 2.25(f) together with the involutiveness of $\mathscr{D}$ implies that $d\omega(X_0, \ldots, X_q) \equiv 0$. Hence $d\omega \in \mathscr{I}(\mathscr{D})$, and $\mathscr{I}(\mathscr{D})$ is a differential ideal. Conversely, suppose that $\mathscr{I}(\mathscr{D})$ is a differential ideal. Let Y_0 and Y_1 be vector fields lying in $\mathscr{D}$, and let $m \in M$. By 2.28(b), there are independent 1-forms $\omega_1, \ldots, \omega_{d-p}$ generating $\mathscr{I}(\mathscr{D})$ on a neighborhood U of m. Extend these forms to M by multiplying by a C^∞ function which is 1 on a neighborhood of m and has support in U. We shall denote the extended forms similarly by $\omega_1, \ldots, \omega_{d-p}$. By 2.25(f),

(1) $$\omega_i[Y_0, Y_1] = -d\omega_i(Y_0, Y_1) + Y_0\omega_i(Y_1) - Y_1\omega_i(Y_0)$$

for $i = 1, \ldots, d-p$. The right-hand side of (1) is identically zero on M since $\mathscr{I}(\mathscr{D})$ is a differential ideal and since $\omega_i \in \mathscr{I}(\mathscr{D})$. Thus $\omega_i([Y_0, Y_1])(m) = 0$ for $i = 1, \ldots, d-p$. Now $\mathscr{D}(m)$ is the subspace of M_m whose annihilator is the subspace of M_m^* spanned by the collection $\{\omega_i(m)\colon i = 1, \ldots, d-p\}$. Thus $[Y_0, Y_1](m) \in \mathscr{D}(m)$, and $\mathscr{D}$ is involutive.

2.31 Definition A submanifold (N, ψ) of M is an *integral manifold of an ideal* $\mathscr{I} \subset E^*(M)$ if for every $\omega \in \mathscr{I}$, $\delta\psi(\omega) \equiv 0$. A connected integral manifold of an ideal $\mathscr{I}$ is *maximal* if its image is not a proper subset of the image of any other connected integral manifold of the ideal.

It now follows immediately that we have the following version of the Frobenius theorem 1.64 in terms of differential ideals.

2.32 Theorem *Let $\mathscr{I} \subset E^*(M)$ be a differential ideal locally generated by $d - p$ independent 1-forms. Let $m \in M$. Then there exists a unique maximal, connected, integral manifold of $\mathscr{I}$ through m, and this integral manifold has dimension p.*

2.33 We shall now consider a technique which in certain situations enables one to find a map by looking for its graph as an integral manifold of a differential ideal. This will have important applications in the theory of Lie groups in Chapter 3, and also has important applications in such areas as isometric imbeddings in Riemannian geometry, for example, see [2, §10.8].

Suppose that $f: N^c \to M^d$ is C^∞ and that $\{\omega_i\}$ is some collection of forms on M. Let π_1 and π_2 denote the natural projections of $N \times M$ onto N and M respectively. For each i, we define a form μ_i on $N \times M$ by setting

$$\mu_i = \delta\pi_1\delta f(\omega_i) - \delta\pi_2(\omega_i). \tag{1}$$

Let $\mathscr{I}$ be the ideal in $E^*(N \times M)$ generated by the μ_i.

Now the *graph* of f is the submanifold (N,g) of $N \times M$ where

$$g(n) = (n, f(n)). \tag{2}$$

We claim that the graph is an integral manifold of the ideal $\mathscr{I}$. For this, it suffices to show that $\delta g(\mu_i) = 0$ for each i. Now, $\pi_1 \circ g = \text{id}$, and $\pi_2 \circ g = f$, and thus it follows that

$$\delta g(\mu_i) = \delta(\pi_1 \circ g)\delta f(\omega_i) - \delta(\pi_2 \circ g)(\omega_i) = \delta f(\omega_i) - \delta f(\omega_i) = 0.$$

Thus starting with a map $f: N \to M$ and a collection of forms on M, we have observed that the graph of f is an integral manifold of a certain ideal of forms on $N \times M$. Now, suppose that we start with the manifold M^d, and suppose that there exists a basis $\omega_1, \ldots, \omega_d$ of 1-forms on M, that is, $\{\omega_i(m)\}$ is a basis of M_m^* for each $m \in M$. (Such a basis does not generally exist, but it does exist in a number of interesting situations for which the following technique is quite useful.) Suppose also that we have a manifold N^c and a collection $\alpha_1, \ldots, \alpha_d$ of 1-forms on N and that we wish to find a map $f: N \to M$ such that

$$\delta f(\omega_i) = \alpha_i \tag{3}$$

for $i = 1, \ldots, d$. If the map f exists, then, as we have observed, its graph will be an integral manifold of a certain ideal of forms. So we attempt to find the graph from the ideal. We define forms μ_i on $N \times M$ by setting

$$\mu_i = \delta\pi_1(\alpha_i) - \delta\pi_2(\omega_i), \tag{4}$$

and we let $\mathscr{I}$ be the ideal in $E^*(N \times M)$ which they generate. If $\mathscr{I}$ happens to be a differential ideal, then we can obtain the desired map f (at least locally) from an integral manifold of $\mathscr{I}$. That is, one gets at f by looking for its graph as an integral manifold of a suitable differential ideal. So suppose that $\mathscr{I}$ is a differential ideal, and let $(n_0,m_0) \in N \times M$. Then since $\mathscr{I}$ is locally (in fact globally) generated by d independent 1-forms, the Frobenius theorem guarantees that there is a maximal, connected, integral manifold I of $\mathscr{I}$ of dimension c through (n_0,m_0). Let $q \in I$. We claim that $d\pi_1 \mid I_q$ is 1:1. For suppose that $v \in I_q$ and that $d\pi_1(v) = 0$. Then since $\mu_i(v) = 0$, it follows from (4) that $\omega_i(d\pi_2(v)) = 0$ for $i = 1, \ldots, d$; and this implies that $d\pi_2(v) = 0$ since the ω_i form a basis of 1-forms on M. But now if both $d\pi_1(v) = 0$ and $d\pi_2(v) = 0$, then $v = 0$. Hence $d\pi_1 \mid I_q$ is 1:1. Thus $\pi_1 \mid I: I \to N$ is locally a diffeomorphism. So there exist neighborhoods V of (n_0,m_0) in I and U of n_0 such that $\pi_1 \mid V: V \to U$ is a diffeomorphism. We define $f: U \to M$ by setting

$$f = \pi_2 \circ (\pi_1 \mid V)^{-1}. \tag{5}$$

Then $f(n_0) = m_0$, the graph of f is an open submanifold of I, and moreover,

$$\delta f(\omega_i) = \alpha_i \mid U. \tag{6}$$

For let $v \in U_n$ for some $n \in U$. Then by (4),

$$\begin{aligned} 0 &= \mu_i(d(\pi_1 \mid V)^{-1}(v)) \\ &= \alpha_i(v) - \omega_i(d\pi_2 \circ d(\pi_1 \mid V)^{-1}(v)) = \alpha_i(v) - \delta f(\omega_i)(v), \end{aligned}$$

which proves (6). Thus we have obtained the desired map locally. That is, given $n_0 \in N$ and an arbitrary choice of $m_0 \in M$, then under the assumption that the ideal $\mathscr{I}$ generated by (4) is a differential ideal, we have found an open neighborhood U of n_0 and a C^∞ map $f: U \to M$ such that $f(n_0) = m_0$ and such that $\delta f(\omega_i) = \alpha_i \mid U$. Moreover, there is a unique such map. More precisely, if $m_0 \in M$, and if U is any connected open neighborhood of n_0 in N for which there exists a C^∞ map $f: U \to M$ such that $f(n_0) = m_0$ and such that $\delta f(\omega_i) = \alpha_i \mid U$ for $i = 1, \ldots, d$, then there is a unique such map on U. For let $\tilde{f}$ be any other such map. Let $(U,\tilde{g})$ and (U,g) be the graphs of $\tilde{f}$ and f over U respectively. Thus

$$\tilde{g}(n) = (n,\tilde{f}(n)) \quad \text{and} \quad g(n) = (n,f(n)) \tag{7}$$

for $n \in U$. Then, according to our remarks near the beginning of this section, not only is (U,g) an integral manifold of $\mathscr{I}$ through (n_0,m_0), but so is $(U,\tilde{g})$. Now, the subset of U on which g and $\tilde{g}$ agree is non-empty since it contains n_0, and is closed by continuity, and is moreover open. For let $\tilde{g}(n) = g(n)$. Then it follows from the uniqueness of integral manifolds that there exist sufficiently small neighborhoods W and $\tilde{W}$ of n so that

$$g(W) = \tilde{g}(\tilde{W}). \tag{8}$$

It follows then from (7) that $W = \tilde{W}$ and that

$$g \mid W = \tilde{g} \mid W. \tag{9}$$

Hence, since U is connected, $g = \tilde{g}$ on U, which implies that $f = \tilde{f}$ on U. This proves uniqueness.

In suitable situations one can conclude that $\pi_1 \mid I$ is a covering of N; and then if N is simply connected, one can conclude that there exists a unique C^∞ map $f\colon N \to M$ such that $f(n_0) = m_0$ and such that (3) holds. In Chapter 3 we shall make several concrete applications of this method for proving existence and uniqueness of certain maps. This technique was originally used by E. Cartan for the problem of finding local isometric imbeddings of Riemannian manifolds in Euclidean space.

We summarize the results of this section in the following.

2.34 Theorem *Let N^c and M^d be differentiable manifolds, and let π_1 and π_2 be the canonical projections of $N \times M$ onto N and M respectively. Suppose that there exists a basis $\{\omega_i\colon i = 1, \ldots, d\}$ for the 1-forms on M.*

(a) *If $f\colon N \to M$ is C^∞, then the graph of f is an integral manifold of the ideal of forms on $N \times M$ generated by*

(1) $$\{\delta\pi_1\delta f(\omega_i) - \delta\pi_2(\omega_i)\colon i = 1, \ldots, d\}.$$

(b) *If $\{\alpha_i\colon i = 1, \ldots, d\}$ are 1-forms on N, and if the ideal of forms on $N \times M$ generated by*

(2) $$\{\delta\pi_1(\alpha_i) - \delta\pi_2(\omega_i)\colon i = 1, \ldots, d\}$$

is a differential ideal, then given $n_0 \in N$ and $m_0 \in M$ there exists a neighborhood U of n_0 and a C^∞ map $f\colon U \to M$ such that $f(n_0) = m_0$ and such that

(3) $$\delta f(\omega_i) = \alpha_i \mid U \qquad (i = 1, \ldots, d).$$

Moreover, if U is any connected open set containing n_0 for which there exists a C^∞ map $f\colon U \to M$ satisfying both $f(n_0) = m_0$ and equation (3), then there exists a unique such map on U.

EXERCISES

1 Supply the proofs of 2.2(a)–(e).

2 (a) Show that homogeneous tensors are generally not decomposable.

(b) Show that if $\dim V \leq 3$, then every homogeneous element in $\Lambda(V)$ is decomposable.

(c) Let $\dim V > 3$. Give an example of an indecomposable homogeneous element of $\Lambda(V)$.

(d) Let α be a differential form. Is $\alpha \wedge \alpha \equiv 0$?

3 Supply the proofs of 2.6(a)–(c).

4 Derive the formulas (2), (3), and (4) of 2.10.

5 Prove that $L_X f = Xf$ whenever $f \in C^\infty(M)$ and X is a C^∞ vector field on M.

6 Let X and Y be C^∞ vector fields on M with corresponding local 1-parameter groups X_t and Y_t. Let $m \in M$, and let

$$\beta(t) = Y_{-\sqrt{t}} X_{-\sqrt{t}} Y_{\sqrt{t}} X_{\sqrt{t}}(m)$$

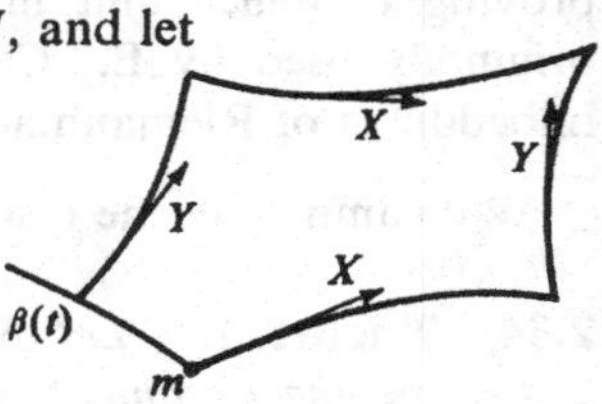

for t in $(-\varepsilon, \varepsilon)$, for a sufficiently small ε. Prove that

$$(1) \qquad [X,Y]|_m f = \lim_{t \to 0} \frac{f(\beta(t)) - f(\beta(0))}{t}.$$

If $\beta(t)$ were a smooth curve at $t = 0$, then the right-hand side of (1) would simply be the effect of the tangent vector to β at $t = 0$ applied to the function f. However, because of the $\sqrt{t}$, β is not generally smooth at $t = 0$. Thus the existence of the limit in (1) is part of the problem, and the problem asserts that this curve β, even though it is not smooth at $t = 0$, defines a tangent vector in M_m in the usual way, and this vector is precisely $[X, Y]|_m$.

7 Prove 2.25(e).

8 Let M be connected, and let $\pi\colon M \times N \to N$ be the natural projection. Prove that a p-form ω on $M \times N$ is $\delta\pi(\alpha)$ for some p-form α on N if and only if $i(X)\omega = 0$ and $L_X \omega = 0$ for every vector field X on $M \times N$ for which $d\pi(X(m,n)) = 0$ at each point $(m,n) \in M \times N$.

9 Prove that the elements $v_1, \ldots, v_r$ of the vector space V are linearly independent if and only if $v_1 \wedge \cdots \wedge v_r \neq 0$.

10 Prove that linearly independent sets $\{v_1, \ldots, v_r\}$ and $\{w_1, \ldots, w_r\}$ are bases of the same r-dimensional subspace of a vector space V if and only if $v_1 \wedge \cdots \wedge v_r = c w_1 \wedge \cdots \wedge w_r$ where necessarily $c \neq 0$; and in this case, $c = \det A$ where $A = (a_{ij})$ and $v_i = \sum a_{ij} w_j$.

11 Let $\mathscr{I}$ be an ideal of forms on M locally generated by r independent 1-forms. Say $\mathscr{I}$ is generated by $\omega_1, \ldots, \omega_r$ on U. Then the condition that $\mathscr{I}$ be a differential ideal is equivalent to each of:

(a) $d\omega_i = \sum_j \omega_{ij} \wedge \omega_j$ for some 1-forms ω_{ij} (for each such $(U, \omega_1, \ldots, \omega_r)$).

(b) If $\omega = \omega_1 \wedge \cdots \wedge \omega_r$, then $d\omega = \alpha \wedge \omega$ for some 1-form α (for each such $(U, \omega_1, \ldots, \omega_r)$).

12 Let V be an n-dimensional vector space, and let l be a linear transformation on V. Since $\Lambda_n(V)$ is one-dimensional, the linear transformation which l induces on $\Lambda_n(V)$ is simply multiplication by a constant. Define the *determinant of* l to be this constant. If A is an $n \times n$ matrix, let $v_1, \ldots, v_n$ be a basis of V, and let l be the linear transformation on V whose matrix with respect to this basis is A. Then define the *determinant of* A to be the determinant of l. Prove that the determinant of A does not depend on the basis $v_1, \ldots, v_n$ chosen. Using this definition, derive the standard properties of determinants. For example, derive the expansion

$$\det A = \sum_{\pi} (\operatorname{sgn} \pi) a_{1\pi(1)} \cdots a_{n\pi(n)}$$

where $A = (a_{ij})$, $\operatorname{sgn} \pi$ is the sign of the permutation π, and π runs over all permutations on n letters. Prove also that the determinant of the product of two matrices is the product of their determinants.

13 Let V be an n-dimensional real inner product space. We extend the inner product from V to all of $\Lambda(V)$ by setting the inner product of elements which are homogeneous of different degrees equal to zero, and by setting

$$\langle w_1 \wedge \cdots \wedge w_p, v_1 \wedge \cdots \wedge v_p \rangle = \det\langle w_i, v_j \rangle \tag{1}$$

and then extending bilinearly to all of $\Lambda_p(V)$. Prove that if $e_1, \ldots, e_n$ is an orthonormal basis of V, then the corresponding basis 2.6(1) of $\Lambda(V)$ is an orthonormal basis for $\Lambda(V)$.

Since $\Lambda_n(V)$ is one-dimensional, $\Lambda_n(V) - \{0\}$ has two components. An *orientation on* V is a choice of a component of $\Lambda_n(V) - \{0\}$. If V is an oriented inner product space, then there is a linear transformation

$$*: \Lambda(V) \rightarrow \Lambda(V), \tag{2}$$

called *star*, which is well-defined by the requirement that for *any* orthonormal basis $e_1, \ldots, e_n$ of V (in particular, for any re-ordering of a given basis),

$$\begin{gathered} *(1) = \pm e_1 \wedge \cdots \wedge e_n, \qquad *(e_1 \wedge \cdots \wedge e_n) = \pm 1, \\ *(e_1 \wedge \cdots \wedge e_p) = \pm e_{p+1} \wedge \cdots \wedge e_n, \end{gathered} \tag{3}$$

where one takes "+" if $e_1 \wedge \cdots \wedge e_n$ lies in the component of $\Lambda_n(V) - \{0\}$ determined by the orientation and "−" otherwise. Observe that

$$*: \Lambda_p(V) \rightarrow \Lambda_{n-p}(V). \tag{4}$$

Prove that on $\Lambda_p(V)$,

$$(5) \qquad ** = (-1)^{p(n-p)}.$$

Also prove that for arbitrary v, $w \in \Lambda_p(V)$, their inner product is given by

$$(6) \qquad \langle v,w \rangle = *(w \wedge *v) = *(v \wedge *w).$$

14 Let V be a real inner product space, as in Exercise 13. Let $\gamma: \Lambda_{p+1}(V) \to \Lambda_p(V)$ be the adjoint of left exterior multiplication by $\xi \in V$. That is,

$$\langle \gamma(v),w \rangle = \langle v, \xi \wedge w \rangle$$

for $v \in \Lambda_{p+1}(V)$ and $w \in \Lambda_p(V)$. Prove that

$$\gamma(v) = (-1)^{np} * (\xi \wedge (*v)).$$

15 Let $\xi \in V$. Prove that the composition

$$\Lambda_p(V) \xrightarrow{\xi \wedge} \Lambda_{p+1}(V) \xrightarrow{\xi \wedge} \Lambda_{p+2}(V)$$

of left exterior multiplication by ξ with itself is an exact sequence; that is, the image of the first map is the kernel of the second.

16 **Cartan Lemma** Let $p \leq d$, and let $\omega_1, \ldots, \omega_p$ be 1-forms on M^d which are linearly independent pointwise. Let $\theta_1, \ldots, \theta_p$ be 1-forms on M such that

$$\sum_{i=1}^{p} \theta_i \wedge \omega_i = 0.$$

Prove that there exist C^∞ functions A_{ij} on M with $A_{ij} = A_{ji}$ such that

$$\theta_i = \sum_{j=1}^{p} A_{ij}\omega_j \qquad (i = 1, \ldots, p).$$

3

LIE GROUPS

Lie groups are without doubt the most important special class of differentiable manifolds. Lie groups are differentiable manifolds which are also groups and in which the group operations are smooth. Well-known examples include the general linear group, the unitary group, the orthogonal group, and the special linear group.

In this chapter we shall set the foundations for the study of Lie groups. Of central importance for Lie theory is the relationship between a Lie group and its Lie algebra of left invariant vector fields. We shall study the correspondence between subgroups and subalgebras and between homomorphisms of Lie groups and homomorphisms of their Lie algebras. We shall study the properties of the exponential mapping which is a generalization to arbitrary Lie groups of the exponentiation of matrices and which provides a key link between a Lie group and its Lie algebra. We shall investigate the adjoint representation, shall prove the closed subgroup theorem, and shall consider basic properties and examples of homogeneous manifolds. Along the way we shall derive many properties of the classical linear groups.

LIE GROUPS AND THEIR LIE ALGEBRAS

3.1 Definition A *Lie group* G is a differentiable manifold which is also endowed with a group structure such that the map $G \times G \to G$ defined by $(\sigma,\tau) \mapsto \sigma\tau^{-1}$ is C^∞.

Throughout this chapter, G and H will denote Lie groups, and we shall universally use e to denote the identity element of a Lie group.

3.2 Remarks

(a) Let G be a Lie group. Then the map $\tau \mapsto \tau^{-1}$ is C^∞ since it is the composition $\tau \mapsto (e,\tau) \mapsto \tau^{-1}$ of C^∞ maps. Also, the map $(\sigma,\tau) \mapsto \sigma\tau$ of $G \times G \to G$ is C^∞ since it is the composition $(\sigma,\tau) \mapsto (\sigma,\tau^{-1}) \mapsto \sigma\tau$ of C^∞ maps.

(b) The identity component of a Lie group is itself a Lie group; and the components of a Lie group are mutually diffeomorphic.

(c) Since we have included second countability in the definition 1.4 of differentiable manifold, Lie groups for us will always be second countable. In particular, they can have at most countably many components. It should be observed, however, that if second countability is dropped from the definition of Lie group, then connected Lie groups (hence also those with countably many components) would still be second countable. We leave the proof of this as an exercise. You will need to use the fact that a connected Lie group is the union of powers of any neighborhood of its identity (see 3.18).

3.3 Examples of Lie Groups

(a) The Euclidean space $\mathbb{R}^n$ is a Lie group under vector addition.

(b) The non-zero complex numbers $\mathbb{C}^*$ form a Lie group under multiplication.

(c) The unit circle $S^1 \subset \mathbb{C}^*$ is a Lie group with the multiplication induced from $\mathbb{C}^*$.

(d) The product $G \times H$ of two Lie groups is itself a Lie group with the product manifold structure and the direct product group structure; that is, $(\sigma_1, \tau_1)(\sigma_2, \tau_2) = (\sigma_1\sigma_2, \tau_1\tau_2)$.

(e) The n-torus T^n (n an integer > 0) is the Lie group which is the product of the Lie group S^1 with itself n times.

(f) The manifold $Gl(n,\mathbb{R})$ of all $n \times n$ non-singular real matrices is a Lie group under matrix multiplication.

(g) The set of all non-singular super-triangular $n \times n$ real matrices (all entries below the diagonal are zero) is a Lie group under matrix multiplication.

(h) Let $\mathbb{R}^*$ denote the non-zero real numbers, and let K be the product manifold $\mathbb{R}^* \times \mathbb{R}$. With a group structure on K defined by

$$(s,t)(s_1,t_1) = (ss_1, st_1 + t),$$

K becomes a Lie group. This Lie group is the *group of affine motions of* $\mathbb{R}$, for if we identify the element (s,t) of K with the affine motion $x \mapsto sx + t$, then the multiplication in K is composition of affine motions.

(i) Let K be the product manifold $Gl(n,\mathbb{R}) \times \mathbb{R}^n$. We define a group structure on K by setting $(A,v)(A_1, v_1) = (AA_1, Av_1 + v)$. With this group structure, K becomes a Lie group. This Lie group is the *group of affine motions of* $\mathbb{R}^n$, for if we identify the element (A,v) of K with the affine motion $x \mapsto Ax + v$ of $\mathbb{R}^n$, then the multiplication in K is composition of affine motions.

3.4 Definition A *Lie algebra* $\mathfrak{g}$ over $\mathbb{R}$ is a real vector space $\mathfrak{g}$ together with a bilinear operator $[\ ,\ ]\colon \mathfrak{g} \times \mathfrak{g} \to \mathfrak{g}$ (called the *bracket*) such that for all $x, y, z \in \mathfrak{g}$,

(a) $[x,y] = -[y,x]$. *(anti-commutativity)*

(b) $[[x,y],z] + [[y,z],x] + [[z,x],y] = 0$. *(Jacobi identity)*

The importance of the concept of Lie algebra is that there is a special finite dimensional Lie algebra intimately associated with each Lie group, and that properties of the Lie group are reflected in properties of its Lie algebra. We shall see, for example, that the connected, simply connected Lie groups are completely determined (up to isomorphism) by their Lie algebras. The study of these Lie groups then reduces in large part to a study of their Lie algebras.

3.5 Examples of Lie Algebras

(a) The vector space of all smooth vector fields on the manifold M forms a Lie algebra under the Lie bracket operation on vector fields.

(b) Any vector space becomes a Lie algebra if all brackets are set equal to 0. Such a Lie algebra is called *abelian*.

(c) The vector space $\mathfrak{gl}(n,\mathbb{R})$ of all $n \times n$ real matrices forms a Lie algebra if we set

$$[A,B] = AB - BA.$$

(d) A 2-dimensional vector space with basis x, y becomes a Lie algebra if we set

$$[x,x] = [y,y] = 0 \qquad \text{and} \qquad [x,y] = y,$$

and extend bilinearly.

(e) $\mathbb{R}^3$ with the bilinear operation $X \times Y$ of the vector cross product is a Lie algebra.

The proof that the above are Lie algebras is left to the reader as an exercise.

3.6 Definitions Let $\sigma \in G$. *Left translation by* σ and *right translation by* σ are respectively the diffeomorphisms l_σ and r_σ of G defined by

$$l_\sigma(\tau) = \sigma\tau,$$
$$r_\sigma(\tau) = \tau\sigma \tag{1}$$

for all $\tau \in G$. If V is a subset of G, we denote $r_\sigma(V)$ and $l_\sigma(V)$ by $V\sigma$ and σV respectively. A vector field X (not assumed apriori to be smooth) on G is called *left invariant* if for each $\sigma \in G$, X is l_σ-related to itself; that is,

$$dl_\sigma \circ X = X \circ l_\sigma. \tag{2}$$

The set of all left invariant vector fields on a Lie group G will be denoted by the corresponding lowercase German letter $\mathfrak{g}$.

3.7 Proposition *Let G be a Lie group and $\mathfrak{g}$ its set of left invariant vector fields.*

(a) *$\mathfrak{g}$ is a real vector space, and the map $\alpha\colon \mathfrak{g} \to G_e$ defined by $\alpha(X) = X(e)$ is an isomorphism of $\mathfrak{g}$ with the tangent space G_e to G at the identity. Consequently,* $\dim \mathfrak{g} = \dim G_e = \dim G$.

(b) *Left invariant vector fields are smooth.*

(c) *The Lie bracket of two left invariant vector fields is itself a left invariant vector field.*

(d) *$\mathfrak{g}$ forms a Lie algebra under the Lie bracket operation on vector fields.*

PROOF That $\mathfrak{g}$ is a real vector space and that α is linear is clear. α is injective, for if $\alpha(X) = \alpha(Y)$, then for each $\sigma \in G$

(1) $$X(\sigma) = dl_\sigma(X(e)) = dl_\sigma(Y(e)) = Y(\sigma);$$

hence $X = Y$. Moreover, α is surjective; for if $x \in G_e$, we let $X(\sigma) = dl_\sigma(x)$ for each $\sigma \in G$. Then $\alpha(X) = x$, and X is left invariant since

$$X(\tau\sigma) = dl_{\tau\sigma}(x) = dl_\tau\, dl_\sigma(x) = dl_\tau(X(\sigma))$$

for all σ and τ in G. This proves part (a).

For (b), let $X \in \mathfrak{g}$, and let $f \in C^\infty(G)$. We need only show that $Xf \in C^\infty(G)$. Now,

(2) $$Xf(\sigma) = X_\sigma f = dl_\sigma(X_e)f = X_e(f \circ l_\sigma).$$

Thus we need to show that $\sigma \mapsto X_e(f \circ l_\sigma)$ is a C^∞ function on G. We do this by exhibiting this function as a suitable composition of C^∞ maps. Let $\varphi\colon G \times G \to G$ denote group multiplication, $\varphi(\sigma,\tau) = \sigma\tau$. And let i_e^1 and i_σ^2 be the maps of $G \to G \times G$ defined by

(3) $$\begin{aligned} i_e^1(\tau) &= (\tau,e),\\ i_\sigma^2(\tau) &= (\sigma,\tau). \end{aligned}$$

Let Y be any C^∞ vector field on G such that $Y(e) = X(e)$. Then $(0, Y)$ is a smooth vector field on $G \times G$, and $[(0,Y)(f \circ \varphi)] \circ i_e^1$ is a C^∞ function on G. Using part (d) of Exercise 24, Chapter 1, we obtain

$$\begin{aligned} [(0,Y)(f \circ \varphi)] \circ i_e^1(\sigma) &= (0,Y)_{(\sigma,e)}(f \circ \varphi)\\ &= 0_\sigma(f \circ \varphi \circ i_e^1) + Y_e(f \circ \varphi \circ i_\sigma^2)\\ &= X_e(f \circ \varphi \circ i_\sigma^2) = X_e(f \circ l_\sigma). \end{aligned}$$

Thus $\sigma \mapsto X_e(f \circ l_\sigma)$ is a smooth function on G, which proves part (b).

Since, by (b), left invariant vector fields are smooth, their Lie brackets are defined. Now, if X is l_σ-related to itself, and Y is l_σ-related to itself, then according to 1.55, $[X,Y]$ is l_σ-related to $[X,Y]$. Thus the Lie bracket of two left invariant vector fields is again a left invariant vector field. Part (d) is now an immediate consequence of 1.45.

3.8 Definition We define the *Lie algebra of the Lie group G* to be the Lie algebra $\mathfrak{g}$ of left invariant vector fields on G. Alternatively, we could take as the Lie algebra of G the tangent space G_e at the identity with Lie algebra structure induced by requiring the vector space isomorphism 3.7(a) of $\mathfrak{g}$ with G_e to be an isomorphism of Lie algebras. It will be convenient at times to consider the Lie algebra from this alternate point of view. In particular, we shall see (3.10 and 3.37) that the Lie algebras of the classical groups admit particularly nice interpretations in terms of their tangent spaces at the identity.

3.9 Remarks on Vector Spaces as Manifolds In 1.5(b) we observe that any finite dimensional real vector space V is, in a natural way, a differentiable manifold. Let $\{e_i\}$ be a basis of V, and let $\{r_i\}$ be the dual basis. Now let $p \in V$. Then there is a natural identification of the tangent space V_p to V at p with V itself, given by

(1) $$\sum a_i \frac{\partial}{\partial r_i}\bigg|_p \leftrightarrow \sum a_i e_i.$$

It is easily seen that this identification is independent of the basis chosen. We shall often make use of this identification in describing tangent vectors to a vector space as elements of the vector space itself. In particular, if $\sigma(t)$ is a smooth curve in V, then

(2) $$\dot\sigma(t_0) = \lim_{t\to t_0} \frac{\sigma(t) - \sigma(t_0)}{t - t_0}.$$

3.10 Examples of Lie Groups and Their Lie Algebras

(a) The real line $\mathbb{R}$ is a Lie group under addition. The left invariant vector fields are simply the constant vector fields $\{\lambda(d/dr) \colon \lambda \in \mathbb{R}\}$. The bracket of any two such vector fields is 0.

(b) *The General Linear Group.* The set $\mathfrak{gl}(n,\mathbb{R})$ of all $n \times n$ real matrices is a real vector space of dimension n^2. Matrices are added and multiplied by scalars componentwise. As we have already remarked, $\mathfrak{gl}(n,\mathbb{R})$ becomes a Lie algebra if we set $[A,B] = AB - BA$.

The general linear group $Gl(n,\mathbb{R})$ inherits its manifold structure as the open subset of $\mathfrak{gl}(n,\mathbb{R})$ where the determinant function does not vanish, and is a Lie group under matrix multiplication. Let x_{ij} be the global coordinate function on $\mathfrak{gl}(n,\mathbb{R})$ which assigns to each matrix its ijth entry. Then if $\sigma, \tau \in Gl(n,\mathbb{R})$, $x_{ij}(\sigma\tau^{-1})$ is a rational function of $\{x_{kl}(\sigma)\}$ and $\{x_{kl}(\tau)\}$ with non-zero denominator, which proves that the map $(\sigma,\tau) \to \sigma\tau^{-1}$ is C^∞.

Now let $\mathfrak{g}$ be the Lie algebra of $Gl(n,\mathbb{R})$. Let $\alpha\colon \mathfrak{gl}(n,\mathbb{R})_e \to \mathfrak{gl}(n,\mathbb{R})$ be the canonical identification of the tangent space to $\mathfrak{gl}(n,\mathbb{R})$ at the identity matrix e with $\mathfrak{gl}(n,\mathbb{R})$ itself. Thus for $v \in \mathfrak{gl}(n,\mathbb{R})_e$,

(1) $$\alpha(v)_{ij} = v(x_{ij}).$$

Then since $Gl(n,\mathbb{R})_e = \mathfrak{gl}(n,\mathbb{R})_e$, we have a natural map $\beta\colon \mathfrak{g} \to \mathfrak{gl}(n,\mathbb{R})$ defined by

(2) $$\beta(X) = \alpha(X(e)).$$

We claim that *β is a Lie algebra isomorphism*, and in this way we consider $\mathfrak{gl}(n,\mathbb{R})$ to be the Lie algebra of $Gl(n,\mathbb{R})$. β is clearly a vector space isomorphism. We need only show that

(3) $$\beta([X,Y]) = [\beta(X),\beta(Y)]$$

whenever X, $Y \in \mathfrak{g}$. Now,

(4) $$(x_{ij} \circ l_\sigma)(\tau) = x_{ij}(\sigma\tau) = \sum_k x_{ik}(\sigma)x_{kj}(\tau).$$

Since Y is a left invariant vector field,

(5) $$\begin{aligned}(Y(x_{ij}))(\sigma) &= dl_\sigma(Y_e)(x_{ij}) = Y_e(x_{ij} \circ l_\sigma)\\ &= \sum_k x_{ik}(\sigma)Y_e(x_{kj}) = \sum_k x_{ik}(\sigma)\alpha(Y_e)_{kj}\\ &= \sum_k x_{ik}(\sigma)\beta(Y)_{kj}.\end{aligned}$$

Using (5), we now compute the ijth component of $\beta([X,Y])$:

(6) $$\begin{aligned}\beta([X,Y])_{ij} &= [X,Y]_e(x_{ij}) = X_e(Y(x_{ij})) - Y_e(X(x_{ij}))\\ &= \sum_k \{X_e(x_{ik})\beta(Y)_{kj} - Y_e(x_{ik})\beta(X)_{kj}\}\\ &= \sum_k \{\beta(X)_{ik}\beta(Y)_{kj} - \beta(Y)_{ik}\beta(X)_{kj}\}\\ &= [\beta(X),\beta(Y)]_{ij}.\end{aligned}$$

Hence β is a Lie algebra isomorphism.

(c) Let V be an n-dimensional real vector space. Let $\mathrm{End}(V)$ denote the set of all linear operators on V (the set of endomorphisms of V), and let $\mathrm{Aut}(V) \subset \mathrm{End}(V)$ denote the subset of non-singular operators (the automorphisms). $\mathrm{End}(V)$ is a real vector space of dimension n^2, and it becomes a Lie algebra if we set

(7) $$[l_1,l_2] = l_1 \circ l_2 - l_2 \circ l_1.$$

A basis of V determines a diffeomorphism of End(V) with gl(n,$\mathbb{R}$) sending Aut(V) onto $Gl(n,\mathbb{R})$. It follows that Aut(V) inherits a manifold structure as an open subset of End(V) and is a Lie group under composition. Under the natural identification of End(V) with $\text{End}(V)_e = \text{Aut}(V)_e$ (where e denotes the identity transformation on V), End(V) inherits a Lie algebra structure from the Lie algebra of Aut(V). This induced Lie algebra structure is precisely the one described in (7).

(d) Let gl(n,$\mathbb{C}$) denote the set of all $n \times n$ complex matrices, and let $Gl(n,\mathbb{C}) \subset \text{gl}(n,\mathbb{C})$ be the subset of non-singular ones. $Gl(n,\mathbb{C})$ is known as the *complex general linear group*. gl(n,$\mathbb{C}$) is a $2n^2$-dimensional real vector space, with a basis consisting of the matrices δ_{ij} and $\sqrt{-1}\delta_{ij}$ $(i,j = 1, \ldots, n)$ where δ_{ij} is the matrix all of whose entries are zero except for a 1 in the ijth spot; and gl(n,$\mathbb{C}$) forms a Lie algebra if we set $[A,B] = AB - BA$. $Gl(n,\mathbb{C})$ inherits a manifold structure as an open subset of gl(n,$\mathbb{C}$) and is a Lie group under matrix multiplication. With considerations entirely analogous to those in Example (b), one sees that the natural identification of the Lie algebra of $Gl(n,\mathbb{C})$ with gl(n,$\mathbb{C}$) is a Lie algebra isomorphism. Thus we may consider gl(n,$\mathbb{C}$) as the Lie algebra of $Gl(n,\mathbb{C})$.

(e) Similarly, in analogy with Example (c), if V is a complex n-dimensional vector space, and if End(V) denotes the set of complex linear transformations of V, and Aut(V) $\subset$ End(V) denotes the non-singular ones, then Aut(V) is a $2n^2$-dimensional Lie group with Lie algebra End(V).

3.11 Definition A form ω on G is called *left invariant* if

(1) $$\delta l_\sigma \omega = \omega$$

for each $\sigma \in G$. As in the case of left invariant vector fields, it is not necessary to assume that left invariant forms are smooth, for smoothness is a consequence of the left invariance. We shall denote the vector space of left invariant p-forms on G by $E^p_{l\,\text{inv}}(G)$, and we let

(2) $$E^*_{l\,\text{inv}}(G) = \sum_{p=0}^{\dim G} E^p_{l\,\text{inv}}(G).$$

Left invariant 1-forms are also known as *Maurer-Cartan forms*.

The following proposition is the analog for left invariant forms of 3.7. The proofs are straightforward, and we leave them to the reader as an exercise.

3.12 Proposition

(a) *Left invariant forms are smooth.*

(b) $E^*_{l\,\mathrm{inv}}(G)$ *is a subalgebra of the algebra* $E^*(G)$ *of all smooth forms on* G, *and the map* $\omega \to \omega(e)$ *is an algebra isomorphism of* $E^*_{l\,\mathrm{inv}}(G)$ *onto* $\Lambda(G^*_e)$. *In particular, this map gives a natural isomorphism of* $E^1_{l\,\mathrm{inv}}(G)$ *with* G^*_e *and hence with* $\mathfrak{g}^*$. (In this way we shall consider $E^1_{l\,\mathrm{inv}}(G)$ as the dual space of the Lie algebra of G.)

(c) *If* ω *is a left invariant* 1*-form and* X *a left invariant vector field, then* $\omega(X)$ *is a constant function on* G, *and this constant is precisely the effect* ω *has on* X *when* ω *is considered as an element of the dual space of* $\mathfrak{g}$ *as in part* (b). (We shall consider $\omega(X)$ either as a constant *function* on G or as the corresponding *real number*, the particular choice depending on the context.)

(d) *If* $\omega \in E^1_{l\,\mathrm{inv}}(G)$ *and* X, $Y \in \mathfrak{g}$, *then it follows from* 2.25(f) *that*

(1) $$d\omega(X,Y) = -\omega[X,Y].$$

(e) *Let* $\{X_1, \ldots, X_d\}$ *be a basis of* $\mathfrak{g}$ *with dual basis* $\{\omega_1, \ldots, \omega_d\}$ *for* $E^1_{l\,\mathrm{inv}}(G)$. *Then there exist constants* c_{ijk} *such that*

(2) $$[X_i, X_j] = \sum_{k=1}^{d} c_{ijk} X_k.$$

(The c_{ijk} are called the *structural constants* of G with respect to the basis $\{X_i\}$ of $\mathfrak{g}$.) *They satisfy*

(3) $$c_{ijk} + c_{jik} = 0,$$
$$\sum_r (c_{ijr}c_{rks} + c_{jkr}c_{ris} + c_{kir}c_{rjs}) = 0.$$

The exterior derivatives of the ω_i *are given by the Maurer-Cartan equations*

(4) $$d\omega_i = \sum_{j<k} c_{jki}\,\omega_k \wedge \omega_j.$$

HOMOMORPHISMS

3.13 Definitions A map $\varphi\colon G \to H$ is a (*Lie group*) *homomorphism* if φ is both C^∞ and a group homomorphism of the abstract groups. We call φ an *isomorphism* if, in addition, φ is a diffeomorphism. An isomorphism of a Lie group with itself is called an *automorphism*. If $H = \mathrm{Aut}(V)$ for some vector space V, or if $H = Gl(n,\mathbb{C})$ or $Gl(n,\mathbb{R})$, then a homomorphism $\varphi\colon G \to H$ is called a *representation of the Lie group G*.

If $\mathfrak{g}$ and $\mathfrak{h}$ are Lie algebras, a map $\psi\colon \mathfrak{g} \to \mathfrak{h}$ is a *(Lie algebra) homomorphism* if it is linear and preserves brackets ($\psi[X,Y] = [\psi(X),\psi(Y)]$ for all $X, Y \in \mathfrak{g}$). If, in addition, ψ is 1:1 and onto, then ψ is an *isomorphism.* An isomorphism of $\mathfrak{g}$ with itself is called an *automorphism.* If $\mathfrak{h} = \mathrm{End}(V)$ for some vector space V, or if $\mathfrak{h} = \mathfrak{gl}(n,\mathbb{C})$ or $\mathfrak{gl}(n,\mathbb{R})$, then a homomorphism $\psi\colon \mathfrak{g} \to \mathfrak{h}$ is called a *representation of the Lie algebra* $\mathfrak{g}$.

Let $\varphi\colon G \to H$ be a homomorphism. Then since φ maps the identity of G to the identity of H, the differential $d\varphi$ of φ is a linear transformation of G_e into H_e. By means of the natural identifications of the tangent spaces at the identities with the Lie algebras, this linear transformation $d\varphi$ of G_e into H_e induces a linear transformation of $\mathfrak{g}$ into $\mathfrak{h}$ which we shall denote also by $d\varphi$. Thus

$$d\varphi\colon \mathfrak{g} \to \mathfrak{h}, \tag{1}$$

where if $X \in \mathfrak{g}$, then $d\varphi(X)$ is the unique left invariant vector field on H such that

$$d\varphi(X)(e) = d\varphi(X(e)). \tag{2}$$

3.14 Theorem *Let G and H be Lie groups with Lie algebras $\mathfrak{g}$ and $\mathfrak{h}$ respectively, and let $\varphi\colon G \to H$ be a homomorphism. Then*

(a) *X and $d\varphi(X)$ are φ-related for each $X \in \mathfrak{g}$.*

(b) *$d\varphi\colon \mathfrak{g} \to \mathfrak{h}$ is a Lie algebra homomorphism.*

PROOF Let $\tilde{X} = d\varphi(X)$. Then $\tilde{X}$ and X are φ-related. For since φ is a homomorphism, $l_{\varphi(\sigma)} \circ \varphi = \varphi \circ l_\sigma$; hence

$$\begin{aligned}(1)\quad \tilde{X}(\varphi(\sigma)) &= dl_{\varphi(\sigma)}\tilde{X}(e) = dl_{\varphi(\sigma)}\, d\varphi(X(e)) \\ &= d(l_{\varphi(\sigma)} \circ \varphi)X(e) = d(\varphi \circ l_\sigma)X(e) = d\varphi(X(\sigma)).\end{aligned}$$

This proves part (a). Now let $X, Y \in \mathfrak{g}$. Then for part (b), we must show that

$$\widetilde{[X,Y]} = [\tilde{X}, \tilde{Y}]. \tag{2}$$

By 1.55, $[X,Y]$ is φ-related to the left invariant vector field $[\tilde{X}, \tilde{Y}]$. In particular,

$$[\tilde{X}, \tilde{Y}](e) = d\varphi([X,Y](e)). \tag{3}$$

But $\widetilde{[X,Y]}$, by Definition 3.13(2), is the unique left invariant vector field on H whose value at the identity is $d\varphi([X,Y](e))$. Thus (2) holds, and the theorem is proved.

3.15 Effect of Homomorphisms on Left Invariant Forms Let $\varphi\colon G \to H$ be a homomorphism. Then $\delta\varphi$ pulls left invariant forms on H back to left invariant forms on G, since whenever ω is a left invariant form on H,

(1)
$$\delta l_\sigma \delta\varphi(\omega) = \delta(\varphi \circ l_\sigma)\omega = \delta(l_{\varphi(\sigma)} \circ \varphi)\omega$$
$$= \delta\varphi\delta l_{\varphi(\sigma)}(\omega) = \delta\varphi(\omega).$$

Moreover, the mapping $\delta\varphi\colon E^1_{l\,\mathrm{inv}}(H) \to E^1_{l\,\mathrm{inv}}(G)$, considered as a mapping of the dual space of $\mathfrak{h}$ to the dual space of $\mathfrak{g}$, is precisely the transpose of $d\varphi\colon \mathfrak{g} \to \mathfrak{h}$; that is,

(2)
$$\big(\delta\varphi(\omega)\big)(X) = \omega\big(d\varphi(X)\big) \qquad \big(\omega \in E^1_{l\,\mathrm{inv}}(H); \quad X \in \mathfrak{g}\big).$$

Recall that if $\{\omega_1, \ldots, \omega_d\}$ is a basis of $E^1_{l\,\mathrm{inv}}(H)$, then there are structural constants $\{c_{ijk}\}$ (see 3.12(4)) such that

(3)
$$d\omega_i = \sum_{j<k} c_{jki}\omega_k \wedge \omega_j.$$

Thus since d and $\delta\varphi$ commute,

(4)
$$d\big(\delta\varphi(\omega_i)\big) = \sum_{j<k} c_{jki}\, \delta\varphi(\omega_k) \wedge \delta\varphi(\omega_j).$$

Now let π_1 and π_2 be the canonical projections of $G \times H$ onto G and H respectively. Then the ideal $\mathscr{I}$ of forms on $G \times H$ generated by the collection of independent 1-forms

(5)
$$\{\delta\pi_1\delta\varphi(\omega_i) - \delta\pi_2(\omega_i)\colon i = 1, \ldots, d\}$$

is a differential ideal; for using (4), we obtain for each i,

(6)
$$d\big(\delta\pi_1\, \delta\varphi(\omega_i) - \delta\pi_2(\omega_i)\big)$$
$$= \sum_{j<k} c_{jki}\big(\delta\pi_1\delta\varphi(\omega_k) \wedge \delta\pi_1\delta\varphi(\omega_j) - \delta\pi_2(\omega_k) \wedge \delta\pi_2(\omega_j)\big)$$
$$= \sum_{j<k} c_{jki}[\{\delta\pi_1\, \delta\varphi(\omega_k) - \delta\pi_2(\omega_k)\} \wedge \delta\pi_1\, \delta\varphi(\omega_j)$$
$$+ \delta\pi_2(\omega_k) \wedge \{\delta\pi_1\, \delta\varphi(\omega_j) - \delta\pi_2(\omega_j)\}],$$

which belongs to the ideal $\mathscr{I}$. Moreover, observe that the 1-forms (5) generating the differential ideal $\mathscr{I}$ are themselves left invariant 1-forms on $G \times H$. This follows immediately from applying $\delta l_{(\sigma,\xi)}$ to the forms (5), where $(\sigma,\xi) \in G \times H$, and from using the fact that $\pi_1 \circ l_{(\sigma,\xi)} = l_\sigma \circ \pi_1$ and $\pi_2 \circ l_{(\sigma,\xi)} = l_\xi \circ \pi_2$. Note also that the basis $\{\omega_i\}$ of $E^1_{l\,\mathrm{inv}}(H)$ is a basis of the 1-forms on H in the sense of 2.33.

To carry out the above construction of a differential ideal on $G \times H$, it is sufficient to start simply with a homomorphism of the Lie algebras rather than a homomorphism of the Lie groups. Let G and H be Lie groups, and let $\psi\colon \mathfrak{g} \to \mathfrak{h}$ be a homomorphism of their Lie algebras. ψ has a transpose $\psi^*\colon E^1_{l\,\mathrm{inv}}(H) \to E^1_{l\,\mathrm{inv}}(G)$, namely,

(7)
$$\big(\psi^*(\omega)\big)(X) = \omega\big(\psi(X)\big) \qquad \big(\omega \in E^1_{l\,\mathrm{inv}}(H); X \in \mathfrak{g}\big).$$

Let $\{\omega_i\}$ be a basis of $E^1_{l\,\mathrm{inv}}(H)$. Then we claim that the ideal $\mathscr{I}$ of forms on $G \times H$ generated by the collection of independent 1-forms

$$\{\delta\pi_1(\psi^*(\omega_i)) - \delta\pi_2(\omega_i) : i = 1, \ldots, d\} \tag{8}$$

is a differential ideal. This follows from a computation similar to (6) together with the observation that

$$d(\psi^*(\omega_i)) = \sum_{j<k} c_{jki}\psi^*(\omega_k) \wedge \psi^*(\omega_j). \tag{9}$$

To prove (9), it suffices to prove that both sides have the same effect on an arbitrary pair X, Y of left invariant vector fields on G:

$$\begin{aligned} d(\psi^*(\omega_i))(X,Y) &= -\psi^*(\omega_i)[X,Y] = -\omega_i[\psi(X),\psi(Y)] \\ &= d\omega_i(\psi(X),\psi(Y)) \\ &= \sum_{j<k} c_{jki}\omega_k \wedge \omega_j(\psi(X),\psi(Y)) \\ &= \sum_{j<k} c_{jki}\psi^*(\omega_k) \wedge \psi^*(\omega_j)(X,Y). \end{aligned} \tag{10}$$

Here the first and third equalities follow from 3.12(1). Finally, observe that, in this case also, the forms (8) are left invariant on $G \times H$.

We shall make use of these ideals on $G \times H$ in proving existence and uniqueness of homomorphisms in various situations.

3.16 Theorem *Let G be connected, and let φ and ψ be homomorphisms of G into H such that the Lie algebra homomorphisms $d\varphi$ and $d\psi$ of $\mathfrak{g}$ into $\mathfrak{h}$ are identical. Then $\varphi = \psi$.*

PROOF Since $d\varphi = d\psi$, we have

$$\delta\varphi = \delta\psi : E^1_{l\,\mathrm{inv}}(H) \to E^1_{l\,\mathrm{inv}}(G). \tag{1}$$

Thus we have two C^∞ maps φ and ψ of the connected manifold G into H, both agreeing at $e \in G$, and both having the same effect of pulling back a basis of 1-forms from H. Thus since the ideal of forms on $G \times H$ generated by the 1-forms 3.15(5) is a differential ideal, it follows from 2.34(b) that $\varphi = \psi$.

LIE SUBGROUPS

3.17 Definitions (H,φ) is a *Lie subgroup* of the Lie group G if

(a) H is a Lie group;

(b) (H,φ) is a submanifold of G;

(c) $\varphi : H \to G$ is a group homomorphism.

(H,φ) is called a *closed subgroup* of G if, in addition, $\varphi(H)$ is a closed subset of G.

Let $\mathfrak{g}$ be a Lie algebra. A subspace $\mathfrak{h} \subset \mathfrak{g}$ is a *subalgebra* if $[X, Y] \in \mathfrak{h}$ whenever X, $Y \in \mathfrak{h}$. A subalgebra $\mathfrak{h} \subseteq \mathfrak{g}$ clearly forms a Lie algebra under the bracket induced from $\mathfrak{g}$.

Let (H,φ) be a Lie subgroup of G, and let $\mathfrak{h}$ and $\mathfrak{g}$ be their respective Lie algebras. Then $d\varphi$ gives an isomorphism of $\mathfrak{h}$ with the subalgebra $d\varphi(\mathfrak{h})$ of $\mathfrak{g}$.

Various theorems assert the existence of unique subgroups satisfying certain conditions, and as in the case of submanifolds (1.33 is pertinent), uniqueness needs a little explanation in view of our definition of subgroup. We will consider two subgroups (H,φ) and (H_1,φ_1) of G equivalent if there exists a Lie group isomorphism $\alpha\colon H \to H_1$ such that $\varphi_1 \circ \alpha = \varphi$. This is an equivalence relation on the Lie subgroups of G, and uniqueness for Lie subgroups means uniqueness up to this equivalence. As in the case of submanifolds (see 1.33), each equivalence class of Lie subgroups of G has a unique representative of the form (A,i), where A is a subset of G which is an abstract subgroup of G and which has a manifold structure (not necessarily with the relative topology) making A into a Lie group such that the inclusion $i\colon A \to G$ yields a submanifold and hence a Lie subgroup of G. When we wish to consider a subgroup of G in this latter form (i.e. as an actual subset of G) we usually drop any reference to a mapping and speak simply of the Lie subgroup A of G, the inclusion map being understood. Also, we usually identify the Lie algebra $\mathfrak{a}$ of A with $di(\mathfrak{a})$ and simply speak of the Lie algebra of A as a subalgebra of the Lie algebra of G. In particular, we identify a left invariant vector field X on A with the left invariant vector field $di(X)$ on G with which it is inclusion-related.

We are now going to prove one of the fundamental theorems of Lie group theory, which asserts that there is a 1:1 correspondence between connected Lie subgroups of a Lie group and subalgebras of its Lie algebra. But first, we will need the following proposition.

3.18 Proposition *Let G be a connected Lie group, and let U be a neighborhood of e. Then*

$$G = \bigcup_{n=1}^{\infty} U^n \tag{1}$$

where U^n consists of all n-fold products of elements of U. (We say that U *generates* G.)

PROOF Let V be an open subset of U containing e such that $V = V^{-1}$ (where $V^{-1} = \{\sigma^{-1}\colon \sigma \in V\}$). For example, $V = U \cap U^{-1}$ will do. Let

$$H = \bigcup_{n=1}^{\infty} V^n \subset \bigcup_{n=1}^{\infty} U^n. \tag{2}$$

Then H is an abstract subgroup of G and is an open subset of G since $\sigma \in H$ implies $\sigma V \subset H$. Thus each coset mod H is open in G. Now, H is the complement in G of the union of all the cosets mod H different from H itself. Therefore H is also a closed subset of G. Since G is connected, and H is also non-empty, H must be all of G. This together with (2) implies (1).

3.19 Theorem *Let G be a Lie group with Lie algebra $\mathfrak{g}$, and let $\tilde{\mathfrak{h}} \subset \mathfrak{g}$ be a subalgebra. Then there is a unique connected Lie subgroup (H,φ) of G such that $d\varphi(\mathfrak{h}) = \tilde{\mathfrak{h}}$.*

PROOF We define a distribution $\mathscr{D}$ on G by setting

(1) $$\mathscr{D}(\sigma) = \{X(\sigma): X \in \tilde{\mathfrak{h}}\}$$

for each $\sigma \in G$. Let $\dim \tilde{\mathfrak{h}} = d$. Then $\mathscr{D}$ is of dimension d. $\mathscr{D}$ is smooth since $\mathscr{D}$ is globally spanned by a basis $X_1, \ldots, X_d$ for $\tilde{\mathfrak{h}}$. Moreover, $\mathscr{D}$ is involutive, since if X and Y are vector fields lying in $\mathscr{D}$, then there are C^∞ functions $\{a_i\}$ and $\{b_i\}$ on G such that $X = \sum a_i X_i$ and $Y = \sum b_i X_i$, and thus

(2) $$[X,Y] = \sum_{i,j=1}^{d} \{a_i b_j [X_i,X_j] + a_i X_i(b_j) X_j - b_j X_j(a_i) X_i\},$$

which again is a vector field in $\mathscr{D}$ since $\tilde{\mathfrak{h}}$ is a subalgebra of $\mathfrak{g}$.

Let (H,φ) be a maximal connected integral manifold of $\mathscr{D}$ through e (see 1.64). Let $\sigma \in \varphi(H)$. Since $\mathscr{D}$ is invariant under left translations, $(H, l_{\sigma^{-1}} \circ \varphi)$ is also an integral manifold of $\mathscr{D}$ through e. Thus by the maximality, $l_{\sigma^{-1}} \circ \varphi(H) \subset \varphi(H)$. Therefore if $\sigma \in \varphi(H)$ and $\tau \in \varphi(H)$, then also $\sigma^{-1}\tau \in \varphi(H)$. It follows that $\varphi(H)$ is an abstract subgroup of G. Thus we can induce a group structure on H so that $\varphi: H \to G$ is a homomorphism of abstract groups. It remains only to check that H is a Lie group, that is, the map $\alpha: H \times H \to H$ where $\alpha(\sigma,\tau) = \sigma\tau^{-1}$ is C^∞. Now, the map $\beta: H \times H \to G$ sending $(\sigma,\tau) \mapsto \varphi(\sigma)\varphi(\tau)^{-1}$ is C^∞, and we have the following commutative diagram:

(3)

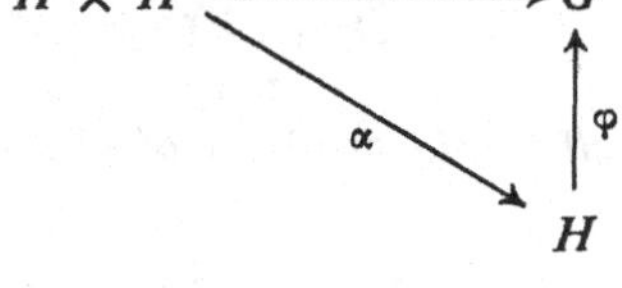

Since (H,φ) is an integral manifold of an involutive distribution on G, it follows from 1.62 that α is C^∞. Thus (H,φ) is a Lie subgroup of G; and if $\mathfrak{h}$ is the Lie algebra of H, clearly $d\varphi(\mathfrak{h}) = \tilde{\mathfrak{h}}$.

For uniqueness, let (K,ψ) be another connected Lie subgroup of G with $d\psi(\mathfrak{k}) = \tilde{\mathfrak{h}}$. Then (K,ψ) must also be an integral manifold of $\mathscr{D}$ through e. By the maximality of (H,φ), $\psi(K) \subset \varphi(H)$, and there is uniquely determined a map η such that $\varphi \circ \eta = \psi$:

(4)

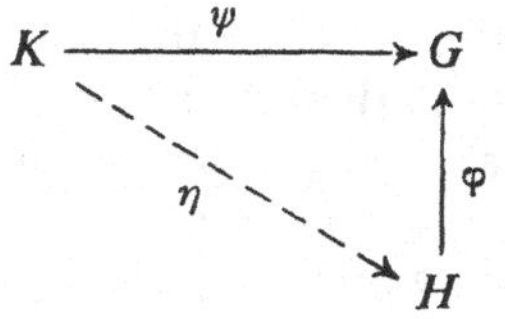

η is smooth by 1.62, and consequently is an injective Lie group homomorphism. Moreover, η is everywhere non-singular; in particular, η is a diffeomorphism on a neighborhood of the identity, and therefore is surjective by 3.18. Thus η is a Lie group isomorphism, and the subgroups (K,ψ) and (H,φ) are equivalent. This proves uniqueness.

Corollary (a) *There is a* 1:1 *correspondence between connected Lie subgroups of a Lie group and subalgebras of its Lie algebra.*

Corollary (b) *Let (H,φ) be a Lie subgroup of G. Then if $\tilde{H}$ is a component of H, $(\tilde{H}, \varphi \mid \tilde{H})$ is a maximal connected integral manifold of the involutive distribution on G determined by the subalgebra $d\varphi(\mathfrak{h})$ of $\mathfrak{g}$.*

We can restate the content of part of the proof of Theorem 3.19 in terms of differential ideals on G as follows (cf. Exercise 6):

Corollary (c) *Suppose that the ideal $\mathscr{I}$ generated by a collection $\{\omega_1, \ldots, \omega_{c-d}\}$ of independent left invariant* 1*-forms on a Lie group G^c is a differential ideal. Then the maximal connected integral manifold I^d of $\mathscr{I}$ through $e \in G$ is a Lie subgroup of G.*

The following theorem should be compared with the situation for submanifolds (see 1.33).

3.20 Theorem *If an abstract subgroup A of a Lie group G has a manifold structure (that is, a second countable locally Euclidean topology together with a differentiable structure) which makes (A,i) into a submanifold of G, where i is the inclusion map, then it has a unique such manifold structure, and in this manifold structure, A is a Lie group, and hence (A,i) is a Lie subgroup of G.*

PROOF We show first that in any such manifold structure, A is a Lie group. Let $\mathscr{D}$ be the distribution on G determined by left translations of the tangent space to A at the identity. We claim that (A,i) is an integral manifold of $\mathscr{D}$ through the identity $e \in G$.

We caution that this is not evident and that its proof requires a little care. The problem is that if $\gamma(t)$ is a smooth curve in A, and if σ lies in A, then $l_\sigma \circ \gamma(t)$ is a smooth curve in G which lies in A but apriori may no longer be smooth in A. Thus it could be that there are elements of $\mathscr{D}(\sigma)$ which are not tangent vectors to A. We must show that this cannot occur. Let $\dim A = k$. If for some $\sigma \in A$ the tangent space A_σ is not contained in $\mathscr{D}(\sigma)$, then we can find $k + 1$ curves which are smooth in G, lie in A, and have independent tangent vectors at σ. Translating to the identity, we have curves $\gamma_1(t), \ldots, \gamma_{k+1}(t)$ which are smooth in G, lie in A, pass through e at $t = 0$, and have independent tangent vectors there. Notice that this is not yet a contradiction. For example, there are two smooth curves in the plane $\mathbb{R}^2$ with independent tangent vectors at the origin and lying in the one-dimensional figure-8 submanifold (see 1.31). Now consider the map

$$(5) \qquad (t_1, \ldots, t_{k+1}) \mapsto \gamma_1(t_1) \cdots \gamma_{k+1}(t_{k+1}).$$

This map is non-singular at the origin, thus is a diffeomorphism of a neighborhood of the origin in $\mathbb{R}^{k+1}$ into G, and has its image contained in A since the individual curves γ_i lie in A, and A is a subgroup. The map (5) can be extended to a diffeomorphism of a neighborhood of the origin in $\mathbb{R}^n$ with a neighborhood U of the identity in G, where $n = \dim G$. Composing the inverse of this diffeomorphism with the inclusion $i: A \to G$, we obtain a C^∞ immersion of the open submanifold $U \cap A$ of A into $\mathbb{R}^n$ with image containing an open set in $\mathbb{R}^{k+1} \subset \mathbb{R}^n$. This now contradicts the fact that A is second countable and has dimension k. (See Exercise 6, Chapter 1.) Thus $A_\sigma = D(\sigma)$ for each $\sigma \in A$ and consequently (A,i) is an integral manifold of $\mathscr{D}$.

It follows that for any $\sigma \in G$, (A,l_σ) is an integral manifold of $\mathscr{D}$ through σ. Thus by 1.59, $\mathscr{D}$ is an involutive distribution; and now it follows from an argument as in 3.19(3) that the map $(\sigma,\tau) \to \sigma\tau^{-1}$ of $A \times A \to A$ is C^∞, which proves that A is a Lie group. Now suppose that we have two manifold structures on A for which (A,i) is a submanifold. Denote A with these manifold structures by A_1 and A_2 respectively. Then (A_1,i) and (A_2,i) are both integral manifolds of involutive distributions on G; hence 1.62 applied to the commutative diagram

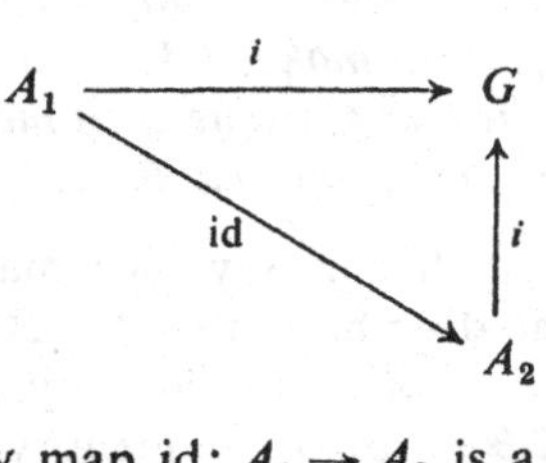

implies that the identity map id: $A_1 \to A_2$ is a diffeomorphism. Thus there is a unique such manifold structure on A.

Hence if a subset of a Lie group can be made into a Lie subgroup under the inclusion map, it can be made so in one and only one way—the group structure, topology, and differentiable structure are uniquely determined. Consequently, it is unambiguous to assert that a subset A of a Lie group G is a Lie subgroup of G. This means that A is an abstract subgroup of G and has a manifold structure (and hence a unique one) making A into a Lie group and (A,i) into a Lie subgroup of G.

We are now in a position to describe exactly in which situation a Lie subgroup $A \subseteq G$ has the relative topology.

3.21 Theorem *Let (H^d,φ) be a Lie subgroup of G^c. Then φ is an imbedding (that is, φ is a homeomorphism of H with $\varphi(H)$ in the relative topology) if and only if (H,φ) is a closed subgroup of G (that is, $\varphi(H)$ is closed in G).*

PROOF Assume that $\varphi(H)$ is closed in G. It suffices to prove that there is some non-empty open set $V \subset H$ such that $\varphi \mid V$ is a homeomorphism of V into $\varphi(H)$ where $\varphi(H)$ has the relative topology. For then it follows from the fact that φ commutes with left translations ($\varphi \circ l_\sigma = l_{\varphi(\sigma)} \circ \varphi$) that φ is an imbedding on all of H. By 3.19, Corollary (b), and by 1.60, there exists a cubic-centered coordinate system (U,τ) about $e \in G$ such that $\varphi(H) \cap U$ consists of a union (at most countable) of slices of the form

(1) $$\tau_i = \text{constant} \qquad \text{for all } i \in \{d+1, \ldots, c\},$$

including at least the slice through e. Let C be a closed subset of U containing e whose image under τ is a cube, and let S be the slice of C given by

(2) $$\tau_1 = 0, \ldots, \tau_d = 0.$$

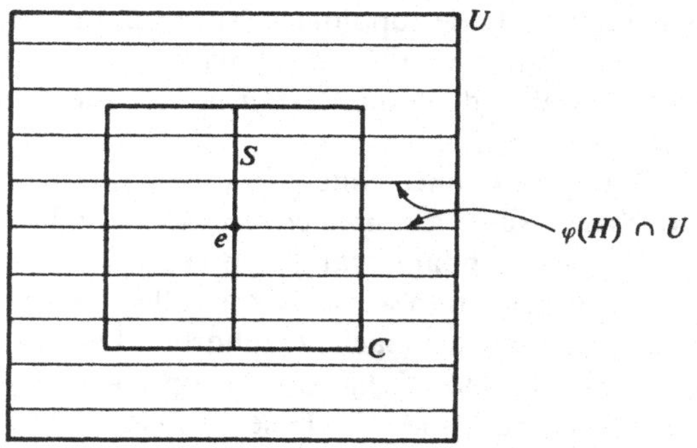

Then $\tau(\varphi(H) \cap S)$ is a non-empty, closed, countable subset of $\mathbb{R}^{c-d}$. Now, any closed countable subset of Euclidean space must have an isolated point. Otherwise, such a set with the induced metric would be a complete metric space representable as a countable union of nowhere dense sets, which is impossible according to the Baire category theorem [27]. Thus there is an isolated slice, call it S_0, among the collection of slices $\varphi(H) \cap U$; and thus $\varphi^{-1}(S_0)$ is an open subset of H on which φ gives an imbedding into $\varphi(H)$.

Conversely, suppose that φ is an imbedding. Let $\{\sigma_i\}$ be a sequence of points in $\varphi(H)$ converging to the point $\sigma \in G$. Since φ is an imbedding, there exists a cubic coordinate system (U,τ) about the identity $e \in G$ such that $\varphi(H) \cap U$ consists of a single slice—call it S. Choose neighborhoods $V \subset W \subset U$ of the identity, cubic relative to τ, so that $V^{-1}V \subset \overline{W} \subset U$. Since $\sigma_i \to \sigma$, there is an N sufficiently large so that $\sigma_n \in \sigma V$ for $n \geq N$. Therefore $\sigma_N{}^{-1}\sigma_n \in \overline{W}$ for $n \geq N$. Also $\sigma_N{}^{-1}\sigma_n \in \varphi(H)$. Thus $\sigma_N{}^{-1}\sigma_n \in S \cap \overline{W}$ and converges to $\sigma_N{}^{-1}\sigma$ which therefore must also lie in $S \cap \overline{W}$. Therefore $\sigma_N{}^{-1}\sigma \in \varphi(H)$. Hence $\sigma \in \varphi(H)$, and thus $\varphi(H)$ is closed.

COVERINGS

We shall assume that the reader has some familiarity with the notions of homotopy, fundamental group, simple connectivity, and covering space. In Theorem 3.23 we state three of the fundamental facts on covering spaces which we shall need. The proofs are sketched in Exercise 7 at the end of this chapter. Details of the proofs and additional facts on covering spaces may be found, for example, in [27] or [28].

3.22 Definition Before stating Theorem 3.23, we recall a few definitions and establish some notation. We let $\pi_1(X,x_0)$ denote the fundamental group of the topological space X with base point $x_0 \in X$. We let $\pi\colon (X,x_0) \to (Y,y_0)$ denote a mapping of the topological space X into the topological space Y sending the base point x_0 in X to the base point y_0 in Y, and we denote the corresponding induced homomorphism of the fundamental groups by $\pi_*\colon \pi_1(X,x_0) \to \pi_1(Y,y_0)$.

We shall assume connectedness in the definition of "simply connected." That is, we shall assume a *simply connected* space X to be a connected topological space whose fundamental group is trivial.

A continuous surjection $\pi\colon X \to Y$ is called a *covering* if X is a connected, locally pathwise connected (each neighborhood of a point contains a pathwise connected neighborhood of that point) topological space, and if each point $y \in Y$ has a neighborhood V whose inverse image under π is a disjoint union of open sets in X each homeomorphic with V under π. Such neighborhoods V in Y are called *evenly covered.* If $\pi\colon X \to Y$ is a covering, then Y is called the *base* of the covering, and X is called the *covering space*.

A topological space Y is called *semi-locally 1-connected* if each point $y \in Y$ has a neighborhood U such that each loop based at y and lying in U is homotopic in Y, through loops based at y, to the constant loop.

3.23 Theorem

(a) *Let* $\pi\colon (X,x_0) \to (Y,y_0)$ *be a covering. Let* Z *be a pathwise connected and locally pathwise connected topological space, and let* $\alpha\colon (Z,z_0) \to (Y,y_0)$ *be a continuous map such that* $\alpha_*(\pi_1(Z,z_0)) \subset \pi_*(\pi_1(X,x_0))$. *Then there exists a unique continuous map* $\tilde{\alpha}\colon (Z,z_0) \to (X,x_0)$ *such that* $\pi \circ \tilde{\alpha} = \alpha$.

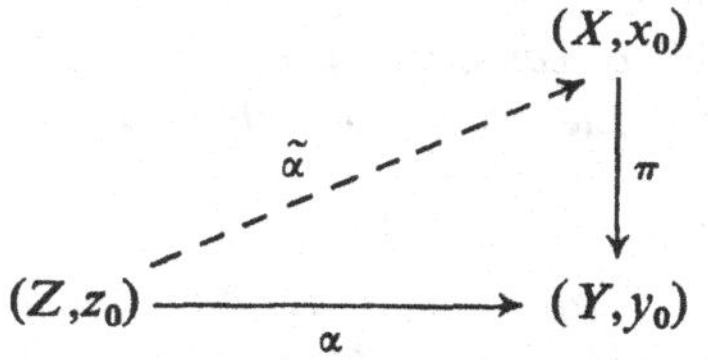

(b) *If* X *is a pathwise connected, locally pathwise connected, and semi-locally 1-connected topological space, then* X *has a simply connected covering space.*

(c) *If* $\pi\colon X \to Y$ *is a covering and* Y *is simply connected, then* π *is a homeomorphism.*

3.24 Simply Connected Covering Group Let $\pi\colon \tilde{M} \to M$ be a covering of a differentiable manifold M. Then $\tilde{M}$ is automatically a locally Euclidean, second countable (cf. Exercise 8) Hausdorff space, and there is a unique differentiable structure on $\tilde{M}$ for which the covering map π is C^∞ and non-singular. This differentiable structure is obtained simply by requiring that the local homeomorphisms obtained from π over evenly covered open sets be diffeomorphisms.

Now let G be a connected Lie group. By 3.23(b), G has a covering $\pi\colon \tilde{G} \to G$ with $\tilde{G}$ simply connected. As we have observed, $\tilde{G}$ has a unique differentiable structure for which π is C^∞ and non-singular. We shall now show that a group structure can be induced in $\tilde{G}$ making $\tilde{G}$ into a Lie group and making π into a Lie group homomorphism. Consider the map $\alpha\colon \tilde{G} \times \tilde{G} \to G$ such that $\alpha(\tilde{\sigma},\tilde{\tau}) = \pi(\tilde{\sigma})\pi(\tilde{\tau})^{-1}$. Choose $\tilde{e} \in \pi^{-1}(e)$. Since $\tilde{G} \times \tilde{G}$ is simply connected, it follows from 3.23(a) that a unique mapping $\tilde{\alpha}\colon \tilde{G} \times \tilde{G} \to \tilde{G}$ exists such that $\pi \circ \tilde{\alpha} = \alpha$ and such that $\tilde{\alpha}(\tilde{e},\tilde{e}) = \tilde{e}$. For $\tilde{\sigma}$ and $\tilde{\tau}$ in $\tilde{G}$ we define

(1) $$\tilde{\tau}^{-1} = \tilde{\alpha}(\tilde{e},\tilde{\tau}), \qquad \tilde{\sigma}\tilde{\tau} = \tilde{\alpha}(\tilde{\sigma},\tilde{\tau}^{-1}).$$

It follows easily from applications of the uniqueness part of 3.23(a) that $\tilde{\sigma}\tilde{e} = \tilde{e}\tilde{\sigma} = \tilde{\sigma}$, for the maps $\tilde{\sigma} \mapsto \tilde{\sigma}\tilde{e}$, $\tilde{\sigma} \mapsto \tilde{e}\tilde{\sigma}$, and $\tilde{\sigma} \mapsto \tilde{\sigma}$ of $\tilde{G} \to \tilde{G}$ all make the following diagram commute, and they all send $\tilde{e}$ to $\tilde{e}$.

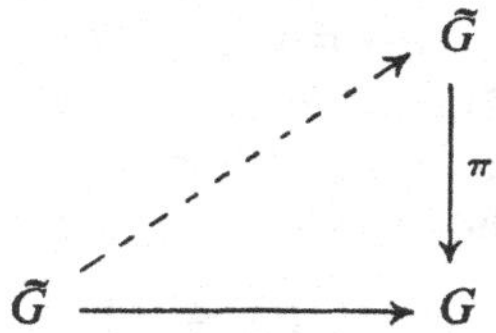

Thus since $\tilde{G}$ is simply connected, it follows from the uniqueness in 3.23(a) that these maps are all identical. Similarly, if follows that $\tilde{\sigma}\tilde{\sigma}^{-1} = \tilde{\sigma}^{-1}\tilde{\sigma} = \tilde{e}$, and that $(\tilde{\sigma}\tilde{\tau})\tilde{\gamma} = \tilde{\sigma}(\tilde{\tau}\tilde{\gamma})$ for all $\tilde{\sigma}$, $\tilde{\tau}$, $\tilde{\gamma} \in \tilde{G}$. Thus (1) makes $\tilde{G}$ into an abstract group; and since $\tilde{\alpha}$ is smooth, $\tilde{G}$ is a Lie group. Also, (1) implies that $\pi(\tilde{\tau}^{-1}) = \pi(\tilde{\tau})^{-1}$ and that $\pi(\tilde{\sigma}\tilde{\tau}) = \pi(\tilde{\sigma})\pi(\tilde{\tau})$; hence $\pi: \tilde{G} \to G$ is a Lie group homomorphism. Thus we have proved the following.

3.25 Theorem *Each connected Lie group has a simply connected covering space which is itself a Lie group such that the covering map is a Lie group homomorphism.*

3.26 Proposition *Let G and H be connected Lie groups, and let $\varphi: G \to H$ be a homomorphism. Then φ is a covering map if and only if $d\varphi: G_e \to H_e$ is an isomorphism.*

PROOF Suppose first that φ is a covering. Then $d\varphi \mid G_e$ must be injective. Otherwise, since φ is a homomorphism, $d\varphi$ would have a nontrivial kernel at each point of G, and these kernels would form an involutive distribution on G, whose integral manifolds are collapsed to points under φ. Thus φ would nowhere be locally one-to-one, contradicting the covering property. $d\varphi \mid G_e$ must also be surjective, for otherwise (G,φ) would locally be a proper submanifold of H, again contradicting the local homeomorphism property of a covering. Thus $d\varphi: G_e \to H_e$ is an isomorphism.

Conversely, suppose that $d\varphi: G_e \to H_e$ is an isomorphism. Then φ is everywhere a local diffeomorphism. Thus, by 3.18, φ maps G onto H since φ is a homomorphism and since $\varphi(G)$ contains a neighborhood of the identity in the connected Lie group H. G certainly is pathwise and locally pathwise connected, so to prove that φ is a covering map, it only remains to show that each point of H is contained in an evenly covered neighborhood. Let

$$D = \ker \varphi = \varphi^{-1}(e). \tag{1}$$

Then since φ is a local diffeomorphism, D is a discrete normal subgroup of G, where "discrete" means that if $\sigma \in D$, then there is an open set V in G such that $V \cap D = \sigma$. Since $(\sigma,\tau) \mapsto \sigma^{-1}\tau$ is a continuous (in fact C^∞) map of $G \times G \to G$, there exists a neighborhood V of e in G such that

$$(V^{-1}V) \cap D = e. \tag{2}$$

We claim that $\varphi(V)$ is a neighborhood of e in H, evenly covered by φ. First, $\varphi \mid V$ is 1:1, since if $\sigma, \tau \in V$ and $\varphi(\sigma) = \varphi(\tau)$, then $\varphi(\sigma^{-1}\tau) = e$, whence $\sigma^{-1}\tau = e$ by (2), and thus $\sigma = \tau$. Moreover, $d\varphi$ is an isomorphism at each point of G. Thus $\varphi \mid V$ is a diffeomorphism of V with the open neighborhood $\varphi(V)$ of e in H. Now, we claim that

$$\varphi^{-1}(\varphi(V)) = \bigcup_{\theta \in D} V\theta. \tag{3}$$

The inclusion $\supset$ is obvious. To prove $\subset$, suppose that $\varphi(\sigma) \in \varphi(V)$. Then there exists $\tau \in V$ such that $\varphi(\tau) = \varphi(\sigma)$. Hence $\tau^{-1}\sigma \in D$ and $\sigma \in V\tau^{-1}\sigma$, proving (3). Finally, $V\theta_1 \cap V\theta_2 = \varnothing$ if $\theta_1 \neq \theta_2 \in D$, for if $\sigma \in V\theta_1 \cap V\theta_2$ then $\sigma = \tau\theta_1 = \eta\theta_2$ for some $\tau, \eta \in V$. Then $\tau^{-1}\eta \in D$ and $\tau^{-1}\eta \in V^{-1}V$; thus, by (2), $\tau = \eta$, which implies that $\theta_1 = \theta_2$. Thus $\varphi(V)$ is an evenly covered neighborhood of $e \in H$. It follows that $\varphi(\sigma V)$ is an open neighborhood of $\varphi(\sigma)$ in H, evenly covered by the disjoint union of the open sets $\sigma V\theta$ for $\theta \in D$. Thus φ is a covering.

SIMPLY CONNECTED LIE GROUPS

3.27 Theorem *Let G and H be Lie groups with Lie algebras $\mathfrak{g}$ and $\mathfrak{h}$ respectively and with G simply connected. Let $\psi\colon \mathfrak{g} \to \mathfrak{h}$ be a homomorphism. Then there exists a unique homomorphism $\varphi\colon G \to H$ such that $d\varphi = \psi$.*

PROOF Uniqueness was proved in 3.16.

Let $\{\omega_i\}$ be a basis of the left invariant 1-forms on H, and let ψ^* be the transpose of ψ (see 3.15(7)). Then according to 3.15, the 1-forms

(1) $$\{\delta\pi_1(\psi^*(\omega_i)) - \delta\pi_2(\omega_i)\}$$

on $G \times H$ (where π_1 and π_2 are the canonical projections of $G \times H$ onto G and H respectively) are left invariant, and the ideal $\mathscr{I}$ which they generate is a differential ideal. Thus by 3.19, Corollary (c), the maximal connected integral manifold I of $\mathscr{I}$ through $(e,e) \in G \times H$ is a Lie subgroup of $G \times H$ with dimension equal to the dimension of G. Now, we know from 2.33 that $(\pi_1 \mid I)\colon I \to G$ is non-singular, and therefore by 3.26 is a covering homomorphism. G was assumed to be simply connected; hence by 3.23(c) and the inverse function theorem, $\pi_1 \mid I\colon I \to G$ is an isomorphism. We define $\varphi\colon G \to H$ by setting

(2) $$\varphi = \pi_2 \circ (\pi_1 \mid I)^{-1}.$$

Then φ is a Lie group homomorphism, and according to 2.33(6),

$$\delta\varphi(\omega_i) = \psi^*(\omega_i).$$

Thus $d\varphi = \psi$, and the theorem is proved.

Corollary *If simply connected Lie groups G and H have isomorphic Lie algebras, then G and H are isomorphic.*

There is a theorem [12, p. 199] due to Ado, which we will not prove, that every Lie algebra has a faithful (1:1) representation in $\mathfrak{gl}(n,\mathbb{R})$ for some n. As a consequence, if $\mathfrak{g}$ is a Lie algebra, then there is a Lie group, in particular a simply connected one, with Lie algebra $\mathfrak{g}$. In view of this we have:

3.28 Theorem *There is a one-to-one correspondence between isomorphism classes of Lie algebras and isomorphism classes of simply connected Lie groups.*

EXPONENTIAL MAP

3.29 Definition A homomorphism $\varphi\colon \mathbb{R} \to G$ is called a *1-parameter subgroup* of G.

3.30 Definition Let G be a Lie group, and let $\mathfrak{g}$ be its Lie algebra. Let $X \in \mathfrak{g}$. Then

(1) $$\lambda \frac{d}{dr} \mapsto \lambda X$$

is a homomorphism of the Lie algebra of $\mathbb{R}$ into $\mathfrak{g}$. Since the real line is simply connected, there exists, by 3.27, a unique 1-parameter subgroup

(2) $$\exp_X\colon \mathbb{R} \to G$$

such that

(3) $$d \exp_X\left(\lambda \frac{d}{dr}\right) = \lambda X.$$

In other words, $t \mapsto \exp_X(t)$ is the unique 1-parameter subgroup of G whose tangent vector at 0 is $X(e)$. We define the *exponential map*

(4) $$\exp\colon \mathfrak{g} \to G$$

by setting

(5) $$\exp(X) = \exp_X(1).$$

The reason for this terminology will become apparent in 3.35(13), where we show that the exponential map for the general linear group is actually given by exponentiation of matrices.

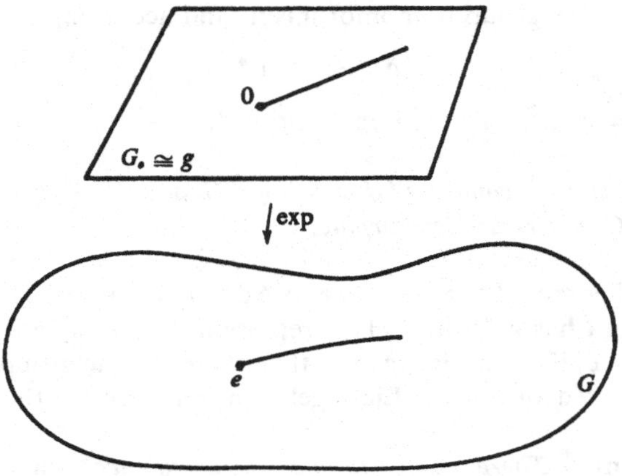

3.31 Theorem *Let X belong to the Lie algebra $\mathfrak{g}$ of the Lie group G. Then*

(a) $\exp(tX) = \exp_X(t)$ *for each* $t \in \mathbb{R}$.

(b) $\exp(t_1 + t_2)X = (\exp t_1X)(\exp t_2X)$ *for all* $t_1, t_2 \in \mathbb{R}$.

(c) $\exp(-tX) = (\exp tX)^{-1}$ *for each* $t \in \mathbb{R}$.

(d) $\exp\colon \mathfrak{g} \to G$ *is* C^∞ *and* $d\exp\colon \mathfrak{g}_0 \to G_e$ *is the identity map (with the usual identifications), so* exp *gives a diffeomorphism of a neighborhood of* 0 *in* $\mathfrak{g}$ *onto a neighborhood of* e *in* G.

(e) $l_\sigma \circ \exp_X$ *is the unique integral curve of* X *which takes the value* σ *at* 0. *As a particular consequence, left invariant vector fields are always complete.*

(f) *The* 1*-parameter group of diffeomorphisms* X_t *associated with the left invariant vector field* X *is given by*

$$X_t = r_{\exp_X(t)}.$$

PROOF By 3.14, (d/dr) and $d\exp_X(d/dr)$, which is X, are $\exp_X$ related. Thus $\exp_X$ is an integral curve of X and is the unique one for which $\exp_X(0) = e$. Since X is left invariant, $l_\sigma \circ \exp_X$ is also an integral curve of X and is the unique one taking the value σ at 0. Thus part (e) is proved; and (f) is an immediate consequence of (e). Now define maps φ and ψ of $\mathbb{R}$ into G by setting

$$\text{(1)} \qquad \psi(t) = \exp_{sX}(t) \quad \text{and} \quad \varphi(t) = \exp_X(st)$$

where $s \in \mathbb{R}$. We have observed that ψ is the unique integral curve of sX such that $\psi(0) = e$. Now,

$$d\varphi\left(\frac{d}{dr}\bigg|_t\right) = d\exp_X\left(s\frac{d}{dr}\bigg|_{st}\right) = sX\big|_{\exp_X(st)}.$$

Thus φ also is an integral curve of sX such that $\varphi(0) = e$. By the uniqueness (1.48(c)) of integral curves, $\varphi = \psi$. Thus

$$\text{(2)} \qquad \exp_{sX}(t) = \exp_X(st) \qquad (s, t \in \mathbb{R};\ X \in \mathfrak{g}).$$

Setting $t = 1$ and changing s to t, we obtain part (a). Since $\exp_X$ is a homomorphism of $\mathbb{R}$ into G, parts (b) and (c) follow immediately from (a). For part (d), we define a vector field V on $G \times \mathfrak{g}$ by setting

$$V(\sigma,X) = (X(\sigma),0) \in G_\sigma \oplus \mathfrak{g}_X.$$

Then V is a smooth vector field, and according to part (e), the integral curve of V through (σ,X) is

$$t \mapsto (\sigma \exp tX, X),$$

or in other words, the local 1-parameter group of transformations associated with the vector field V is given by

$$V_t(\sigma,X) = (\sigma \exp tX, X).$$

In particular, V is complete; hence V_1 is defined and smooth on all of $G \times \mathfrak{g}$. Now let $\pi: G \times \mathfrak{g}$ be the projection onto G. Then

$$\exp X = \pi \circ V_1(e, X).$$

Thus we have exhibited exp as the composition of C^∞ mappings, so exp is C^∞. That $d\exp: \mathfrak{g}_0 \to G_e$ is the identity map is immediate, for tX is a curve in $\mathfrak{g}$ whose tangent vector at $t = 0$ is X, and by part (a), $\exp tX$ is a curve in G whose tangent vector at $t = 0$ is $X(e)$.

3.32 Theorem *Let $\varphi: H \to G$ be a homomorphism, Then the following diagram is commutative:*

$$(1) \qquad \begin{array}{ccc} H & \xrightarrow{\ \varphi\ } & G \\ \uparrow{\scriptstyle \exp} & & \uparrow{\scriptstyle \exp} \\ \mathfrak{h} & \xrightarrow{\ d\varphi\ } & \mathfrak{g} \end{array}$$

PROOF Let $X \in \mathfrak{h}$. Then $t \mapsto \varphi(\exp tX)$ is a smooth curve in G whose tangent at 0 is $d\varphi(X(e))$. This is also a 1-parameter subgroup of G since φ is a homomorphism. But $t \mapsto \exp t(d\varphi(X))$ is the unique 1-parameter subgroup of G whose tangent at 0 is $(d\varphi(X))(e)$. Thus

$$(2) \qquad \varphi(\exp tX) = \exp t(d\varphi(X)),$$

whence

$$(3) \qquad \varphi(\exp X) = \exp(d\varphi(X)).$$

3.33 Proposition *Let (H, φ) be a Lie subgroup of G, and let $X \in \mathfrak{g}$. If $X \in d\varphi(\mathfrak{h})$, then $\exp tX \in \varphi(H)$ for all t. Conversely, if $\exp tX \in \varphi(H)$ for t in some open interval, then $X \in d\varphi(\mathfrak{h})$.*

PROOF If $X \in d\varphi(\mathfrak{h})$, then $\exp tX \in \varphi(H)$ for all t, according to 3.32. On the other hand, if $\exp tX \in \varphi(H)$ for t in some interval I, then for $t \in I$, the map $t \mapsto \exp tX$ can be expressed as a composition $\varphi \circ \alpha$ where α, according to 1.62, is a smooth map of I into H. Let $t_0 \in I$, and let $\tilde{X}$ be the left invariant vector field on H determined by $\dot{\alpha}(t_0)$. Then $d\varphi(\tilde{X}) = X$.

3.34 Theorem *Let A be an abstract subgroup of a Lie group G, and let $\mathfrak{a}$ be a subspace of $\mathfrak{g}$. Let U be a neighborhood of 0 in $\mathfrak{g}$ diffeomorphic under the exponential map with a neighborhood V of the identity in G. Suppose that*

$$(1) \qquad \exp(U \cap \mathfrak{a}) = A \cap V.$$

Then A with the relative topology is a Lie subgroup of G, $\mathfrak{a}$ is a subalgebra of $\mathfrak{g}$, and $\mathfrak{a}$ is the Lie algebra of A.

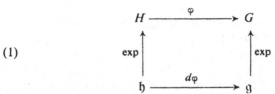

PROOF We need only show that in the relative topology, the subgroup A has a differentiable structure such that (A,i) is a submanifold of G, where i is the inclusion map. For then, according to 3.20, with this manifold structure (and no other) A is a Lie subgroup of G, and it follows from 3.33 that the Lie algebra of A is $\mathfrak{a}$. Let

$$\varphi = \exp \big| U \cap \mathfrak{a} \colon U \cap \mathfrak{a} \to A \cap V.$$

Then the desired differentiable structure on A is the maximal collection of smoothly overlapping coordinate systems containing the collection

$$\{(A \cap \sigma V, \varphi^{-1} \circ l_{\sigma^{-1}}) \colon \sigma \in A\}.$$

3.35 Example We shall see that the exponential map

(1) $$\exp\colon \mathfrak{gl}(n,\mathbb{C}) \to Gl(n,\mathbb{C})$$

for the complex general linear group is given by exponentiation of matrices. We shall now let I rather than e denote the identity matrix in $Gl(n,\mathbb{C})$. Let

(2) $$e^A = I + A + \frac{A^2}{2!} + \cdots + \frac{A^j}{j!} + \cdots$$

for $A \in \mathfrak{gl}(n,\mathbb{C})$. To see that this makes sense, that is, to see that the right-hand side of (2) converges, observe that in fact the right-hand side of (2) converges uniformly for A in a bounded region of $\mathfrak{gl}(n,\mathbb{C})$. For given a bounded region Ω of $\mathfrak{gl}(n,\mathbb{C})$, there is a $\mu > 0$ such that for any matrix A in this region, $|x_{ik}(A)| \leq \mu$ for each component $x_{ik}(A)$ of the matrix A. It follows easily by induction that $|x_{ik}(A^j)| \leq n^{(j-1)}\mu^j$. Thus, by the Weierstrass M-test, each of the series

(3) $$\sum_{j=0}^{\infty} \frac{x_{ik}(A^j)}{j!} \qquad (1 \leq i \leq n;\ 1 \leq k \leq n)$$

converges uniformly for A in Ω. Thus the right-hand side of (2) converges uniformly for A in Ω.

Now let $S_j(A)$ be the jth partial sum of the series (2), that is,

(4) $$S_j(A) = I + A + \frac{A^2}{2!} + \cdots + \frac{A^j}{j!},$$

and let $B \in \mathfrak{gl}(n,\mathbb{C})$. Then since $C \mapsto BC$ is a continuous map of $\mathfrak{gl}(n,\mathbb{C})$ into itself, it follows that

(5) $$B\left(\lim_{j\to\infty} S_j(A)\right) = \lim_{j\to\infty} (BS_j(A)).$$

In particular, if $B \in Gl(n,\mathbb{C})$, then $B\left(\lim_{j\to\infty} S_j(A)\right)B^{-1} = \lim_{j\to\infty} BS_j(A)B^{-1}$, from which it follows that

(6) $$Be^AB^{-1} = e^{BAB^{-1}}.$$

There exists a $B \in Gl(n,\mathbb{C})$ such that BAB^{-1} is super-triangular; that is, all entries below the diagonal are zero. (Simply take B to be the inverse of the matrix whose columns $v_1, \ldots, v_n$ are determined as follows: Let l be the linear transformation of $\mathbb{C}^n$ whose matrix with respect to the canonical basis is A, let v_1 be an eigenvector of l, and inductively let v_{i+1} be an eigenvector of $\pi_i \circ l \mid W_i$ where W_i is any complement of the subspace V_i of C^n spanned by $v_1, \ldots, v_i$ and where π_i is projection of $\mathbb{C}^n = V_i \oplus W_i$ onto W_i.) If the diagonal entries of BAB^{-1} are $\lambda_1, \ldots, \lambda_n$, then $e^{BAB^{-1}}$ is also super-triangular, with diagonal entries $e^{\lambda_1}, \ldots, e^{\lambda_n}$. In particular, $\det e^{BAB^{-1}} \neq 0$, so it follows from (6) that

$$e^A \in Gl(n,\mathbb{C}) \quad \text{for each} \quad A \in \mathfrak{gl}(n,\mathbb{C}). \tag{7}$$

Since the trace of a matrix is the sum of its eigenvalues, and the determinant is the product of its eigenvalues, we have also shown that

$$\det e^A = e^{\text{trace } A}. \tag{8}$$

Next we shall prove that

$$e^{A+B} = e^A e^B \quad \text{if} \quad AB = BA. \tag{9}$$

By definition, $e^{A+B} = \lim_{j\to\infty} S_j(A + B)$. It follows from the fact that matrix multiplication gives a continuous map of $\mathfrak{gl}(n,\mathbb{C}) \times \mathfrak{gl}(n,\mathbb{C})$ into $\mathfrak{gl}(n,\mathbb{C})$ that $e^A e^B = \lim_{j\to\infty} S_j(A) \cdot S_j(B)$. So to prove (9), it suffices to show that

$$\lim_{j\to\infty} \{S_j(A)S_j(B) - S_j(A + B)\} = 0. \tag{10}$$

Now,

$$S_j(A)S_j(B) - S_j(A + B) = \sum \frac{B^l A^k}{l!k!} \tag{11}$$

where the sum is over all integers l and k for which $1 \leq l \leq j$, $1 \leq k \leq j$, and $j + 1 \leq l + k \leq 2j$. If $\mu > 1$ is an upper bound for the absolute value of each of the entries of A and B, then it follows that each entry of the right-hand side of (11) is bounded in absolute value by

$$\sum \frac{n^{(l+k-1)}\mu^{(l+k)}}{l!k!} \leq \frac{(n\mu)^{2j}(j^2)}{[j/2]!} \tag{12}$$

where $[j/2]$ is the greatest integer less than or equal to $j/2$. The estimate on the right-hand side of (12) goes to zero as $j \to \infty$, which proves (10) and hence proves (9).

Consider the map $t \mapsto e^{tA}$ of $\mathbb{R}$ into $Gl(n,\mathbb{C})$. It is smooth since the real and imaginary components of each entry of e^{tA} are power series in t with infinite radii of convergence. Its tangent vector at 0 is A (simply differentiate the power series term by term), and this map is a homomorphism by (9).

Thus $t \mapsto e^{tA}$ is the unique 1-parameter subgroup of $Gl(n,\mathbb{C})$ whose tangent vector at 0 is A. So the exponential map (1) for $Gl(n,\mathbb{C})$ is given by exponentiation of matrices:

$$\exp(A) = e^A \qquad (A \in \mathfrak{gl}(n,\mathbb{C})). \tag{13}$$

If $A \in \mathfrak{gl}(n,\mathbb{R})$, then $e^A \in Gl(n,\mathbb{R})$, and $A \to e^A$ is the exponential map for the real general linear group. Similarly, the exponential map

$$\exp\colon \operatorname{End}(V) \to \operatorname{Aut}(V), \tag{14}$$

where V is a real or complex vector space, is given by exponentiation of endomorphisms. If $l \in \operatorname{End}(V)$, then

$$\exp(l) = e^l = 1 + l + \frac{l^2}{2!} + \cdots + \frac{l^j}{j!} + \cdots \tag{15}$$

where l^j means "l composed with itself j times," and where 1 is the identity transformation.

3.36 Remarks If $\varphi\colon G \to \operatorname{Aut}(V)$ is a representation, and $X \in \mathfrak{g}$, then it follows from 3.35(15) and 3.32 that

$$\varphi(\exp X) = 1 + d\varphi(X) + \frac{(d\varphi(X))^2}{2!} + \cdots, \tag{1}$$

and it follows from 3.32 and 3.9(2) that

$$d\varphi(X) = \lim_{t\to 0} \frac{\varphi(\exp tX) - 1}{t} = \left.\frac{d}{dt}\right|_{t=0} (\varphi(\exp tX)). \tag{2}$$

3.37 Subgroups of $Gl(n,\mathbb{C})$ Using 3.34 and 3.35, we shall now obtain some of the classical Lie subgroups of $Gl(n,\mathbb{C})$ by exponentiating subalgebras of $\mathfrak{gl}(n,\mathbb{C})$. We already have one example, namely, $Gl(n,\mathbb{R})$ is a closed Lie subgroup of $Gl(n,\mathbb{C})$ with Lie algebra $\mathfrak{gl}(n,\mathbb{R}) \subset \mathfrak{gl}(n,\mathbb{C})$. Let $A \in \mathfrak{gl}(n,\mathbb{C})$. Then the *transpose* of A is the matrix A^t such that $(A^t)_{ij} = A_{ji}$; and the *complex conjugate* of A is the matrix $\bar{A}$ whose ijth entry $(\bar{A})_{ij}$ is the complex conjugate $\overline{A_{ij}}$ of A_{ij}. And n, of course, is an integer ≥ 1.

(a) *Unitary group* $\qquad U(n) = \{A \in Gl(n,\mathbb{C})\colon A^{-1} = \overline{A^t}\}$.

(b) *Special linear group* $\qquad Sl(n,\mathbb{C}) = \{A \in Gl(n,\mathbb{C})\colon \det A = 1\}$.

(c) *Complex orthogonal group* $\qquad O(n,\mathbb{C}) = \{A \in Gl(n,\mathbb{C})\colon A^{-1} = A^t\}$.

Each of these is an abstract subgroup and a closed subset of $Gl(n,\mathbb{C})$. We shall use 3.34 to show that they are Lie subgroups. Their Lie algebras will be

(a′) *Skew-hermitian matrices* $\qquad \mathfrak{u}(n) = \{A \in \mathfrak{gl}(n,\mathbb{C})\colon \bar{A} + A^t = 0\}$.

(b′) *Matrices of trace* 0 $\qquad \mathfrak{sl}(n,\mathbb{C}) = \{A \in \mathfrak{gl}(n,\mathbb{C})\colon \operatorname{trace} A = 0\}$.

(c′) *Skew-symmetric matrices* $\qquad \mathfrak{o}(n,\mathbb{C}) = \{A \in \mathfrak{gl}(n,\mathbb{C})\colon A + A^t = 0\}$.

Each of these examples is a subalgebra of $\mathfrak{gl}(n,\mathbb{C})$. To apply 3.34, let U be a neighborhood of 0 in $\mathfrak{gl}(n,\mathbb{C})$, diffeomorphic under the exponential map with a neighborhood V of the identity in $Gl(n,\mathbb{C})$. We can assume, in addition, that if $A \in U$, then $\bar{A}$, A^t, and $-A$ belong to U, and $|\text{trace } A| < 2\pi$. For let W be a neighborhood of 0 in $\mathfrak{gl}(n,\mathbb{C})$ that is small enough for the exponential map to be a diffeomorphism and also small enough for the trace condition to be satisfied, and then let $U = W \cap \overline{W} \cap W^t \cap (-W)$. We shall also assume that $\exp(U \cap \mathfrak{gl}(n,\mathbb{R})) = Gl(n,\mathbb{R}) \cap V$.

If $A \in U \cap \mathfrak{u}(n)$, then $(\overline{e^A})^t = e^{\bar{A}^t} = e^{-A}$; hence $(\overline{e^A})^t e^A = e^{-A}e^A = e^0 = I$, which implies that $e^A \in U(n) \cap V$. Conversely, suppose that $A \in U$ and that $e^A \in U(n) \cap V$. Then $e^{-A} = (e^A)^{-1} = (\overline{e^A})^t = e^{\bar{A}^t}$, which implies that $-A = \bar{A}^t$ since $-A$ and $\bar{A}^t$ also belong to U and since the exponential map is 1:1 on U. Thus $A \in U \cap \mathfrak{u}(n)$. It follows from 3.34 that $U(n)$ is a closed Lie subgroup of $Gl(n,\mathbb{C})$ with Lie algebra $\mathfrak{u}(n)$. A similar argument shows that $O(n,\mathbb{C})$ is a closed Lie subgroup of $Gl(n,\mathbb{C})$ with Lie algebra $\mathfrak{o}(n,\mathbb{C})$.

If $A \in \mathfrak{sl}(n,\mathbb{C})$, then by 3.35(8), $\det e^A = 1$; hence $e^A \in Sl(n,\mathbb{C})$. Conversely, if $\det e^A = 1$, then according to 3.35(8), trace $A = (2\pi i)j$ for some integer j. If in addition $A \in U$, then trace $A = 0$. Thus 3.34 implies that $Sl(n,\mathbb{C})$ is a closed Lie subgroup of $Gl(n,\mathbb{C})$ with Lie algebra $\mathfrak{sl}(n,\mathbb{C})$.

It follows immediately from 3.34 and from considerations as given above that the *special unitary group* $SU(n)$, which is by definition $U(n) \cap Sl(n,\mathbb{C})$, is a closed Lie subgroup of $Gl(n,\mathbb{C})$ with Lie algebra $\mathfrak{su}(n)$ the subalgebra of $\mathfrak{gl}(n,\mathbb{C})$ consisting of skew-hermitian matrices of trace 0; the *real special linear group* $Sl(n,\mathbb{R})$, which is by definition $Sl(n,\mathbb{C}) \cap Gl(n,\mathbb{R})$, is a closed Lie subgroup of $Gl(n,\mathbb{R})$ with Lie algebra $\mathfrak{sl}(n,\mathbb{R})$ the real matrices of trace 0; the *orthogonal group* $O(n)$, which is by definition $U(n) \cap Gl(n,\mathbb{R})$, is a closed Lie subgroup of $Gl(n,\mathbb{R})$ with Lie algebra $\mathfrak{o}(n)$ the real skew-symmetric matrices; and the *special orthogonal group* $SO(n)$, which is by definition $O(n) \cap Sl(n,\mathbb{R})$, is a closed Lie subgroup of $Gl(n,\mathbb{R})$ also with Lie algebra $\mathfrak{o}(n)$ the real skew-symmetric matrices.

Each of these examples is a closed subgroup of either $Gl(n,\mathbb{R})$ or $Gl(n,\mathbb{C})$, and it follows from either 3.34 or 3.21 that in each case the Lie topology is the relative topology.

The unitary group is compact since not only is it closed, but also it is bounded in $Gl(n,\mathbb{C})$, for $A^t\bar{A} = I$ implies that $\sum_j |A_{ij}|^2 = 1$ for each i, which implies that $|A_{ij}|^2 \leq 1$. It follows that $SU(n)$, $O(n)$, and $SO(n)$ are also compact.

The dimensions of these Lie groups are easily computed from their Lie algebras. $U(n)$ has dimension n^2; $Sl(n,\mathbb{C})$ has dimension $2n^2 - 2$; $O(n,\mathbb{C})$ has dimension $n(n-1)$; $SU(n)$ has dimension $n^2 - 1$; $Sl(n,\mathbb{R})$ has dimension $n^2 - 1$; $O(n)$ and $SO(n)$ both have dimension $n(n-1)/2$.

CONTINUOUS HOMOMORPHISMS

3.38 Theorem *Let $\varphi\colon \mathbb{R} \to G$ be a continuous homomorphism of the real line $\mathbb{R}$ into the Lie group G. Then φ is C^∞.*

PROOF It suffices to prove that φ is C^∞ on a neighborhood of 0, for then it follows from composition with suitable left translations that φ is C^∞ everywhere. Let V be a neighborhood of $e \in G$ diffeomorphic with a neighborhood U of $0 \in \mathfrak{g}$ under the exponential map. We can assume that U is starlike, that is, $tX \in U$ for $0 \leq t \leq 1$ whenever $X \in U$, and we let

$$U' = \{X/2\colon X \in U\}.$$

Choose $t_0 > 0$ small enough so that $\varphi(t) \in \exp(U')$ for $|t| \leq t_0$. Let n be a positive integer. Then there are uniquely determined elements X and Y of U' such that $\exp X = \varphi(t_0/n)$ and $\exp Y = \varphi(t_0)$. We claim that $nX = Y$. Since

$$\exp(nX) = \varphi(t_0) = \exp(Y),$$

and since exp is injective on U', we need only show that $nX \in U'$. Now, $X \in U'$, so let $1 \leq j < n$, and assume that $jX \in U'$. We will prove that $(j+1)X \in U'$. Now, $(j+1)X \in U$, and $\exp((j+1)X) = \varphi((j+1)t_0/n)$ which by assumption belongs to $\exp(U')$. Since exp is injective on U, it follows that $(j+1)X \in U'$. Hence $nX \in U'$ and $nX = Y$, as claimed. Now, let m be an integer with $0 < |m| \leq n$. If m is positive, then $\varphi(mt_0/n) = \varphi(t_0/n)^m = \exp(Y/n)^m = \exp(mY/n)$. If m is negative, then also $\varphi(mt_0/n) = \varphi((-m)t_0/n)^{-1} = \exp((-m)Y/n)^{-1} = \exp(mY/n)$. It follows by continuity that $\varphi(t) = \exp(tY/t_0)$ for $|t| \leq t_0$. Thus φ is C^∞.

3.39 Theorem *Let $\varphi\colon H \to G$ be a continuous homomorphism of Lie groups. Then φ is C^∞.*

PROOF Let H be of dimension d, and let $X_1, \ldots, X_d$ be a basis of $\mathfrak{h}$. The map $\alpha\colon \mathbb{R}^d \to H$ defined by

$$(1) \qquad \alpha(t_1, \ldots, t_d) = (\exp t_1X_1) \cdots (\exp t_dX_d)$$

is C^∞, and is non-singular at $0 \in \mathbb{R}^d$ by 3.31(d); so there is a neighborhood V of $0 \in \mathbb{R}^d$ diffeomorphic under α with a neighborhood U of e in H. Now, $t \mapsto \varphi(\exp tX_i)$ is a continuous homomorphism of $\mathbb{R}$ into G, and so is C^∞ by 3.38. Thus $\varphi \circ \alpha$ is C^∞, and therefore $\varphi \mid U$, which can be expressed as $(\varphi \circ \alpha) \circ \alpha^{-1} \mid U$, is C^∞. Since $\varphi \mid \sigma U = l_{\varphi(\sigma)} \circ \varphi \circ l_{\sigma^{-1}} \mid \sigma U$, φ is C^∞ on all of H.

3.40 Definition A *topological group G* is an abstract group G which has a topology such that the map $(\sigma,\tau) \mapsto \sigma\tau^{-1}$ of $G \times G \to G$ is continuous.

3.41 Corollary to Theorem 3.39 *A second countable locally Euclidean topological group can have at most one differentiable structure making it into a Lie group.*

PROOF The identity map would give a diffeomorphism of any two such differentiable structures.

One of the outstanding problems in the theory of Lie groups was that of deciding whether every connected locally Euclidean topological group has a differentiable structure which makes it into a Lie group. The problem was posed by Hilbert in his famous address to the International Congress of Mathematics in 1900, and was solved with an affirmative answer by Gleason together with Montgomery and Zippen in 1952; see [20].

We have dealt exclusively with the C^∞ structure on Lie groups. It can be shown [23] that the C^∞ structure on a Lie group contains an analytic structure, that is, a collection of coordinate systems which overlap analytically (the compositions $\varphi_\alpha \circ \varphi_\beta^{-1}$ of coordinate maps being locally represented by convergent power series). In this context, the analog of 3.39 would state that every continuous homomorphism of Lie groups is analytic, from which it follows that the C^∞ structure on a Lie group contains a unique analytic structure.

CLOSED SUBGROUPS

3.42 Theorem *Let G be a Lie group, and let A be a closed abstract subgroup of G. Then A has a unique manifold structure which makes A into a Lie subgroup of G.*

According to 3.21, the topology in this manifold structure on A must be the relative topology.

PROOF Uniqueness has been proved in 3.20. Let

$$\mathfrak{a} = \{X \in \mathfrak{g} \colon \exp tX \in A \text{ for all } t \in \mathbb{R}\}. \tag{1}$$

The idea of the proof is to show that $\mathfrak{a}$ is a subspace of $\mathfrak{g}$, and to apply 3.34. It is clear from the definition (1) that if $X \in \mathfrak{a}$, then also $tX \in \mathfrak{a}$ for any $t \in \mathbb{R}$. Now, let $X, Y \in \mathfrak{a}$. Suppose that

$$\lim_{n\to\infty}\left(\exp\frac{t}{n}X\exp\frac{t}{n}Y\right)^n = \exp(t(X+Y)). \tag{2}$$

Since A is closed, the left-hand side of (2) belongs to A. Thus (2) implies that $(X + Y) \in \mathfrak{a}$. Assuming (2) for the moment, we have proved that $\mathfrak{a}$ is a subspace of $\mathfrak{g}$ and shall use this to complete the proof of the theorem. We shall return to the proof of (2) in a separate lemma.

The theorem will follow from 3.34 once we have shown that there exists a neighborhood U of 0 in $\mathfrak{g}$, diffeomorphic under the exponential map with a neighborhood V of e in G such that

$$\exp(U \cap \mathfrak{a}) = V \cap A. \tag{3}$$

Suppose, on the contrary, that no such U exists. Then there is a neighborhood W of 0 in $\mathfrak{a}$ and a sequence $\{\sigma_k\} \subset A$ such that $\sigma_k \to e$ in G and

$$\sigma_k \notin \exp(W). \tag{4}$$

Let $\mathfrak{b}$ be any complementary subspace to $\mathfrak{a}$ in $\mathfrak{g}$. It follows from 3.31(d) that the map $\alpha : \mathfrak{a} \times \mathfrak{b} \to G$ defined by $\alpha(X, Y) = \exp X \exp Y$ is C^∞ and non-singular at (0,0). Thus there are neighborhoods $W_\mathfrak{a} \subset W$ of 0 in $\mathfrak{a}$ and $W_\mathfrak{b}$ of 0 in $\mathfrak{b}$ such that $\alpha \mid W_\mathfrak{a} \times W_\mathfrak{b}$ is a diffeomorphism of $W_\mathfrak{a} \times W_\mathfrak{b}$ with a neighborhood $\tilde{V}$ of e in G. If we can choose $W_\mathfrak{b}$ small enough that

$$A \cap \exp(W_\mathfrak{b} - \{0\}) = \varnothing, \tag{5}$$

then we can reach a contradiction as follows. For k large enough, $\sigma_k \in \tilde{V}$, and thus $\sigma_k = \exp X_k \exp Y_k$ for $X_k \in W_\mathfrak{a}$ and $Y_k \in W_\mathfrak{b}$, where $Y_k \neq 0$ by (4). But $\sigma_k \in A$, as also $\exp X_k \in A$; so $\exp Y_k \in A$, contradicting (5).

Finally, we show that $W_\mathfrak{b}$ can be chosen satisfying (5). Again we argue by contradiction. Suppose that there is a sequence $\{Y_i\} \subset W_\mathfrak{b}$ such that $Y_i \neq 0$, $Y_i \to 0$, and $\exp Y_i \in A$. Then there is a subsequence $\{Y_j\}$ and a sequence $\{t_j\}$ of positive real numbers with $t_j \to 0$ such that the sequence $\{Y_j/t_j\}$ converges to some $Y \neq 0$ in $W_\mathfrak{b}$ (choose a norm on $\mathfrak{b}$ whose unit sphere lies in $W_\mathfrak{b}$, and let t_i be the norm of Y_i). For $t > 0$, let $n_j(t)$ be the largest integer less than or equal to (t/t_j). Then

$$\frac{t}{t_j} - 1 < n_j(t) \leq \frac{t}{t_j},$$

so $\lim_{j\to\infty} t_j n_j(t) = t$. Thus

$$\exp tY = \exp\left(\lim_{j\to\infty} n_j(t) Y_j\right) = \lim_{j\to\infty} (\exp Y_j)^{n_j(t)},$$

which belongs to A. Since $\exp(-tY) = (\exp tY)^{-1}$, then $\exp tY \in A$ for all real numbers t, which implies that $Y \in \mathfrak{a}$. This contradicts the fact that Y is a non-zero element of $W_\mathfrak{b}$.

To complete the proof we need the following lemma, which clearly implies (2).

Lemma *Let G be a Lie group with Lie algebra $\mathfrak{g}$. If X, $Y \in \mathfrak{g}$, then for t sufficiently small,*

$$\text{(6)} \qquad \exp tX \exp tY = \exp\{t(X + Y) + O(t^2)\},$$

where $O(t^2)$ denotes a $\mathfrak{g}$-valued C^∞ function of t such that $(1/t^2)O(t^2)$ is bounded at $t = 0$.

PROOF For t small enough, there is a C^∞ curve $Z(t)$ in $\mathfrak{g}$ such that

$$\text{(7)} \qquad \exp tX \exp tY = \exp Z(t).$$

Since the tangent vector to the curve

$$t \mapsto \exp tX \exp tY$$

at $t = 0$ is $X(e) + Y(e)$ (see Exercise 11), it follows that the tangent vector to $Z(t)$ at $t = 0$ is $X + Y$, so the Taylor expansion with integral remainder of $Z(t)$ about $t = 0$ has the form

$$\text{(8)} \qquad Z(t) = t(X + Y) + O(t^2),$$

where $O(t^2)$ is a C^∞ $\mathfrak{g}$-valued function of t such that $(1/t^2)O(t^2)$ is bounded at $t = 0$. From (7) and (8) we obtain (6).

3.43 Theorem *Let $\psi\colon G \to K$ be a homomorphism of Lie groups. If $A = \ker \psi$ and $\mathfrak{a} = \ker d\psi$, then A is a closed Lie subgroup of G with Lie algebra $\mathfrak{a}$.*

PROOF A is a closed abstract subgroup of G, hence, by 3.42, is a Lie subgroup of G. If $X \in \mathfrak{g}$, then according to 3.33 (with $H = A$ and φ the inclusion map), X belongs to the Lie algebra of A if and only if $\exp tX \in A$ for all $t \in \mathbb{R}$, and this occurs if and only if $\psi(\exp tX) = e$ for all $t \in \mathbb{R}$. By 3.32, this latter condition is equivalent to having $\exp t\, d\psi(X) = e$ for all $t \in \mathbb{R}$, and this occurs if and only if $d\psi(X) = 0$, that is, if and only if $X \in \mathfrak{a}$.

THE ADJOINT REPRESENTATION

3.44 Definitions Let M be a manifold, and let G be a Lie group. A C^∞ map $\mu\colon G \times M \to M$ such that

$$\text{(1)} \qquad \mu(\sigma\tau,m) = \mu(\sigma,\mu(\tau,m)), \qquad \mu(e,m) = m$$

for all $\sigma, \tau \in G$ and $m \in M$ is called an *action of G on M on the left*. If $\mu\colon G \times M \to M$ is an action of G on M on the left, then for a fixed $\sigma \in G$ the map $m \mapsto \mu(\sigma,m)$ is a diffeomorphism of M which we shall denote by μ_σ. Similarly, a C^∞ map $\mu\colon M \times G \to M$ such that

$$\text{(2)} \qquad \mu(m,\sigma\tau) = \mu(\mu(m,\sigma),\tau), \qquad \mu(m,e) = m$$

for all $\sigma, \tau \in G$ and $m \in M$ is called an *action of G on M on the right*.

3.45 Theorem *Let $\mu: G \times M \to M$ be an action of G on M on the left. Assume that $m_0 \in M$ is a fixed point, that is, $\mu_\sigma(m_0) = m_0$ for each $\sigma \in G$. Then the map*

(1) $$\psi: G \to \operatorname{Aut}(M_{m_0})$$

defined by

(2) $$\psi(\sigma) = d\mu_\sigma \mid M_{m_0}$$

is a representation of G.

PROOF ψ is a homomorphism for

(3) $$\psi(\sigma\tau) = d\mu_{\sigma\tau} \mid M_{m_0} = d(\mu_\sigma \circ \mu_\tau) \mid M_{m_0} = \psi(\sigma) \circ \psi(\tau).$$

It remains only to prove that ψ is C^∞. For this, it suffices to prove that ψ composed with an arbitrary coordinate function on $\operatorname{Aut}(M_{m_0})$ is C^∞. Now, one gets a coordinate system on $\operatorname{Aut}(M_{m_0})$ by choosing a basis for M_{m_0} and then by using this basis to identify $\operatorname{Aut}(M_{m_0})$ with non-singular matrices. One gets the matrix associated with an element of $\operatorname{Aut}(M_{m_0})$ by applying this element to the basis of M_{m_0} and then applying the dual basis. So it suffices to prove that if $v_0 \in M_{m_0}$ and if $\alpha \in M^*_{m_0}$, then

(4) $$\sigma \mapsto \alpha(d\mu_\sigma(v_0))$$

is a C^∞ function on G. For (4), it suffices to prove that

(5) $$\sigma \mapsto d\mu_\sigma(v_0)$$

is a C^∞ map of G into M_{m_0}, or equivalently that (5) is a C^∞ map of G into $T(M)$. But (5) is exactly the composition of C^∞ maps

$$G \to T(G) \times T(M) \to T(G \times M) \to T(M)$$

in which the first map sends $\sigma \mapsto ((\sigma,0),(m_0,v_0))$, the second map is the canonical diffeomorphism of $T(G) \times T(M)$ with $T(G \times M)$, and the third map is $d\mu$. Thus ψ is C^∞.

3.46 The Adjoint Representation A Lie group G acts on itself on the left by inner automorphisms:

(1) $$a: G \times G \to G, \qquad a(\sigma,\tau) = \sigma\tau\sigma^{-1} = a_\sigma(\tau).$$

The identity is a fixed point of this action. Hence, by 3.45, the map

(2) $$\sigma \mapsto da_\sigma \mid G_e \cong \mathfrak{g}$$

is a representation of G into $\operatorname{Aut}(\mathfrak{g})$. This is called the *adjoint representation* and is denoted by

(3) $$\operatorname{Ad}: G \to \operatorname{Aut}(\mathfrak{g}).$$

We let the differential of the adjoint representation be denoted by ad,

(4)
$$d(\mathrm{Ad}) = \mathrm{ad},$$

and we denote $\mathrm{Ad}(\sigma)$ by Ad_σ and $\mathrm{ad}(X)$ by ad_X. Thus by 3.32 we have a commutative diagram

(5)
$$\begin{array}{ccc} G & \xrightarrow{\mathrm{Ad}} & \mathrm{Aut}(\mathfrak{g}) \\ \uparrow{\scriptstyle \exp} & & \uparrow{\scriptstyle \exp} \\ \mathfrak{g} & \xrightarrow{\mathrm{ad}} & \mathrm{End}(\mathfrak{g}) \end{array}$$

Also applying 3.32 to the automorphism a_σ of G, we obtain the commutative diagram

(6)
$$\begin{array}{ccc} G & \xrightarrow{a_\sigma} & G \\ \uparrow{\scriptstyle \exp} & & \uparrow{\scriptstyle \exp} \\ \mathfrak{g} & \xrightarrow{\mathrm{Ad}_\sigma} & \mathfrak{g} \end{array}$$

In other words,

(7)
$$\exp t\mathrm{Ad}_\sigma(X) = \sigma(\exp tX)\sigma^{-1}.$$

In the special case in which $G = \mathrm{Aut}(V)$, the above diagrams become

(8)
$$\begin{array}{ccc} \mathrm{Aut}(V) & \xrightarrow{\mathrm{Ad}} & \mathrm{Aut}(\mathrm{End}\ V) \\ \uparrow{\scriptstyle \exp} & & \uparrow{\scriptstyle \exp} \\ \mathrm{End}(V) & \xrightarrow{\mathrm{ad}} & \mathrm{End}(\mathrm{End}\ V) \end{array}$$

$$\begin{array}{ccc} \mathrm{Aut}(V) & \xrightarrow{a_B} & \mathrm{Aut}(V) \\ \uparrow{\scriptstyle \exp} & & \uparrow{\scriptstyle \exp} \\ \mathrm{End}(V) & \xrightarrow{\mathrm{Ad}_B} & \mathrm{End}(V) \end{array}$$

where $B \in \mathrm{Aut}(V)$. If in addition $C \in \mathrm{End}(V)$, then

(9)
$$\mathrm{Ad}_B(C) = B \circ C \circ B^{-1}.$$

For from 3.36(2), using 3.35(15) and 3.35(6), we obtain

$$\begin{aligned} \mathrm{Ad}_B(C) &= \frac{d}{dt}\bigg|_{t=0} \big(a_B(\exp tC)\big) = \frac{d}{dt}\bigg|_{t=0} (B \circ e^{tC} \circ B^{-1}) \\ &= \frac{d}{dt}\bigg|_{t=0} e^{tB \circ C \circ B^{-1}} = B \circ C \circ B^{-1}. \end{aligned}$$

Similarly, in the case in which $G = Gl(n,\mathbb{R})$(or $Gl(n,\mathbb{C})$) and $B \in Gl(n,\mathbb{R})$ and $C \in \mathfrak{gl}(n,\mathbb{R})$, then

(10) $$\mathrm{Ad}_B(C) = BCB^{-1}.$$

3.47 Proposition *Let G be a Lie group with Lie algebra $\mathfrak{g}$, and let $X, Y \in \mathfrak{g}$. Then*

(1) $$\mathrm{ad}_X Y = [X, Y].$$

PROOF By 3.36(2),

(2) $$\begin{aligned}\mathrm{ad}_X Y &= \left(\frac{d}{dt}\bigg|_{t=0} \mathrm{Ad}(\exp tX)\right)Y \\ &= \frac{d}{dt}\bigg|_{t=0} \mathrm{Ad}_{\exp tX}(Y) = \frac{d}{dt}\bigg|_{t=0} d(a_{\exp tX})(Y).\end{aligned}$$

Thus if X_t denotes, as usual, the 1-parameter group of diffeomorphisms associated with X, then

(3) $$\begin{aligned}\mathrm{ad}_X Y(e) &= \frac{d}{dt}\bigg|_{t=0} d(r_{\exp(-tX)})\big(d(l_{\exp tX})(Y(e))\big) \\ &= \frac{d}{dt}\bigg|_{t=0} d(r_{\exp(-tX)})(Y|_{\exp tX}) \\ &= \frac{d}{dt}\bigg|_{t=0} d(X_{-t})(Y_{X_t(e)}) \\ &= (L_X Y)(e) = [X,Y](e).\end{aligned}$$

The last equality follows from 2.25(b). Since $\mathrm{ad}_X Y$ and $[X, Y]$ are both left invariant, (1) is an immediate consequence of (3).

3.48 Theorem *Let $A \subseteq G$ be a connected Lie subgroup of a connected Lie group G. Then A is a normal subgroup of G if and only if the Lie algebra $\mathfrak{a}$ of A is an ideal in $\mathfrak{g}$.*

PROOF Assume that $\mathfrak{a}$ is an ideal in $\mathfrak{g}$. Let $Y \in \mathfrak{a}$, let $X \in \mathfrak{g}$, and let $\sigma = \exp X$. Then

(1) $$\begin{aligned}\sigma(\exp Y)\sigma^{-1} &= \exp \mathrm{Ad}_\sigma(Y) && (3.46(7)) \\ &= \exp((\exp \mathrm{ad}_X)(Y)) && (3.46(5)) \\ &= \exp\left(Y + [X,Y] + \frac{(\mathrm{ad})^2 X}{2!}(Y) + \cdots\right).\end{aligned}$$

(3.35(15) and 3.47)

Under the assumption that $\mathfrak{a}$ is an ideal in $\mathfrak{g}$, the series in the last term in (1) converges to an element of $\mathfrak{a}$, so that

$$(2) \qquad \sigma(\exp Y)\sigma^{-1} \in A.$$

Now, by 3.18 and 3.31(d), A is generated by elements of the form $\exp Y$, and G is generated by elements of the form $\exp X$. This, together with (2), implies that A is a normal subgroup of G.

Conversely, assume that A is normal in G. Let s and t be real numbers, let $Y \in \mathfrak{a}$ and $X \in \mathfrak{g}$, and let $\sigma = \exp tX$. Then, according to (1),

$$(3) \qquad \sigma(\exp sY)\sigma^{-1} = \exp \mathrm{Ad}_\sigma(sY) = \exp s\{(\exp \mathrm{ad}_{tX})(Y)\}.$$

Since A is normal, $\sigma(\exp sY)\sigma^{-1} \in A$; so it follows from (3) and 3.33 that $(\exp \mathrm{ad}_{tX})(Y) \in \mathfrak{a}$ for all $t \in \mathbb{R}$. Now

$$(4) \qquad (\exp \mathrm{ad}_{tX})(Y) = \big(\exp t(\mathrm{ad}_X)\big)(Y) = Y + t[X,Y] + \frac{t^2}{2!}[X,[X,Y]] + \cdots,$$

which is a smooth curve in $\mathfrak{a}$ whose tangent vector at $t = 0$ is $[X, Y]$. Thus $[X, Y] \in \mathfrak{a}$, and $\mathfrak{a}$ is an ideal in $\mathfrak{g}$.

3.49 Definitions

(a) *Center of* $\mathfrak{g} = \{X \in \mathfrak{g}\colon [X, Y] = 0 \text{ for all } Y \in \mathfrak{g}\}$.

(b) *Center of* $G = \{\sigma \in G\colon \sigma\tau = \tau\sigma \text{ for all } \tau \in G\}$.

3.50 Theorem *Let G be a connected Lie group. Then the center of G is the kernel of the adjoint representation.*

PROOF Let σ belong to the center of G, and let $X \in \mathfrak{g}$. Then

$$(1) \qquad \exp tX = \sigma(\exp tX)\sigma^{-1} = \exp t\mathrm{Ad}_\sigma(X)$$

for all $t \in \mathbb{R}$. Thus $X = \mathrm{Ad}_\sigma(X)$; so $\sigma \in \ker(\mathrm{Ad})$. Conversely, let $\sigma \in \ker(\mathrm{Ad})$. Then (1) again holds, so σ commutes with every element in a neighborhood of e in G. Since G is connected, σ commutes with every element of G. Thus σ lies in the center of G.

Corollary (a) *Let G be a connected Lie group. Then the center of G is a closed Lie subgroup of G with Lie algebra the center of $\mathfrak{g}$.*

PROOF This corollary is an immediate consequence of 3.50 and 3.43.

Corollary (b) *A connected Lie group G is abelian if and only if its Lie algebra $\mathfrak{g}$ is abelian.*

3.51 Proposition *Let X, Y belong to the Lie algebra $\mathfrak{g}$ of a Lie group G. Then*

$$[X, Y] = 0 \Rightarrow \exp(X + Y) = \exp X \exp Y.$$

PROOF The subspace $\mathfrak{a}$ of $\mathfrak{g}$ spanned by X and Y is a subalgebra of $\mathfrak{g}$, and the corresponding connected Lie subgroup A of G is abelian. Let

$$\alpha(t) = \exp tX \exp tY.$$

Then α is a C^∞ map of $\mathbb{R}$ into G and is a homomorphism since A is abelian. The tangent vector to α at 0 is $X(e) + Y(e)$ (see Exercise 11), so that

$$\exp tX \exp tY = \exp t(X + Y)$$

for all t.

3.52 Remark Recall that in 3.27 we stated Ado's theorem which asserts that any Lie algebra $\mathfrak{g}$ has a faithful representation in some $\mathfrak{gl}(n,\mathbb{R})$ from which it follows that there is a Lie group with Lie algebra $\mathfrak{g}$. If $\mathfrak{g}$ has trivial center, we can get such a faithful representation from the adjoint representation. Simply define $\text{ad}: \mathfrak{g} \to \text{End}(\mathfrak{g})$ by $\text{ad}_X(Y) = [X,Y]$, and check that this is a Lie algebra homomorphism. If $\mathfrak{g}$ has trivial center, ad is one-to-one; so we have a faithful representation of $\mathfrak{g}$ in $\text{End}(\mathfrak{g})$. This together with 3.19 gives a simple proof that every Lie algebra with trivial center is the Lie algebra of some Lie group.

AUTOMORPHISMS AND DERIVATIONS OF BILINEAR OPERATIONS AND FORMS

3.53 Definitions Let V be a finite dimensional real or complex vector space. A *bilinear operation* on V is a linear map $\psi: V \otimes V \to V$ where the tensor product is taken over the real (resp. complex) numbers when V is a real (resp. complex) vector space. We shall use the notation

$$\psi(v \otimes w) = \{v,w\}.$$

(a) We let

$$A_\psi(V) = \{\alpha \in \text{Aut}(V): \alpha\{v,w\} = \{\alpha(v),\alpha(w)\} \text{ for all } v,\, w \in V\}.$$

The elements of $A_\psi(V)$ are the automorphisms of V which *preserve the bilinear operation* ψ.

(b) We let

$$\mathfrak{d}_\psi = \{l \in \text{End}(V): l\{v,w\} = \{l(v),w\} + \{v,l(w)\} \text{ for all } v,\, w \in V\}.$$

Elements of $\mathfrak{d}_\psi$ are called *derivations of the bilinear operation* ψ.

3.54 Theorem *$A_\psi(V)$ is a closed Lie subgroup of* $\text{Aut}(V)$ *with Lie algebra* $\mathfrak{d}_\psi$.

PROOF It is easy to check that $\mathfrak{d}_\psi$ is a subalgebra of End(V) and that $A_\psi(V)$ is a closed abstract subgroup of Aut(V). According to Theorem 3.42, $A_\psi(V)$ is a closed Lie subgroup of Aut(V). Let $\mathfrak{a}$ be the Lie algebra of $A_\psi(V)$. We need only show that $\mathfrak{a} = \mathfrak{d}_\psi$.

If $l \in \mathfrak{a}$, then $\exp tl \in A_\psi(V)$, so that

$$(1) \qquad (\exp tl)\{v,w\} = \{(\exp tl)(v), (\exp tl)(w)\}.$$

Both sides of (1) are smooth curves in V. Taking their derivatives at $t = 0$, we obtain

$$(2) \qquad l\{v,w\} = \{l(v),w\} + \{v,l(w)\},$$

which proves that $l \in \mathfrak{d}_\psi$. (Observe that a curve $\varphi(t) \otimes \psi(t)$ in $V \otimes V$ has derivative (or tangent vector) at $t = 0$ equal to $\dot\varphi(0) \otimes \psi(0) + \varphi(0) \otimes \dot\psi(0)$.)

Conversely, suppose that $l \in \mathfrak{d}_\psi$. To prove that $l \in \mathfrak{a}$, we need only prove that $\exp tl \in A_\psi(V)$ for all t. We let $l \otimes 1$ denote the endomorphism of $V \otimes V$ defined by

$$(l \otimes 1)(v \otimes w) = l(v) \otimes w.$$

The fact that $l \in \mathfrak{d}_\psi$ means that

$$(3) \qquad l\{v,w\} = \{l(v),w\} + \{v,l(w)\}$$

for all v and w in V; and this can be expressed as

$$(4) \qquad l \circ \psi = \psi \circ (l \otimes 1 + 1 \otimes l).$$

It follows from (4) that

$$(5) \qquad l^n \circ \psi = \psi \circ (l \otimes 1 + 1 \otimes l)^n,$$

so that

$$(6) \qquad e^{tl} \circ \psi = \psi \circ e^{t(l \otimes 1 + 1 \otimes l)}.$$

Now, $(l \otimes 1) \circ (1 \otimes l) = (1 \otimes l) \circ (l \otimes 1)$, so we can apply 3.35(9) to the right-hand term in (6). Then, by the fact that $e^{t(l \otimes 1)} = e^{tl} \otimes 1$, equation (6) becomes

$$(7) \qquad e^{tl} \circ \psi = \psi \circ e^{t(l \otimes 1)} \circ e^{t(1 \otimes l)} = \psi \circ (e^{tl} \otimes 1) \circ (1 \otimes e^{tl}) = \psi \circ e^{tl} \otimes e^{tl}.$$

Thus

$$(\exp tl)\{v,w\} = \{(\exp tl)(v), (\exp tl)(w)\}$$

for all v and w in V, which implies that $\exp tl \in A_\psi(V)$ for all $t \in \mathbb{R}$. Thus $\mathfrak{a} = \mathfrak{d}_\psi$, and the theorem is proved.

3.55 Definitions Again we let V be a finite dimensional vector space over a field F where F is either $\mathbb{R}$ or $\mathbb{C}$. A *bilinear form* B on V is a linear map $B\colon V \otimes V \to F$. We shall use the notation $B(v \otimes w) = (v,w)$.

(a) We let

$$A_B(V) = \{\alpha \in \mathrm{Aut}(V)\colon (v,w) = (\alpha(v),\alpha(w)) \text{ for all } v,\, w \in V\}.$$

The elements of $A_B(V)$ are the automorphisms of V which *preserve the bilinear form B.*

(b) We let

$$\mathfrak{d}_B = \{l \in \mathrm{End}(V)\colon (l(v),w) + (v,l(w)) = 0 \text{ for all } v,\, w \in V\}.$$

Elements of $\mathfrak{d}_B$ are called *derivations of the bilinear form B.*

3.56 Theorem *$A_B(V)$ is a closed Lie subgroup of* Aut(V) *with Lie algebra $\mathfrak{d}_B$.*

The proof is similar to the proof of Theorem 3.54, and we leave it to the reader as an exercise.

3.57 Applications of 3.54 and 3.56

(a) Let V be a real vector space with an inner product B. According to Theorem 3.56, the automorphisms of V which preserve the inner product form a closed Lie subgroup $A_B(V) \subset \mathrm{Aut}(V)$ whose Lie algebra is the Lie subalgebra $\mathfrak{d}_B \subset \mathrm{End}(V)$ of derivations of the inner product B. Choose an orthonormal basis for V. Now consider the Lie group isomorphism of Aut(V) with $Gl(n,\mathbb{R})$ and the associated Lie algebra isomorphism of End(V) with $\mathfrak{gl}(n,\mathbb{R})$ determined by associating with each linear transformation on V its matrix relative to this basis. It is easily seen that $A_B(V)$ is mapped onto the orthogonal group $O(n) \subset Gl(n,\mathbb{R})$ and that $\mathfrak{d}_B$ is mapped onto the set $\mathfrak{o}(n)$ of skew-symmetric matrices in $\mathfrak{gl}(n,\mathbb{R})$. This provides another proof of the fact that the orthogonal group $O(n)$ is a closed Lie subgroup of $Gl(n,\mathbb{R})$ with Lie algebra $\mathfrak{o}(n)$.

(b) A Lie algebra $\mathfrak{g}$ has a bilinear operation—the bracket [,]. The automorphisms of $\mathfrak{g}$ which preserve the bracket are precisely the Lie algebra isomorphisms of $\mathfrak{g}$. We shall denote them by $A(\mathfrak{g})$. According to 3.54, $A(\mathfrak{g})$ is a closed Lie subgroup of Aut$(\mathfrak{g})$ with Lie algebra the derivations of [,] which we shall denote by $\mathfrak{d}(\mathfrak{g})$.

Suppose that G is a simply connected Lie group with Lie algebra $\mathfrak{g}$. Let $A(G)$ be the group of all Lie group automorphisms of G. The map

$$\psi\colon A(G) \to A(\mathfrak{g}) \quad \text{defined by} \quad \psi(\alpha) = d\alpha \tag{1}$$

is a homomorphism and is 1:1 by 3.16 and onto by 3.27. Thus we can induce via ψ a manifold structure on $A(G)$, making $A(G)$ into a Lie group with Lie algebra isomorphic with $\mathfrak{d}(\mathfrak{g})$.

HOMOGENEOUS MANIFOLDS

3.58 Theorem *Let H be a closed subgroup of a Lie group G, and let G/H be the set $\{\sigma H: \sigma \in G\}$ of left cosets modulo H. Let $\pi: G \to G/H$ denote the natural projection $\pi(\sigma) = \sigma H$. Then G/H has a unique manifold structure such that*

(a) *π is C^∞.*

(b) *There exist local smooth sections of G/H in G; that is, if $\sigma H \in G/H$, there is a neighborhood W of σH and a C^∞ map $\tau: W \to G$ such that $\pi \circ \tau = \mathrm{id}$.*

PROOF *Existence* We topologize G/H by requiring U in G/H to be open if and only if $\pi^{-1}(U)$ is an open set in G. With this topology, π is an open map since if W is open in G, then

$$\pi^{-1}(\pi(W)) = \bigcup_{h \in H} Wh,$$

which implies that $\pi(W)$ is open in G/H. Moreover, G/H is Hausdorff. To see this, first observe that the set $R \subset G \times G$ consisting of all pairs (σ,τ) for which there exists an $h \in H$ such that $\sigma = \tau h$ is a closed set since $R = \alpha^{-1}(H)$ where α is the continuous map $(\sigma,\tau) \mapsto \tau^{-1}\sigma$ of $G \times G$ into G. Now, if σH and τH are distinct points of G/H, (σ,τ) does not belong to R, so there exist open neighborhoods V of σ and W of τ in G such that $(V \times W) \cap R = \varnothing$. Then $\pi(V)$ and $\pi(W)$ are disjoint open neighborhoods of σH and τH respectively, which proves that G/H is Hausdorff. A countable basis for the topology on G projects under π to a countable basis for the topology on G/H. Thus G/H is second countable.

According to 3.42, the closed subgroup H is a Lie subgroup of G. Suppose that G is of dimension d and that H is of dimension $d - k$. To prove that G/H is locally Euclidean and to obtain a cover of smoothly overlapping coordinate systems on G/H, we first prove that there exists a cubic centered coordinate system (U,φ) about e in G, with coordinate functions $x_1, \ldots, x_d$, such that distinct slices of the form

(1) $\qquad x_i = \text{constant} \qquad \text{for all } i \in \{1, \ldots, k\}$

lie on distinct left cosets of H. Let $\mathscr{D}$ be the distribution on G determined by the Lie algebra of H. Then, by 1.60, there is a cubic centered coordinate system (V,φ) about e in G, with coordinate functions $x_1, \ldots, x_d$, such that the integral manifolds of $\mathscr{D}$ in V are slices of the form (1). Since H is a closed subgroup of G, and therefore as a Lie subgroup has the relative topology, V can be chosen small enough so that

(2) $\qquad V \cap H = \text{the slice } S_0 \text{ through } e.$

Choose neighborhoods U and V_1 of e, cubic relative to the coordinate system (V,φ), such that

(3) $$V_1 V_1 \subset V \qquad \text{and} \qquad U^{-1}U \subset V_1.$$

Now suppose that σ and τ are points of U which lie in the same coset modulo H, so $\sigma \in \tau H$. Then

(4) $$\tau^{-1}\sigma \in V_1 \cap H = V_1 \cap S_0.$$

So $\sigma \in \tau(V_1 \cap S_0)$. Now $\tau(V_1 \cap S_0)$ is an integral manifold of $\mathscr{D}$ which lies in V by the choice of V_1 in (3), and $\tau(V_1 \cap S_0)$ is connected. Therefore $\tau(V_1 \cap S_0)$ lies in a single slice of V. So σ and τ lie in the same slice. Conversely, it is easily seen that a single slice lies on a single coset. Thus (U,φ) is the desired coordinate system.

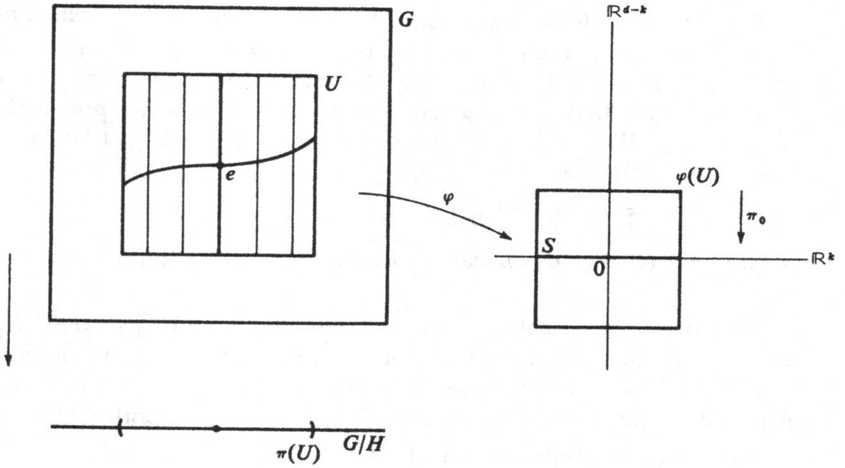

Let S be the slice of $\varphi(U)$ on which $x_{k+1}, \ldots, x_d$ vanish. Let $\tilde{\varphi}^{-1}$ be the map defined by setting

(5) $$\tilde{\varphi}^{-1} = \pi \circ \varphi^{-1} \,|\, S\colon S \to \pi(U).$$

Then $\tilde{\varphi}^{-1}$ is one-to-one by the choice of the coordinate system (U,φ), and is also continuous and an open map; hence it is a homeomorphism. Let $\tilde{\varphi}$ be the inverse,

(6) $$\tilde{\varphi}\colon \pi(U) \to S \subset \mathbb{R}^k.$$

Then $(\pi(U),\tilde{\varphi})$ is a coordinate system about the identity coset in G/H. We obtain coordinate systems about other points of G/H by left translations. Indeed, if $\sigma \in G$, we let $\tilde{l}_\sigma$ be the homeomorphism of G/H induced by left translation l_σ on G; that is,

(7) $$\tilde{l}_\sigma(\tau H) = \sigma\tau H.$$

Then for each $\sigma H \in G/H$ we define a map $\tilde{\varphi}_{\sigma H}$ by setting

(8) $$\tilde{\varphi}_{\sigma H} = \tilde{\varphi} \circ \tilde{l}_{\sigma^{-1}} \mid \tilde{l}_\sigma(\pi(U)).$$

Then $(\tilde{l}_\sigma(\pi(U)), \tilde{\varphi}_{\sigma H})$ is a coordinate system about σH. Note that in this notation $\tilde{\varphi}_H$ is just the map $\tilde{\varphi}$. Now, we claim that one obtains a differentiable structure on G/H by maximizing the collection of coordinate systems

(9) $$\{(\tilde{l}_\sigma(\pi(U)), \tilde{\varphi}_{\sigma H}) \colon \sigma \in G\}.$$

One has only to check differentiability on overlaps. So let $(\tilde{l}_{\sigma_1}(\pi(U)), \tilde{\varphi}_{\sigma_1 H})$ and $(\tilde{l}_{\sigma_2}(\pi(U)), \tilde{\varphi}_{\sigma_2 H})$ be two such coordinate systems, and let

(10) $$V = \tilde{\varphi}_{\sigma_1 H}(\tilde{l}_{\sigma_1}(\pi(U)) \cap \tilde{l}_{\sigma_2}(\pi(U))).$$

We must prove that $\tilde{\varphi}_{\sigma_2 H} \circ \tilde{\varphi}^{-1}_{\sigma_1 H} \mid V$ is C^∞. Let $t \in V$. Then since $\tilde{l}_{\sigma_2^{-1}} \circ \tilde{l}_{\sigma_1} \circ \tilde{\varphi}^{-1}(t) \in \pi(U)$, there exists an element $g \in H$ such that $\sigma_2^{-1}\sigma_1\varphi^{-1}(t)g \in U$. It follows that there exists a neighborhood W of t in V such that $\sigma_2^{-1}\sigma_1\varphi^{-1}(W)g \subset U$. It suffices to prove that $\tilde{\varphi}_{\sigma_2 H} \circ \tilde{\varphi}^{-1}_{\sigma_1 H} \mid W$ is C^∞. But we can express $\tilde{\varphi}_{\sigma_2 H} \circ \tilde{\varphi}^{-1}_{\sigma_1 H} \mid W$ as the following composition of C^∞ maps:

(11) $$\tilde{\varphi}_{\sigma_2 H} \circ \tilde{\varphi}^{-1}_{\sigma_1 H} \mid W = \pi_0 \circ \varphi \circ r_g \circ l_{\sigma_2^{-1}\sigma_1} \circ \varphi^{-1} \mid W,$$

where π_0 is the canonical projection of $\varphi(U)$ onto S. Thus $\tilde{\varphi}_{\sigma_2 H} \circ \tilde{\varphi}^{-1}_{\sigma_1 H} \mid V$ is C^∞.

With this differentiable structure on G/H, the projection $\pi\colon G \to G/H$ is C^∞; indeed, π restricted to an open set of the form $l_\sigma(U)$ is nothing other than the composition $\tilde{\varphi}^{-1}_{\sigma H} \circ \pi_0 \circ \varphi \circ l_{\sigma^{-1}} \mid l_\sigma(U)$. Thus result (a) holds; and as for (b), $l_\sigma \circ \varphi^{-1} \circ \tilde{\varphi}_{\sigma H}$ is a local smooth section of G/H in G on the neighborhood $\tilde{l}_\sigma(\pi(U))$ of σH.

Uniqueness Let $(G/H)_1$ denote G/H with another differentiable structure satisfying (a) and (b).

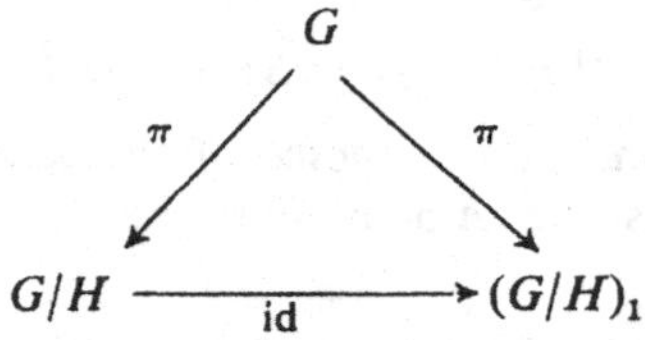

Then the identity map and its inverse are both C^∞ since locally they can be expressed as the composition of local smooth sections into G followed by π. Thus id is a diffeomorphism, which proves uniqueness.

3.59 Definition Manifolds of the form G/H where G is a Lie group, H is a closed subgroup of G, and the manifold structure is the unique one satisfying (a) and (b) of Theorem 3.58 are called *homogeneous manifolds*.

3.60 Remark Observe that f is a C^∞ function on G/H if and only if $f \circ \pi$ is a C^∞ function on G; for if f is C^∞, then certainly $f \circ \pi$ is C^∞. Conversely if $f \circ \pi$ is C^∞, then f, which can locally be represented as the composition of a smooth section of G/H in G with $f \circ \pi$, is also C^∞.

3.61 Definitions Let

(1) $$\eta: G \times M \to M$$

be an action of G on M on the left (cf. 3.44); and, as usual, let

(2) $$\eta_\sigma(m) = \eta(\sigma, m).$$

The action is called *effective* if e is the only element of G for which η_e is the identity map on M. The action is called *transitive* if whenever m and n belong to M there exists a σ in G such that $\eta_\sigma(m) = n$. Let $m_0 \in M$, and let

(3) $$H = \{\sigma \in G: \eta_\sigma(m_0) = m_0\}.$$

H is a closed subgroup of G called the *isotropy group at* m_0. The action η restricted to H gives an action of H on M on the left with a fixed point m_0, so by 3.45 we have a representation

(4) $$\alpha: H \to \mathrm{Aut}(M_{m_0}) \quad \text{where} \quad \alpha(\sigma) = d\eta_\sigma \mid M_{m_0}.$$

The group $\alpha(H)$ of linear transformations of M_{m_0} is called the *linear isotropy group at* m_0.

3.62 Theorem *Let $\eta: G \times M \to M$ be a transitive action of the Lie group G on the manifold M on the left. Let $m_0 \in M$, and let H be the isotropy group at m_0. Define a mapping*

(1) $$\tilde{\beta}: G/H \to M \quad \text{by} \quad \tilde{\beta}(\sigma H) = \eta_\sigma(m_0).$$

Then $\tilde{\beta}$ is a diffeomorphism.

PROOF Observe first that $\tilde{\beta}$ is well-defined since $\eta_{\sigma h}(m_0) = \eta_\sigma(\eta_h(m_0)) = \eta_\sigma(m_0)$ for any $h \in H$. $\tilde{\beta}$ is surjective since G acts transitively, and is injective since if $\tilde{\beta}(\sigma H) = \tilde{\beta}(\tau H)$, then $\eta_{\tau^{-1}\sigma}(m_0) = m_0$, which implies that $\tau^{-1}\sigma$ belongs to H, or in other words that $\sigma H = \tau H$. In view of Exercise 6 of Chapter 1, it suffices to prove that $\tilde{\beta}$ is C^∞ and that $d\tilde{\beta}$ is non-singular at each point.

According to Remark 3.60, $\tilde{\beta}$ is C^∞ if and only if $\tilde{\beta} \circ \pi$ is C^∞ where π is the natural projection of G onto G/H. But

(2) $$\tilde{\beta} \circ \pi = \eta \circ i_{m_0}$$

where i_{m_0} is the C^∞ map of G into $G \times M$ defined by

(3) $$i_{m_0}(\sigma) = (\sigma, m_0).$$

Thus $\tilde{\beta}$ is C^∞.

Now let $\beta = \tilde{\beta} \circ \pi$. Since the kernel of $d\pi \mid G_\sigma$ is $(\sigma H)_\sigma \subset G_\sigma$, in order to prove that $d\tilde{\beta} \mid (G/H)_{\sigma H}$ is non-singular it is sufficient to prove that the kernel of $d\beta \mid G_\sigma$ is also $(\sigma H)_\sigma$. Since for each $\sigma \in G$,

(4) $$\beta = \eta_\sigma \circ \beta \circ l_{\sigma^{-1}},$$

it suffices to prove that the kernel of $d\beta \mid G_e$ is H_e. Certainly $H_e \subset \ker(d\beta \mid G_e)$. So let $x \in \ker(d\beta \mid G_e)$. To show that $x \in H_e$, we need only prove that $\exp tX \in H$ for all $t \in \mathbb{R}$ where X is the left invariant vector field on G determined by x. For this, it is sufficient to prove that the tangent vector to the curve $t \mapsto \beta(\exp tX)$ in M is identically zero, for then $\beta(\exp tX) \equiv m_0$, which implies that $\exp tX \in H$ for all t. The tangent vector to this curve at t is

$$\begin{aligned} d\beta(X_{\exp tX}) &= d(\eta_{\exp tX} \circ \beta \circ l_{\exp(-tX)})(X_{\exp tX}) \\ &= d\eta_{\exp tX} \circ d\beta(X(e)) = d\eta_{\exp tX} \circ d\beta(x) = 0. \end{aligned}$$

Thus $d\tilde{\beta}$ is everywhere non-singular, and the theorem is proved.

3.63 Remarks If G is a Lie group and H a closed subgroup of G, then there is a natural action $\tilde{l}$ of G on the homogeneous manifold G/H on the left, namely,

(1) $$\tilde{l}: G \times G/H \to G/H, \qquad \tilde{l}(\sigma,\tau H) = \sigma\tau H.$$

It is easily checked that $\tilde{l}$ is C^∞ and indeed gives an action of G on G/H on the left. Moreover, it is obvious that $\tilde{l}$ is a transitive action. Now, for each σ in G, $\tilde{l}_\sigma$ (notation as in 3.61(2)) is a diffeomorphism of G/H; and given any two points τH and γH of G/H there is a diffeomorphism $\tilde{l}_{\tau\gamma^{-1}}$ taking γH to τH. The reason that manifolds of the form G/H are called *homogeneous* is that they possess this transitive group of diffeomorphisms. Conversely, Theorem 3.62 shows that if a manifold M has a transitive group of diffeomorphisms in the sense of 3.61(1), then M is diffeomorphic with a homogeneous manifold G/H.

From Theorem 3.62 we get another characterization of the manifold structure on G/H, namely, *the set G/H has a unique manifold structure such that the natural map* (1) *is C^∞.*

3.64 Theorem *Let G be a Lie group and H a closed normal subgroup of G. Then the homogeneous manifold G/H with its natural group structure is a Lie group.*

PROOF One has only to check that the map

(1) $$(\sigma H,\tau H) \mapsto \sigma\tau^{-1}H$$

of $G/H \times G/H \to G/H$ is C^∞. Let $\alpha_\sigma: W_\sigma \to G$ and $\alpha_\tau: W_\tau \to G$ be local sections of G/H in G on neighborhoods W_σ of σH and W_τ of τH

respectively. Then locally the map (1) can be expressed as the following composition of C^∞ maps:

$$\pi \circ \varphi \circ (\alpha_\sigma \times \alpha_\tau), \tag{2}$$

where $\varphi: G \times G \to G$ is the map $\varphi(\sigma,\tau) = \sigma\tau^{-1}$.

3.65 Examples of homogeneous manifolds

(a) Let $\{e_i: i = 1, \ldots, n\}$ be the canonical basis of $\mathbb{R}^n$ where e_i is the n-tuple consisting of all zeros except for a 1 in the ith spot. Each matrix $\sigma \in Gl(n,\mathbb{R})$ uniquely determines a linear transformation on $\mathbb{R}^n$, which we shall also denote by σ, by requiring that

$$\sigma(e_j) = \sum_i \sigma_{ij} e_i. \tag{1}$$

In other words, if we consider the n-tuples of $\mathbb{R}^n$ as $n \times 1$ matrices, then σ acts on $\mathbb{R}^n$ in the natural way by matrix multiplication. The map $(\sigma,v) \mapsto \sigma(v)$ gives an action of $Gl(n,\mathbb{R})$ on $\mathbb{R}^n$ on the left:

$$Gl(n,\mathbb{R}) \times \mathbb{R}^n \to \mathbb{R}^n. \tag{2}$$

Let $\langle\ ,\ \rangle$ denote the standard inner product on $\mathbb{R}^n$ with respect to which the basis $\{e_i\}$ is orthonormal. Then if $\sigma \in Gl(n,\mathbb{R})$,

$$\langle \sigma(v), w\rangle = \langle v, \sigma^t(w)\rangle. \tag{3}$$

If $\sigma \in O(n)$, then $\sigma^t\sigma = I$, so it follows from (3) that

$$\langle \sigma(v),\sigma(v)\rangle = \langle v,\sigma^t\sigma(v)\rangle = \langle v,v\rangle; \tag{4}$$

thus σ preserves lengths of vectors. Thus the action (2) restricted to $O(n) \times S^{n-1}$ factors through the unit sphere S^{n-1}.

$$\begin{array}{ccc} O(n) \times S^{n-1} & \longrightarrow & \mathbb{R}^n \\ & \searrow & \uparrow i \\ & & S^{n-1} \end{array} \tag{5}$$

According to 1.32, the factoring map is smooth, so we have a natural C^∞ action

$$O(n) \times S^{n-1} \to S^{n-1} \tag{6}$$

of the orthogonal group $O(n)$ on the unit sphere S^{n-1} on the left. We claim that the action (6) is transitive. If $v_1 \in S^{n-1}$, let $\{v_1, v_2, \ldots, v_n\}$ be an orthonormal basis of $\mathbb{R}^n$ containing v_1 as the first element. Let

$$v_i = \sum_j \sigma_{ji} e_j. \tag{7}$$

Then the matrix σ whose entries are determined by (7) is orthogonal, and

$$\sigma(e_1) = v_1. \tag{8}$$

It follows that if v and w are any two points of S^{n-1}, there is an orthogonal matrix σ such that $\sigma(v) = w$. Thus the action (6) is transitive.

The set of matrices in $O(n)$ of the form

$$\sigma = \begin{pmatrix} \begin{pmatrix} & & \\ & \tilde{\sigma} & \\ & & \end{pmatrix} & \begin{matrix} 0 \\ \cdot \\ \cdot \\ \cdot \\ 0 \end{matrix} \\ 0 \quad \cdots \quad 0 & 1 \end{pmatrix} \tag{9}$$

forms a closed subgroup of $O(n)$. The matrices $\tilde{\sigma}$ occurring in this subgroup are precisely the matrices in $O(n-1)$. So $O(n-1)$ sits in $O(n)$ as this natural closed subgroup. We claim that $O(n-1)$ is precisely the isotropy group for the action (6) at $e_n \in S^{n-1}$. Clearly the elements of $O(n-1)$ leave e_n fixed. On the other hand, suppose that $\sigma \in O(n)$ and that $\sigma(e_n) = e_n$. Now

$$\sigma(e_n) = \sum_i \sigma_{in} e_i,$$

so $\sigma_{in} = 0$ for $i < n$ and $\sigma_{nn} = 1$. Since σ is orthogonal, $\sigma\sigma^t = I$, and this implies that $\sum_i \sigma_{ni}^2 = 1$. Since $\sigma_{nn} = 1$, we must have $\sigma_{ni} = 0$, $i < n$. Thus $\sigma \in O(n-1)$. It follows from Theorem 3.62 that the map

$$O(n)/O(n-1) \to S^{n-1}, \qquad \sigma O(n-1) \mapsto \sigma(e_n) \tag{10}$$

is a diffeomorphism. *Thus the sphere S^{n-1} is in a natural way diffeomorphic with the homogeneous manifold $O(n)/O(n-1)$.*

By a similar argument one can show that *there is a diffeomorphism of the sphere S^{n-1} with the homogeneous manifold $SO(n)/SO(n-1)$.*

(b) Let $\{e_i : i = 1, \ldots, n\}$ be the canonical complex basis of $\mathbb{C}^n$ where e_i is the n-tuple consisting of all zeros except for a 1 in the ith spot. Just as in Example (a), each matrix $\sigma \in Gl(n,\mathbb{C})$ uniquely determines a linear transformation of $\mathbb{C}^n$, which we still denote by σ, by requiring

$$\sigma(e_j) = \sum_i \sigma_{ij} e_i. \tag{11}$$

So if we consider the complex n-tuples of $\mathbb{C}^n$ as $n \times 1$ matrices, then σ acts by matrix multiplication. The map $(\sigma,v) \mapsto \sigma(v)$ is an action of the real $2n^2$-dimensional Lie group $Gl(n,\mathbb{C})$ on the real $2n$-dimensional manifold $\mathbb{C}^n$ on the left:

$$Gl(n,\mathbb{C}) \times \mathbb{C}^n \to \mathbb{C}^n. \tag{12}$$

Let $\langle\ ,\ \rangle$ denote the standard inner product on $\mathbb{C}^n$, where

$$\left\langle \sum_i a_i e_i, \sum_i b_i e_i \right\rangle = \sum_{i=1}^n a_i \bar{b}_i. \tag{13}$$

Then if $\sigma \in Gl(n,\mathbb{C})$,

$$\langle \sigma(v), w\rangle = \langle v, \bar{\sigma}^t(w)\rangle. \tag{14}$$

If σ is an element of the unitary group $U(n)$, then $\bar{\sigma}^t\sigma = I$, so it follows from (14) that σ preserves lengths of vectors in $\mathbb{C}^n$. If X denotes the unit sphere in $\mathbb{C}^n$, then the action (12) restricted to $U(n) \times X$ factors through X, and so, by 1.32, yields a C^∞ action

$$U(n) \times X \to X \tag{15}$$

of the unitary group $U(n)$ on the unit sphere $X \subset \mathbb{C}^n$ on the left. It follows by an argument similar to that in Example (a) that the action (15) is transitive and that the isotropy group for the action (15) at the point e_n is simply $U(n-1)$, which we consider to be a closed subgroup of $U(n)$ by identifying $\tilde{\sigma} \in U(n-1)$ with

$$\sigma = \begin{pmatrix} \begin{pmatrix} & & \\ & \tilde{\sigma} & \\ & & \end{pmatrix} & \begin{matrix} 0 \\ \vdots \\ 0 \end{matrix} \\ 0 \cdots 0 & 1 \end{pmatrix} \tag{16}$$

in $U(n)$. Thus, by 3.62, X is diffeomorphic with the homogeneous manifold $U(n)/U(n-1)$. But now under the canonical global coordinate system on $\mathbb{C}^n$ (given by the dual basis to the real basis $\{e_1, \ldots, e_n, \sqrt{-1}e_1, \ldots, \sqrt{-1}e_n\}$ of $\mathbb{C}^n$), X is diffeomorphic with $S^{2n-1} \subset \mathbb{R}^{2n}$. *Thus the sphere S^{2n-1} is diffeomorphic with the homogeneous manifold $U(n)/U(n-1)$.*

By a similar argument, *the sphere S^{2n-1} is also diffeomorphic with the homogeneous manifold $SU(n)/SU(n-1)$.* In particular, since $SU(1)$ consists only of the 1×1 identity matrix, S^3 is diffeomorphic with $SU(2)$. Thus S^3 can be given a Lie group structure. It can be shown that S^1 and S^3 are the only spheres possessing Lie group structures [25].

(c) The *real projective space* P^{n-1} is the set of equivalence classes of points in $\mathbb{R}^n - \{0\}$ where $(a_1, \ldots, a_n)$ is equivalent to $(b_1, \ldots, b_n)$ if there is a non-zero real number c such that $ca_i = b_i$ for $i = 1, \ldots, n$. If P^{n-1} is given the largest topology in which the natural projection

$$\pi\colon (\mathbb{R}^n - \{0\}) \to P^{n-1} \tag{17}$$

is continuous, then π restricted to S^{n-1} is a 2-fold covering of P^{n-1}, and it is easily established that P^{n-1} has a unique differentiable structure such that this covering map is locally a diffeomorphism. By an argument similar to that in Example (a), one can show that P^{n-1} is diffeomorphic with the homogeneous space $SO(n)/O(n-1)$, where we consider $O(n-1)$ as a closed subgroup of $SO(n)$ by identifying $\tilde{\sigma} \in O(n-1)$ with the following element σ of $SO(n)$:

$$\sigma = \begin{pmatrix} \begin{pmatrix} & & \\ & \tilde{\sigma} & \\ & & \end{pmatrix} & \begin{matrix} 0 \\ \vdots \\ 0 \end{matrix} \\ 0 \quad \cdots \quad 0 & \det \tilde{\sigma} \end{pmatrix}$$

(d) As a set, the *complex projective space* $\mathbb{C}P^{n-1}$ is the set of equivalence classes of points of $\mathbb{C}^n - \{0\}$ under the equivalence relation in which $(\alpha_1, \ldots, \alpha_n)$ is equivalent to $(\beta_1, \ldots, \beta_n)$ if there is a non-zero complex number γ such that $\gamma\alpha_i = \beta_i$ for $i = 1, \ldots, n$. We make $\mathbb{C}P^{n-1}$ into a real $2(n-1)$ dimensional manifold as follows. The action of the special unitary group on the unit sphere in $\mathbb{C}^n$ preserves these equivalence classes, and each element of $\mathbb{C}P^{n-1}$ has representatives of unit length; thus we have a natural transitive action of $SU(n)$ on the *set* $\mathbb{C}P^{n-1}$:

$$SU(n) \times \mathbb{C}P^{n-1} \to \mathbb{C}P^{n-1}. \tag{18}$$

The subgroup of $SU(n)$ leaving fixed the point of $\mathbb{C}P^{n-1}$ determined by $e_n \in \mathbb{C}^n$ is simply $U(n-1)$, which we consider to be a closed subgroup of $SU(n)$ by identifying $\tilde{\sigma} \in U(n-1)$ with

$$\sigma = \begin{pmatrix} \begin{pmatrix} & & \\ & \tilde{\sigma} & \\ & & \end{pmatrix} & \begin{matrix} 0 \\ \vdots \\ 0 \end{matrix} \\ 0 \quad \cdots \quad 0 & 1/(\det \tilde{\sigma}) \end{pmatrix}$$

in $SU(n)$. It follows that the map

$$\sigma U(n-1) \mapsto \{\sigma(e_n)\}, \tag{19}$$

where $\{\sigma(e_n)\}$ denotes the point of $\mathbb{C}P^{n-1}$ determined by $\sigma(e_n)$, is a well-defined one-to-one map of $SU(n)/U(n-1)$ onto $\mathbb{C}P^{n-1}$. We give $\mathbb{C}P^{n-1}$ the structure of a real $2(n-1)$ dimensional manifold by requiring that the map (19) be a diffeomorphism.

(e) Let V be a real d-dimensional vector space, and let $S_p(V)$ be the set of p-frames in V. That is,

(20) $$S_p(V) = \{\tilde{w} = (w_1, \ldots, w_p)\text{: the } w_1, \ldots, w_p \text{ are linearly independent elements of } V\}.$$

If we choose a basis $v_1, \ldots, v_d$ for V, then $Gl(d,\mathbb{R})$ acts on V by matrix multiplication. We define

(21) $$\eta\colon Gl(d,\mathbb{R}) \times S_p(V) \to S_p(V)$$
$$\text{by} \quad \eta(\sigma, (w_1, \ldots, w_p)) = (\sigma(w_1), \ldots, \sigma(w_p)).$$

Observe that if $\tilde{v}$, $\tilde{w} \in S_p(V)$, then there is a $\sigma \in Gl(d,\mathbb{R})$ such that $\eta(\sigma,\tilde{v}) = \tilde{w}$. Now let $\tilde{s}$ be the element of $S_p(V)$ determined by the first p elements of the basis $v_1, \ldots, v_d$; hence $\tilde{s} = (v_1, \ldots, v_p)$. Let H be the subset of $Gl(d,\mathbb{R})$ leaving $\tilde{s}$ fixed. Then

(22) $$H = \left\{\left(\begin{array}{c|c} I & A \\ \hline O & B \end{array}\right) \in Gl(d,\mathbb{R})\right\},$$

where I is the $p \times p$ identity matrix; thus H is a closed subgroup of $Gl(d,\mathbb{R})$. It follows that the map $\sigma H \mapsto \eta(\sigma,\tilde{s})$ is a one-to-one map of the homogeneous manifold $Gl(d,\mathbb{R})/H$ onto the set $S_p(V)$. We give $S_p(V)$ the structure of a $d \cdot p$-dimensional manifold by requiring that this map be a diffeomorphism. It is easily checked that this manifold structure is independent of the basis of V chosen. $S_p(V)$ is called the *Stiefel manifold* of p-frames in V.

(f) Again let V be a real d-dimensional vector space, and now let $M_k(V)$ be the set of all k-dimensional subspaces (k-planes) of V. If we choose a basis $v_1, \ldots, v_d$ for V, then the orthogonal group $O(d)$ acts naturally on V by matrix multiplication; and since non-singular linear transformations map k-planes to k-planes, we have a map

(23) $$\eta\colon O(d) \times M_k(V) \to M_k(V).$$

Observe that if P and Q are k-planes, then there is a $\sigma \in O(d)$ such that $\eta(\sigma,P) = Q$. Now, let P_0 be the k-plane spanned by the first k elements of the basis $v_1, \ldots, v_d$, and let H be the subset of $O(d)$ leaving P_0 fixed. Then

(24) $$H = \left\{\left(\begin{array}{c|c} \sigma & 0 \\ \hline 0 & \tau \end{array}\right) \in O(d)\colon \sigma \in O(k),\ \tau \in O(d-k)\right\},$$

so H is a closed subgroup of $O(d)$ which we can identify with $O(k) \times O(d-k)$. Then the map $\sigma(O(k) \times O(d-k)) \mapsto \eta(\sigma, P_0)$ is a one-to-one map of the homogeneous manifold $O(d)/[O(k) \times O(d-k)]$ onto the set $M_k(V)$. We make $M_k(V)$ into a $(d-k)k$ dimensional manifold by requiring that this map be a diffeomorphism. One can check that this manifold structure on $M_k(V)$ is independent of the basis chosen for V. $M_k(V)$ is known as the *Grassmann manifold* of k-planes in V.

3.66 Proposition *Let H be a closed subgroup of the Lie group G. If H and G/H are connected, then G is connected.*

PROOF Assume that

(1) $$G = U \cup V$$

where U and V are non-empty open subsets of G. Then

(2) $$G/H = \pi(U) \cup \pi(V)$$

where $\pi(U)$ and $\pi(V)$ are non-empty open subsets of G/H. Since G/H is connected, there must be a point σH of G/H such that

(3) $$\sigma H \in \pi(U) \cap \pi(V).$$

Now, (1) implies that

(4) $$\sigma H = (\sigma H \cap U) \cup (\sigma H \cap V)$$

where $(\sigma H \cap U)$ and $(\sigma H \cap V)$ are both open subsets of σH (since H has the relative topology). According to (3), both $(\sigma H \cap U)$ and $(\sigma H \cap V)$ are non-empty; thus, since σH is homeomorphic with H and therefore connected, we have

(5) $$(\sigma H \cap U) \cap (\sigma H \cap V) \neq \varnothing,$$

which implies that

(6) $$U \cap V \neq \varnothing,$$

which proves that G is connected.

3.67 Theorem *Each of the Lie groups $SO(n)$, $SU(n)$, and $U(n)$ is connected for $n \geq 1$, and $O(n)$ has two components ($n \geq 1$).*

PROOF $SO(1)$ and $SU(1)$ are connected since they both consist only of the 1×1 identity matrix, and $U(1)$ is connected since

$$U(1) = \{(\lambda) \colon \lambda \in \mathbb{C}, |\lambda| = 1\}.$$

That $SO(n)$, $SU(n)$, and $U(n)$ are connected for all n now follows from 3.66 by using induction on n and the representation of spheres as homogeneous manifolds given in 3.65.

Since every matrix in $O(n)$ has determinant ± 1, the orthogonal group can be written as the following union of two non-empty disjoint connected open subsets:

$$O(n) = SO(n) \cup \sigma SO(n) \quad \text{where} \quad \sigma = \begin{pmatrix} -1 & & & & 0 \\ & 1 & & & \\ & & \cdot & & \\ & & & \cdot & \\ 0 & & & & 1 \end{pmatrix}$$

Thus $O(n)$ has two components.

3.68 Theorem *$Gl(n,\mathbb{R})$ has two components.*

PROOF Let $Gl(n, \mathbb{R})^+$ be the subset of $Gl(n, \mathbb{R})$ consisting of matrices with positive determinant, and let $Gl(n,\mathbb{R})^-$ be the subset consisting of matrices with negative determinant. $Gl(n,\mathbb{R})^+$ and $Gl(n,\mathbb{R})^-$ are disjoint homeomorphic open subsets of $Gl(n,\mathbb{R})$, so it suffices to prove that $Gl(n,\mathbb{R})^+$ is connected. To do this, we shall show that each element of $Gl(n,\mathbb{R})^+$ can be joined to the identity matrix by a continuous curve.

First we show that each element of $Gl(n,\mathbb{R})$ has a *polar decomposition;* that is, each matrix $\sigma \in Gl(n,\mathbb{R})$ can be expressed in the form

(1) $$\sigma = PR$$

where P is a positive definite symmetric matrix and $R \in O(n)$. (Recall that all the eigenvalues of a symmetric matrix are real, and that a symmetric matrix is *positive definite* if each of its eigenvalues is strictly positive.) Since $(\sigma\sigma^t)^t = \sigma\sigma^t$, $\sigma\sigma^t$ is symmetric. Let a be an eigenvalue of $\sigma\sigma^t$ with eigenvector $v \in \mathbb{R}^n$. Then, according to 3.65(3), if $\langle\ ,\ \rangle$ denotes the standard inner product in $\mathbb{R}^n$,

$$a\langle v,v\rangle = \langle \sigma\sigma^t(v),v\rangle = \langle \sigma^t(v),\sigma^t(v)\rangle.$$

Consequently, $a \geq 0$, and since $\sigma\sigma^t$ is non-singular, we must have $a > 0$. Thus $\sigma\sigma^t$ is a positive definite symmetric matrix. Since $\sigma\sigma^t$ is symmetric, there exists an orthogonal matrix $\beta \in O(n)$ such that the matrix

(2) $$\beta\sigma\sigma^t\beta^t$$

is diagonal (cf. Exercise 22(a)). Since the eigenvalues of $\sigma\sigma^t$ are all positive, the matrix (2) has a square root

(3) $$(\beta\sigma\sigma^t\beta^t)^{1/2};$$

that is, (3) is a diagonal matrix, and each diagonal entry is the positive square root of the corresponding entry in the matrix (2). Let

(4) $$P = \beta^t(\beta\sigma\sigma^t\beta^t)^{1/2}\beta,$$

and let

$$R = P^{-1}\sigma. \tag{5}$$

P is a positive definite symmetric matrix, and R is orthogonal since (4) implies that $P^2 = \sigma\sigma^t$ and therefore that

$$\begin{aligned} RR^t &= P^{-1}\sigma\sigma^t(P^{-1})^t = P^{-1}\sigma\sigma^t(P^t)^{-1} \\ &= P^{-1}\sigma\sigma^t P^{-1} = P^{-1}PPP^{-1} = I. \end{aligned} \tag{6}$$

Thus $\sigma = PR$, as we asserted in (1).

If $\sigma \in Gl(n,\mathbb{R})^+$, then σ has a polar decomposition (1), where now R must have positive determinant; thus $R \in SO(n)$. Let

$$P_t = tI + (1 - t)P \tag{7}$$

for $t \in [0,1]$. Then P_t is positive definite for each t, so the path $t \mapsto P_tR$ is a continuous curve in $Gl(n,\mathbb{R})^+$ joining σ to R. Since $SO(n)$ is connected, and therefore pathwise connected, R can be joined to the identity matrix I by a continuous curve. Thus $Gl(n,\mathbb{R})^+$ is pathwise connected, which completes the proof that $Gl(n,\mathbb{R})$ has two components.

EXERCISES

1 Prove that a connected Lie group is automatically second countable; that is, the assumption of second countability in the definition of connected Lie group is redundant.

2 Show that the examples in 3.3 are Lie groups.

3 Prove that the examples in 3.5 are Lie algebras.

4 Supply a proof for Proposition 3.12.

5 Prove the necessity of the connectedness assumption in 3.16.

6 Let $\mathscr{I}$ be an ideal of forms on G^c generated by a collection $\{\omega_1, \ldots, \omega_{c-d}\}$ of independent left invariant 1-forms. Let $\mathfrak{h} \subseteq \mathfrak{g}$ be the d-dimensional subspace of the Lie algebra of G annihilated by the ω_i ($i = 1, \ldots, c - d$). Prove that $\mathfrak{h}$ is a subalgebra of $\mathfrak{g}$ if and only if $\mathscr{I}$ is a differential ideal.

7 In this exercise we outline the proof of Theorem 3.23.

(a) Let $\pi: (X,x_0) \to (Y,y_0)$ be a covering, and let $\alpha: (Z,z_0) \to (Y,y_0)$ be a continuous map where Z is pathwise and locally pathwise connected, and where $a_*(\pi_1(Z,z_0)) \subset \pi_*(\pi_1(X,x_0))$. Prove that there is a unique continuous map $\tilde{\alpha}: (Z,z_0) \to (X,x_0)$ such that $\pi \circ \tilde{\alpha} = \alpha$.

(*Sketch:* First prove part (a) for the case in which Z is the unit rectangle $[0,1] \times [0,1]$ and $z_0 = (0,0)$. Here the condition on the fundamental groups is trivially satisfied, and the proof follows by partitioning the unit rectangle fine enough so that each subrectangle of the partition maps under α into an evenly covered set in Y. As a particular application of this case, it follows that each path in Y has a unique lift to X once base points are chosen.

For the general case, define the lifting $\tilde{\alpha}$ of α as follows. Let $z \in Z$, and let $\gamma: [0,1] \to Z$ be a path joining z_0 to z; thus $\gamma(0) = z_0$ and $\gamma(1) = z$. Then $\alpha \circ \gamma$ has a unique lift $\tilde{\gamma}$ to X such that $\tilde{\gamma}(0) = x_0$. Set $\tilde{\alpha}(z) = \tilde{\gamma}(1)$. Now use the condition on the fundamental groups and the unique lifting property already established for rectangles to show that $\tilde{\alpha}$ is well-defined, independent of the choice of the path γ from z_0 to z. Finally use the local path-connected property of Z to show that $\tilde{\alpha}$ is continuous.)

(b) If X is a pathwise connected, locally pathwise connected, and semi-locally 1-connected topological space, then X has a simply connected covering space.

(*Sketch:* Fix a base point $x_0 \in X$. We define an equivalence relation on the set of all paths in X with domain $[0,1]$ and with initial point x_0. Two such paths γ_1, γ_2 will be called equivalent if $\gamma_1(1) = \gamma_2(1)$, and the loop $\gamma_2^{-1}\gamma_1$ at x_0 obtained by first traversing γ_1 and then traversing γ_2 in the reverse direction is homotopically trivial at x_0. The point set of the simply connected covering space $\tilde{X}$ of X is the set of such equivalence classes $\{\gamma\}$. The projection map $\pi: \tilde{X} \to X$ is defined by $\pi(\{\gamma\}) = \gamma(1)$. Let $\{\gamma\} \in \tilde{X}$, and let U be an open neighborhood of $\gamma(1)$ in X. Let $U(\{\gamma\})$ consist of all elements $\{\tau\}$ of $\tilde{X}$ possessing a representative path τ which first traverses γ and then remains in U. Prove that the collection of such sets $U(\{\gamma\})$ forms a basis for a topology on $\tilde{X}$ in which $\tilde{X}$ is simply connected, and in which $\pi: \tilde{X} \to X$ becomes a covering.)

(c) A covering of a simply connected space is a homeomorphism. (This is an immediate application of part (a).)

8 Prove that if $\pi: \tilde{M} \to M$ is a covering of a differentiable manifold M, then $\tilde{M}$ is second countable. (It suffices to prove that $\pi_1(M,m)$ is countable. This follows from the observation that M can be covered by a countable number of open sets, each homeomorphic with an open ball in Euclidean space.)

9 Let G and H be Lie groups with G connected. Prove that a C^∞ map $\varphi: G \to H$ sending the identity of G to the identity of H is a homomorphism if $\delta\varphi$ pulls left invariant forms on H back to left invariant forms on G. Show that the assumption of connectedness for G is necessary. (*Hint:* One can construct a differential ideal on $G \times H$ as in 3.15(5) whose maximal connected integral submanifold I through (e,e) is a Lie subgroup of $G \times H$. Factor the graph of φ through I, and show that the natural covering homomorphism of I onto G is 1:1.)

10 Prove that $\begin{pmatrix} -2 & 0 \\ 0 & -1 \end{pmatrix}$ is not e^A for any $A \in \mathfrak{gl}(2,\mathbb{R})$.

11 Let $\sigma(t)$ and $\beta(t)$ be smooth curves in a Lie group G such that $\sigma(0) = \beta(0) = e$, and let $\alpha(t) = \sigma(t)\beta(t)$. Prove that

$$\dot{\alpha}(0) = \dot{\sigma}(0) + \dot{\beta}(0).$$

(*Hint:* Consider the group multiplication map $\eta: G \times G \to G$, that is, $\eta(\gamma,\tau) = \gamma\tau$. Let $v, w \in G_e$, and show that

$$d\eta(v,w) = d\eta((v,0) + (0,w)) = v + w.)$$

12 Let G be a connected Lie group, and let $\varphi: G \to H$ be a homomorphism with a discrete kernel. Prove that the kernel lies in the center of G. Use this fact to prove that the fundamental group of a Lie group is abelian.

13 Prove that Example 3.5(d) is, up to isomorphism, the only 2-dimensional non-abelian Lie algebra. Conclude that Example 3.3(h) is, up to isomorphism, the unique simply connected 2-dimensional non-abelian Lie group.

14 Show that there are matrices $A, B \in \mathfrak{gl}(n,\mathbb{C})$ such that $e^{A+B} \neq e^A e^B$.

15 Prove that the exponential map for $Gl(n,\mathbb{C})$ is surjective. (*Outline:* First, by using the Jordan canonical form, reduce the problem to that of showing that each elementary Jordan matrix (a matrix with a fixed constant λ on the diagonal, with 1's immediately above each diagonal entry, and with zeros elsewhere) in $Gl(j,\mathbb{C})$, for $1 \leq j \leq n$, is the exponential of an element of $\mathfrak{gl}(j,\mathbb{C})$. Let A be an elementary Jordan matrix. Write A as $(\lambda I) \cdot N$ where the diagonal entries of N are 1's. Prove that $\lambda I = e^{A_1}$ and that $N = e^{A_2}$, where $A_1A_2 = A_2A_1$. To find A_2, observe that since sufficiently high powers of the difference $I - N$ vanish, N has a logarithm, namely,

$$\log N = -\sum_{k=1}^{\infty} \frac{(I-N)^k}{k}.$$

Use a formal power series argument to show that $e^{\log N} = N$.)

16 Let G be a Lie group. A vector field Y on G is right invariant if Y is r_σ-related to itself for each $\sigma \in G$. Prove that the set of right invariant vector fields on G forms a Lie algebra under the Lie bracket operation and is naturally isomorphic as a vector space with G_e. Let $\varphi: G \to G$ be the diffeomorphism defined by $\varphi(\sigma) = \sigma^{-1}$. Prove that if X is a left invariant vector field on G, then $d\varphi(X)$ is the right invariant vector field whose value at e is $-X(e)$. Prove that $X \mapsto d\varphi(X)$ gives a Lie algebra isomorphism of the Lie algebra of left invariant vector fields on G with the Lie algebra of right invariant vector fields on G.

17 Find an example of a Lie group G with a Lie subgroup A that is not closed in G.

18 Let G be an n-dimensional abelian connected Lie group. Prove there is an integer k, $0 \leq k \leq n$, such that G is isomorphic with $\mathbb{R}^{n-k} \times T^k$ where T^k is a torus of dimension k. First, prove that up to isomorphism, $\mathbb{R}^n$ is the unique simply connected n-dimensional abelian Lie group. Next, show that if D is a discrete subgroup of $\mathbb{R}^n$, then either $D = \{0\}$ or there is an integer k, $1 \leq k \leq n$, and k linearly independent vectors $v_1, \ldots, v_k$ in $\mathbb{R}^n$ which generate D. To prove this, suppose that $D \neq \{0\}$, and let k be the smallest integer such that D is contained in a k-dimensional subspace V of $\mathbb{R}^n$. Find a basis $w_1, \ldots, w_k$ for V such that the $w_i \in D$. Let

$$A_i = \left\{ \sum_{j=1}^{i} r_j w_j : 0 \leq r_j \leq 1, r_i > 0 \right\}$$

for $i = 1, \ldots, k$. Choose v_i to be an element of $A_i \cap D$ with smallest possible coefficient r_i. Then $\{v_1, \ldots, v_k\}$ will be a linearly independent set generating D.

19 Supply a proof for 3.56.

20 Prove that the group of automorphisms of a connected Lie group G is itself a Lie group. (*Outline:* Let $\tilde{G}$ be the simply connected covering group of G, and let π be the covering homomorphism. Let $D = \ker \pi$. We know from 3.57(b) that the group $A(\tilde{G})$ of automorphisms of $\tilde{G}$ is a Lie group. Show that the group of automorphisms of G is naturally isomorphic with the (closed) subgroup of $A(\tilde{G})$ consisting of those automorphisms of $\tilde{G}$ which map D onto D.)

21 Prove that $U(n)$ is diffeomorphic with $S^1 \times SU(n)$.

22 (a) Deduce from a "super-triangularization" argument similar to that following 3.35(6), but now using unit eigenvectors and orthogonal complements, that if A is a hermitian matrix (that is, A is complex and $\bar{A} = A^t$), then there exists a unitary matrix B such that BAB^{-1} is diagonal. By a similar argument, prove that if A is a real symmetric matrix ($A = A^t$), then there exists a real orthogonal matrix B such that BAB^{-1} is diagonal.

(b) Using part (a) and 3.46(7), prove that the exponential map maps the hermitian matrices one-to-one onto the set of positive definite (all eigenvalues positive) hermitian matrices. Deduce also that the exponential map maps the real symmetric matrices one-to-one onto the set of positive definite symmetric matrices.

23 Use part (b) of Exercise 22 to prove that the polar decomposition 3.68(1) is unique. (*Hint:* If $\sigma = PR = P_1R_1$, first prove that $P^2 = P_1^2$, then apply part (b) of Exercise 22 to P and P_1.)

24 Prove that every matrix $\sigma \in Gl(n,\mathbb{C})$ can uniquely be written as a product $\sigma = PR$, where P is a positive definite hermitian matrix and R is a unitary matrix.

25 Prove that $Gl(n,\mathbb{C})$ is connected.

26 Complete the details of 3.65(b) and (c).

4

INTEGRATION ON MANIFOLDS

We shall consider integration of p-forms over differentiable singular p-chains in n-dimensional manifolds, and integration of n-forms over regular domains in oriented n-dimensional manifolds. For both of these situations we shall prove a version of Stokes' theorem. This is a generalization of the Fundamental Theorem of Calculus and is undoubtedly the single most important theorem in the subject. We shall also consider integration on Riemannian manifolds and on Lie groups. Finally, we shall introduce the de Rham cohomology groups and shall prove the Poincaré lemma, from which we will conclude that the de Rham cohomology groups of Euclidean space are trivial. This lemma will be of central importance for the de Rham theorem, which is stated at the end of this chapter and proved in Chapter 5.

ORIENTATION

4.1 Definitions Let V be a real vector space of dimension n. The notion of an orientation on V was introduced in Exercise 13 of Chapter 2. Recall that the nth exterior power $\Lambda_n(V)$ is 1-dimensional, so that $\Lambda_n(V) - \{0\}$ has two components. An *orientation* on V is a choice of a component of $\Lambda_n(V) - \{0\}$.

Now let M be a connected differentiable manifold of dimension n. We shall call M *orientable* if it is possible to choose in a consistent way an orientation on M_m^* for each $m \in M$. More precisely, let O be the "0-section" of the exterior n-bundle $\Lambda_n^*(M)$; that is,

(1) $$O = \bigcup_{m \in M} \{0 \in \Lambda_n(M_m^*)\}.$$

Then since each $\Lambda_n(M_m^*) - \{0\}$ has exactly two components, it follows easily that $\Lambda_n^*(M) - O$ has at most two components. We say that M is *orientable* if $\Lambda_n^*(M) - O$ has two components; and if M is orientable, an *orientation on M* is a choice of one of the two components of $\Lambda_n^*(M) - O$.

A non-connected manifold M is said to be orientable if each component of M is orientable, and an orientation is a choice of orientation on each component. Let M be oriented, and let $v_1, \ldots, v_n$ be a basis of M_m with dual basis $\delta_1, \ldots, \delta_n$. We say that the (ordered) basis $v_1, \ldots, v_n$ is *oriented* if $\delta_1 \wedge \cdots \wedge \delta_n$ belongs to the orientation.

Let M and N be orientable n-dimensional manifolds, and let ψ: $M \to N$ be a differentiable map. We say that ψ *preserves orientations* if the induced map $\delta\psi$: $\Lambda_n^*(N) \to \Lambda_n^*(M)$ maps the component of $\Lambda_n^*(N) - O$ determining the orientation on N into the component of $\Lambda_n^*(M) - O$ determining the orientation on M. Equivalently, ψ is orientation-preserving if $d\psi$ sends oriented bases of the tangent spaces to M into oriented bases of the tangent spaces to N.

4.2 Proposition *Let M be a differentiable manifold of dimension n. Then the following are equivalent:*

(a) *M is orientable.*

(b) *There is a collection $\Phi = \{(V,\psi)\}$ of coordinate systems on M such that*

$$(1) \qquad M = \bigcup_{(V,\psi)\in\Phi} V \qquad \textit{and} \qquad \det\left(\frac{\partial x_i}{\partial y_j}\right) > 0 \quad \textit{on} \quad U \cap V$$

whenever $(U, x_1, \ldots, x_n)$ and $(V, y_1, \ldots, y_n)$ belong to Φ.

(c) *There is a nowhere-vanishing n-form on M.*

PROOF We can assume, without loss of generality, that M is connected. We prove that (a) $\Rightarrow$ (b) $\Rightarrow$ (c) $\Rightarrow$ (a). Given (a), that M is orientable, choose an orientation on M; that is, we choose one of the two components, call it Λ, of $\Lambda_n^*(M) - O$. Observe that for each $m \in M$, $\Lambda \cap \Lambda_n(M_m^*)$ is precisely one of the two components of $\Lambda_n(M_m^*) - \{0\}$. Now let Φ consist of all of those coordinate systems $(V, y_1, \ldots, y_n)$ on M such that the map of V into $\Lambda_n^*(M)$ defined by

$$(2) \qquad m \mapsto (dy_1 \wedge \cdots \wedge dy_n)(m)$$

has range in Λ. Now, if $(U, x_1, \ldots, x_n)$ and $(V, y_1, \ldots, y_n)$ are any two coordinate systems on M, then for $m \in U \cap V$,

$$(3) \qquad (dx_1 \wedge \cdots \wedge dx_n)(m) = \det\left(\left.\frac{\partial x_i}{\partial y_j}\right|_m\right)(dy_1 \wedge \cdots \wedge dy_n)(m).$$

If these coordinate systems belong to Φ, then necessarily

$$\det\left(\left.\frac{\partial x_i}{\partial y_j}\right|_m\right) > 0$$

for each $m \in U \cap V$. Consequently, (1) is satisfied, and result (b) follows from (a).

Now assume (b). Let $\{\varphi_i\}$ be a partition of unity subordinate to the cover of M given by the coordinate neighborhoods in the collection Φ with φ_i subordinate to $(V_i, x_1^i, \ldots, x_n^i)$. Then

$$\omega = \sum_i \varphi_i \, dx_1^i \wedge \cdots \wedge dx_n^i \tag{4}$$

is a global n-form on M, where $\varphi_i \, dx_1^i \wedge \cdots \wedge dx_n^i$ is defined to be the 0-form outside of V_i. That ω vanishes nowhere follows from the fact that for each m, $\omega(m)$ is a finite sum with positive coefficients of elements of one component of $\Lambda_n(M_m^*) - \{0\}$. Thus (c) follows from (b).

Finally, let ω be a nowhere-vanishing n-form on M, and let

$$\Lambda^+ = \bigcup_{m \in M} \{a\omega(m): a \in \mathbb{R}, a > 0\},$$

$$\Lambda^- = \bigcup_{m \in M} \{a\omega(m): a \in \mathbb{R}, a < 0\}.$$

Then $\Lambda_n^*(M) - O$ is the disjoint union of the two open subsets Λ^+ and Λ^-, so $\Lambda_n^*(M) - O$ is disconnected, and M is orientable.

4.3 Examples

(a) Every Lie group G is orientable, for if $\omega_1, \ldots, \omega_n$ is a basis for the left invariant 1-forms on G, then $\omega_1 \wedge \cdots \wedge \omega_n$ is a global nowhere-vanishing n-form on G.

(b) The *standard orientation* on the Euclidean space $\mathbb{R}^d$ is the one determined by the d-form $dr_1 \wedge \cdots \wedge dr_d$.

(c) Let X be a d-dimensional manifold, and suppose that there exists an immersion $f: X \to \mathbb{R}^{d+1}$. A *normal vector field* along (X, f) is a smooth map $N: X \to T(\mathbb{R}^{d+1})$ such that for each $p \in X$, the vector $N(p)$ lies in $(\mathbb{R}^{d+1})_{f(p)}$ and is orthogonal to the subspace $df(X_p) \subset (\mathbb{R}^{d+1})_{f(p)}$. Such a manifold X is orientable if and only if there is a smooth nowhere-vanishing normal vector field along (X, f). (See Exercise 1.)

(d) As an immediate application of Example (c), the sphere S^n is orientable for each $n \geq 1$.

(e) The real projective space P^n is orientable if and only if n is odd. (See Exercise 2.)

INTEGRATION ON MANIFOLDS

4.4 Integration in the Euclidean Space $\mathbb{R}^n$ We assume that the reader is familiar with some theory of integration in $\mathbb{R}^n$. Since we shall be integrating continuous (in fact usually C^∞) functions over nice subsets of $\mathbb{R}^n$

(polyhedra for example), the theory of the Riemann integral will be quite sufficient. The principal theorem that we need to recall is the change of variables formula. Several versions of this formula and their proofs may be found in [6], [18], or [29]. A version sufficient for our purposes is the following. Let φ be a diffeomorphism of a bounded open set D in $\mathbb{R}^n$ with a bounded open set $\varphi(D)$. Let $J\varphi$ denote the determinant of the Jacobian matrix of φ:

$$J\varphi = \det\left(\frac{\partial\varphi_i}{\partial r_j}\right).$$

Let f be a bounded continuous function on $\varphi(D)$, and let A be a nice subset of D. (A will be polyhedral in most of our applications. Generally, for the Riemann theory, a nice subset would be one which has Jordan content.) Then

$$(1) \qquad \int_{\varphi(A)} f = \int_A f \circ \varphi \, |J\varphi| .$$

4.5 Integration of n-forms in $\mathbb{R}^n$ As usual, we let $r_1, \ldots, r_n$ denote the canonical coordinate system on $\mathbb{R}^n$. The standard orientation is determined by the n-form $dr_1 \wedge \cdots \wedge dr_n$. Now let ω be an n-form on an open set $D \subset \mathbb{R}^n$. Then there is a uniquely determined function f on D such that $\omega = f\, dr_1 \wedge \cdots \wedge dr_n$. Let $A \subset D$. We define

$$(1) \qquad \int_A \omega = \int_A f,$$

provided that the latter exists. We can restate the change of variables formula 4.4(1) in terms of differential forms. Let φ, D, and A be as in 4.4, and let ω be an n-form on $\varphi(D)$. Then

$$(2) \qquad \int_{\varphi(A)} \omega = \pm \int_A \delta\varphi(\omega),$$

where one uses "+" if φ is orientation preserving and "−" if φ is orientation reversing.

4.6 Integration over Chains The first type of integration that we shall consider on general manifolds involves integration of p-forms over differentiable singular p-chains in an n-manifold.

For each $p \geq 1$ we let

$$(1) \qquad \Delta^p = \left\{(a_1, \ldots, a_p) \in \mathbb{R}^p : \sum_{i=1}^p a_i \leq 1, \text{ and each } a_i \geq 0\right\}.$$

Δ^p is called the *standard p-simplex* in $\mathbb{R}^p$. For $p = 0$, we set Δ^0 equal to the 1-point space $\{0\}$; Δ^0 is the *standard* 0-*simplex*. Let M be a manifold. A *differentiable singular p-simplex* σ in M is a map σ of Δ^p into M which extends to be a differentiable (C^∞) map of a neighborhood of Δ^p in $\mathbb{R}^p$ into M.

In this chapter we shall refer to such a σ as simply a p-simplex σ in M. (In Chapter 5 we shall deal, in addition, with continuous singular simplices, and so there we shall need to retain the term "differentiable" for distinction.) A 0-simplex in M consists of a map of the 1-point space $\{0\}$ into M. A *p-chain* c in M (with real coefficients) is a finite linear combination $c = \sum a_i \sigma_i$ of p-simplices σ_i in M where the a_i are real numbers.

For each $p \geq 0$ we define a collection of maps $k_i{}^p: \Delta^p \to \Delta^{p+1}$ for $0 \leq i \leq p + 1$ as follows:

$$\text{for } p = 0, \quad k_0{}^0(0) = 1 \quad \text{and} \quad k_1{}^0(0) = 0;$$

$$(2)\qquad \text{for } p \geq 1, \quad \begin{cases} k_0{}^p(a_1, \ldots, a_p) = \left(1 - \sum_{i=1}^{p} a_i, a_1, \ldots, a_p\right) \quad \text{and} \\ k_i{}^p(a_1, \ldots, a_p) = (a_1, \ldots, a_{i-1}, 0, a_i, \ldots, a_p) \\ \qquad\qquad (1 \leq i \leq p + 1). \end{cases}$$

If σ is a p-simplex in M with $p \geq 1$, we define its *ith face*, $0 \leq i \leq p$, to be the $(p - 1)$ simplex

$$(3)\qquad \sigma^i = \sigma \circ k_i^{p-1},$$

and we define the *boundary* of σ to be the $(p - 1)$ chain

$$(4)\qquad \partial\sigma = \sum_{i=0}^{p} (-1)^i \sigma^i.$$

We extend the boundary operator linearly to chains. Note carefully the use of the superscript to denote faces. Thus the boundary of the p-chain $\sum_{j=1}^{k} a_j \sigma_j$ is

$$\sum_{i=0}^{p} \sum_{j=1}^{k} (-1)^i a_j \sigma_j{}^i.$$

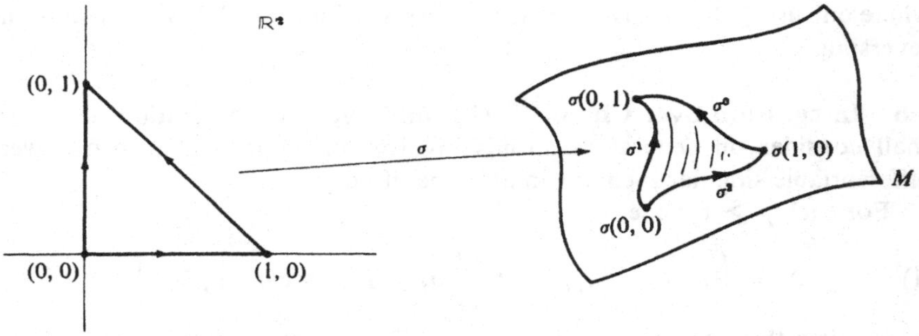

We claim that

$$(5)\qquad k_i^{p+1} \circ k_j{}^p = k_{j+1}^{p+1} \circ k_i{}^p \qquad (p \geq 0;\ i \leq j).$$

For $p = 0$, check the three possible cases separately. For $p \geq 1$, observe that both sides of (5) give the following maps:

$$\begin{array}{lll} 1 \leq i < j & (a_1, \ldots, a_p) \mapsto (a_1, \ldots, a_{i-1}, 0, a_i, \ldots, a_{j-1}, 0, a_j, \ldots, a_p) \\ 1 \leq i = j & (a_1, \ldots, a_p) \mapsto (a_1, \ldots, a_{i-1}, 0, 0, a_i, \ldots, a_p) \\ 0 = i < j & (a_1, \ldots, a_p) \mapsto \left(1 - \sum_{i=1}^{p} a_i, a_1, \ldots, a_{j-1}, 0, a_j, \ldots, a_p\right) \\ 0 = i = j & (a_1, \ldots, a_p) \mapsto \left(0, 1 - \sum_{i=1}^{p} a_i, a_1, \ldots, a_p\right). \end{array} \tag{6}$$

It follows from (5), (3), and (4) that the boundary of the boundary of a chain is always 0. That is,

$$\partial \circ \partial = 0. \tag{7}$$

Now let σ be a p-simplex in M, and let ω be a p-form defined on a neighborhood of the image of σ. In most of our applications we shall deal only with smooth forms, but for the purposes of the following definition it would be quite sufficient for ω to be a *continuous* p-form. First of all, if $p = 0$, then a 0-simplex consists simply of a point in M, and a 0-form is simply a function. In this case, we define the integral of the function ω over the 0-simplex σ to be the value of ω at the point $\sigma(0) \in M$:

$$\int_\sigma \omega = \omega(\sigma(0)). \tag{8}$$

If $p \geq 1$, then since σ extends to be a smooth map of a neighborhood of Δ^p in $\mathbb{R}^p$ into M, ω can be pulled back via σ to a p-form $\delta\sigma(\omega)$ on a neighborhood of Δ^p. In this case, we define the integral of the p-form ω over the p-simplex σ by

$$\int_\sigma \omega = \int_{\Delta^p} \delta\sigma(\omega). \tag{9}$$

We extend these integrals linearly to chains, so that if $c = \sum a_i \sigma_i$, then

$$\int_c \omega = \sum a_i \int_{\sigma_i} \omega. \tag{10}$$

Perhaps the single most important theorem in integration theory on manifolds is Stokes' theorem. This is a generalization of the Fundamental Theorem of Calculus. Observe that in our context the Fundamental Theorem says that if F is a smooth function on the real line, and if σ is a smooth 1-simplex in the real line, then

$$\int_{\partial\sigma} F = \int_\sigma dF. \tag{11}$$

We shall present two versions of Stokes' theorem. The first is in terms of integration of forms over chains.

4.7 Stokes' Theorem I *Let c be a p-chain ($p \geq 1$) in a differentiable manifold M, and let ω be a smooth $(p-1)$ form defined on a neighborhood of the image of c. Then*

$$\int_{\partial c} \omega = \int_c d\omega. \tag{1}$$

PROOF It is sufficient to consider the case in which the p-chain c consists of a single p-simplex σ. Thus we must prove that

$$\int_\sigma d\omega = \int_{\partial\sigma} \omega. \tag{2}$$

It follows immediately from our definitions that (2) is equivalent with

$$\int_{\Delta^p} d(\delta\sigma(\omega)) = \sum_{i=0}^{p} (-1)^i \int_{\Delta^{p-1}} \delta\sigma^i(\omega) = \sum_{i=0}^{p} (-1)^i \int_{\Delta^{p-1}} \delta k_i^{p-1} \circ \delta\sigma(\omega). \tag{3}$$

Observe first of all that the case $p = 1$ reduces directly to the Fundamental Theorem of Calculus:

$$\int_{\Delta^1} \frac{d}{dr}(\omega \circ \sigma)\, dr = \omega(\sigma(1)) - \omega(\sigma(0)), \tag{4}$$

so we now assume that $p \geq 2$. Then the $(p-1)$ form $\delta\sigma(\omega)$ can be expressed as

$$\delta\sigma(\omega) = \sum_{j=1}^{p} a_j\, dr_1 \wedge \cdots \wedge \widehat{dr_j} \wedge \cdots \wedge dr_p, \tag{5}$$

where the circumflex over a term means that the term is to be omitted, and where the a_j are C^∞ functions on a neighborhood of Δ^p in $\mathbb{R}^p$. Since the integral is linear, we may consider the special case in which $\delta\sigma(\omega)$ consists of a single term of the form $a_j\, dr_1 \wedge \cdots \wedge \widehat{dr_j} \wedge \cdots \wedge dr_p$. In this case, the left-hand side of (3) becomes

$$(-1)^{j-1} \int_{\Delta^p} \frac{\partial a_j}{\partial r_j}\, dr_1 \wedge \cdots \wedge dr_p. \tag{6}$$

To evaluate the right-hand side of (3), we observe that for $1 \leq i \leq p$,

$$\delta k_i^{p-1}(r_j) = \begin{cases} r_j & (1 \leq j \leq i-1) \\ 0 & (j = i) \\ r_{j-1} & (i+1 \leq j \leq p), \end{cases} \tag{7}$$

and

$$\delta k_0^{p-1}(r_j) = \begin{cases} 1 - \sum_{i=1}^{p-1} r_i & (j = 1) \\ r_{j-1} & (1 < j \leq p). \end{cases} \tag{8}$$

Applying (7) and (8) to the right-hand term of (3), we obtain

$$\sum_{i=0}^{p}(-1)^i\int_{\Delta^{p-1}}\delta k_i^{p-1}(a_j\, dr_1\wedge\cdots\wedge\widehat{dr_j}\wedge\cdots\wedge dr_p)$$

$$(9)\qquad = (-1)^{j-1}\int_{\Delta^{p-1}} a_j\left(1-\sum_{i=1}^{p-1} r_i, r_1,\ldots,r_{p-1}\right) dr_1\wedge\cdots\wedge dr_{p-1}$$

$$+ (-1)^j\int_{\Delta^{p-1}} a_j(r_1,\ldots,r_{j-1},0,r_j,\ldots,r_{p-1})\, dr_1\wedge\cdots\wedge dr_{p-1}.$$

We shall now apply a change of variables to the first term on the right-hand side of (9). Let φ_j be the diffeomorphism of $\mathbb{R}^{p-1}$ defined by

$$(10)\qquad \varphi_j(r_1,\ldots,r_{p-1}) = \begin{cases} (r_1,\ldots,r_{p-1}) & (j=1)\\ \left(1-\sum_{i=1}^{p-1} r_i, r_2,\ldots,r_{p-1}\right) & (j=2)\\ \left(r_2,\ldots,r_{j-1},1-\sum_{i=1}^{p-1} r_i, r_j,\ldots,r_{p-1}\right) & \\ & (3\le j\le p). \end{cases}$$

Then $\varphi_j(\Delta^{p-1}) = \Delta^{p-1}$, and $|J\varphi_j| = 1$, so that by 4.4(1), the first term on the right-hand side of (9) is equal to

(11)

$$(-1)^{j-1}\int_{\Delta^{p-1}} a_j\left(r_1,\ldots,r_{j-1},1-\sum_{i=1}^{p-1} r_i, r_j,\ldots,r_{p-1}\right) dr_1\wedge\cdots\wedge dr_{p-1}.$$

From (3), (6), (9), and (11) we see that the proof has been reduced to showing that

$$(12)\qquad \int_{\Delta^p}\frac{\partial a_j}{\partial r_j}\, dr_1\wedge\cdots\wedge dr_p$$

$$= \int_{\Delta^{p-1}} a_j\left(r_1,\ldots,r_{j-1},1-\sum_{i=1}^{p-1} r_i, r_j,\ldots,r_{p-1}\right) dr_1\wedge\cdots\wedge dr_{p-1}$$

$$- \int_{\Delta^{p-1}} a_j(r_1,\ldots,r_{j-1},0,r_j,\ldots,r_{p-1})\, dr_1\wedge\cdots\wedge dr_{p-1}.$$

But now (12) is simply the evaluation of the integral of $\partial a_j/\partial r_j$ over Δ^p by iterating first with respect to r_j and applying again the Fundamental Theorem of Calculus.

4.8 Integration on an Oriented Manifold Let M be an n-dimensional oriented manifold. We shall integrate n-forms over regular domains in M. A subset D of M will be called a *regular domain* if for each point $m \in M$ one of the following holds:

(a) There is an open neighborhood of m which is contained in $M - D$.

(b) There is an open neighborhood of m which is contained in D.

(c) There is a centered coordinate system (U,φ) about m such that $\varphi(U \cap D) = \varphi(U) \cap H^n$, where H^n is the half-space of $\mathbb{R}^n$ defined by $r_n \geq 0$.

Points of D of type (b) are called *interior points* and comprise the interior, $\text{Int}(D)$, of D. Points of type (c) are called *boundary points* and comprise the boundary, ∂D, of D. Coordinate systems of type (c) restricted to ∂D overlap differentiably and yield a manifold structure of dimension $n - 1$ on ∂D, making ∂D into an imbedded $(n - 1)$ dimensional submanifold of M.

Let $m \in \partial D$, and let $v \in M_m$. We call v an *outer vector* to D if for each smooth curve $\alpha(t)$ in M with $\dot{\alpha}(0) = v$, one has $\alpha(t) \notin D$ for $0 < t < \varepsilon$, for some $\varepsilon > 0$. The orientation on M induces an orientation on ∂D as follows. Let v be an outer vector to ∂D at m, and let $v_1, \ldots, v_{n-1}$ be a basis of the tangent space $(\partial D)_m$. Then we define $v_1, \ldots, v_{n-1}$ to be an oriented basis of $(\partial D)_m$ if and only if $v, v_1, \ldots, v_{n-1}$ is an oriented basis of M_m. One can easily check that this definition is independent of the outer vector v chosen, and that this defines a smooth orientation on ∂D in the sense of 4.1.

Now let ω be an n-form ($n = \dim M$) with compact support, and let D be a regular domain in M. As a particular case, D could be all of M. We are going to define the integral of ω over D. As in 4.6, it will be sufficient for the purposes of this definition for ω to be a *continuous* n-form. We shall use a partition of unity to reduce the support of ω to certain n-simplices in M over which we can integrate as in 4.6(9). First, we shall choose some n-simplices suitably related to D and ∂D.

An n-simplex σ in M will be called *regular* if σ extends to a diffeomorphism on a neighborhood of Δ^n. When speaking of regular n-simplices, we shall always assume that they have been extended in this way to a neighborhood of Δ^n. An *oriented regular n-simplex* is one in which the map σ preserves orientations. (We always take the standard orientation on $\mathbb{R}^n$.)

Associated with a given regular domain D, we shall consider only oriented regular n-simplices of the following two types:

(α) $\sigma(\Delta^n) \subset \text{Int}(D)$.

(β) $\sigma(\Delta^n) \subset D$ and $\sigma(\Delta^n) \cap \partial D = \sigma^n(\Delta^{n-1})$; that is, precisely the nth face of σ lies in the boundary of D.

Now cover D by open sets U of the following types:

(α') U lies in the interior of an oriented regular n-simplex σ of type (α).

(β') U is the image under a type (β) oriented regular n-simplex σ of an open set V in $\mathbb{R}^n$ which is a neighborhood of a point in the nth face of Δ^n, which intersects the boundary of Δ^n only in that nth face, and whose image under σ is contained in $\sigma(\Delta^n) \cup (M - D)$.

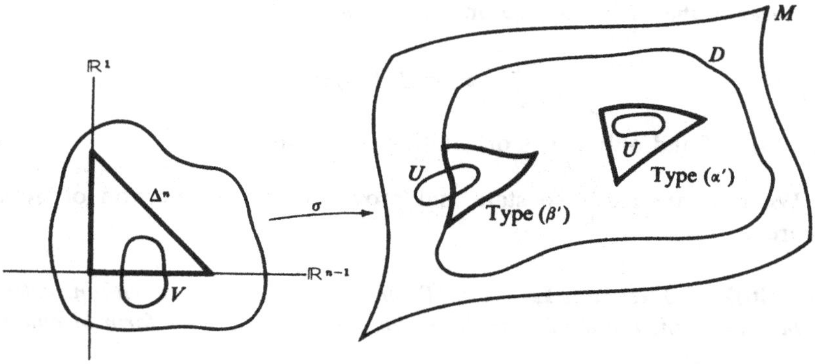

Since supp $\omega \cap D$ is compact, it has a finite cover $U_1, \ldots, U_k$ by open sets of type (α') or (β'). Let the associated oriented regular n-simplices be $\sigma_1, \ldots, \sigma_k$. Let $U = M - (\text{supp}\, \omega \cap D)$, and let $\varphi, \varphi_1, \ldots, \varphi_k$ be a partition of unity subordinate to the cover $U, U_1, \ldots, U_k$ of M. We define the integral of ω over D by

(1) $$\int_D \omega = \sum_{i=1}^{k} \int_{\sigma_i} \varphi_i \omega.$$

We must check that the definition (1) is independent of the cover and the partition of unity chosen. Let $V, V_1, \ldots, V_l$ and $\psi, \psi_1, \ldots, \psi_l$ be another such cover and another such partition of unity respectively, with V_j associated with the oriented regular n-simplex τ_j. Since $\psi = 0$ on supp $\omega \cap D$, it follows that $\sum_{j=1}^{l} \psi_j = 1$ there, so that

(2) $$\sum_{i=1}^{k} \int_{\sigma_i} \varphi_i \omega = \sum_{i=1}^{k} \int_{\sigma_i} \sum_{j=1}^{l} \psi_j \varphi_i \omega = \sum_{i,j} \int_{\sigma_i} \psi_j \varphi_i \omega.$$

Similarly,

(3) $$\sum_{j=1}^{l} \int_{\tau_j} \psi_j \omega = \sum_{i,j} \int_{\tau_j} \psi_j \varphi_i \omega.$$

Now, since $\sigma_i^{-1} \circ \tau_j$ is an orientation-preserving diffeomorphism on the open set (possibly empty) where it is defined, and since $(\text{supp}\, \psi_j \varphi_i \omega) \cap \sigma_i(\Delta^n) = (\text{supp}\, \psi_j \varphi_i \omega) \cap \tau_j(\Delta^n)$, it follows from the change of variables formula 4.5(2) that

(4) $$\int_{\sigma_i} \psi_j \varphi_i \omega = \int_{\Delta^n} \delta\sigma_i(\psi_j \varphi_i \omega) = \int_{\Delta^n} \delta(\sigma_i^{-1} \circ \tau_j)[\delta\sigma_i(\psi_j \varphi_i \omega)]$$
$$= \int_{\Delta^n} \delta\tau_j(\psi_j \varphi_i \omega) = \int_{\tau_j} \psi_j \varphi_i \omega.$$

It follows from (2), (3), and (4) that $\int_D \omega$ is well-defined, independent of the choice of cover and partition of unity.

Observe that if γ is a diffeomorphism of M, then

$$\int_{\gamma(D)} \omega = \pm \int_D \delta\gamma(\omega) \tag{5}$$

with "+" if and only if γ is orientation-preserving.

We now are ready to state and prove our second version of Stokes' theorem.

4.9 Stokes' Theorem II *Let D be a regular domain in an oriented n-dimensional manifold M, and let ω be a smooth $(n-1)$ form of compact support. Then*

$$\int_D d\omega = \int_{\partial D} \omega. \tag{1}$$

PROOF Let $\varphi_1, \ldots, \varphi_k$ and $\sigma_1, \ldots, \sigma_k$ be chosen as in 4.8(1) relative to $(\text{supp}\ \omega) \cap D$. Since $\sum_{i=1}^{k} \varphi_i = 1$ on a neighborhood of $(\text{supp}\ \omega) \cap D$, $d(\sum \varphi_i) = 0$ there. Thus on a neighborhood of $(\text{supp}\ \omega) \cap D$ we have

$$\sum_{i=1}^{k} d(\varphi_i\omega) = \sum_{i=1}^{k} d\varphi_i \wedge \omega + \sum_{i=1}^{k} \varphi_i\, d\omega = d\omega. \tag{2}$$

Now, if σ_i is an n-simplex of type 4.8(α), then

$$\int_{\partial\sigma_i} \varphi_i\omega = 0 = \int_{\partial D} \varphi_i\omega \tag{3}$$

since $\text{supp}\ \varphi_i\omega \subset \text{Int}\ \sigma_i(\Delta^n) \subset \text{Int}\ D$. On the other hand, suppose that σ_i is an n-simplex of type 4.8(β). In this case, $\varphi_i\omega$ is zero on the boundary of σ_i except possibly at points in the interior of the nth face $\sigma_i{}^n$. Now $\sigma_i{}^n$ is an orientation-preserving regular $(n-1)$ simplex in ∂D if n is even, and is orientation-reversing if n is odd. It follows that

$$\int_{\partial\sigma_i} \varphi_i\omega = (-1)^n \int_{\sigma_i{}^n} \varphi_i\omega = (-1)^n(-1)^n \int_{\partial D} \varphi_i\omega = \int_{\partial D} \varphi_i\omega. \tag{4}$$

From (2), (3), (4), and from our first version (4.7) of Stokes' theorem, we see that

$$\begin{aligned} \int_D d\omega &= \sum_i \int_D d(\varphi_i\omega) = \sum_i \int_{\sigma_i} d(\varphi_i\omega) \\ &= \sum_i \int_{\partial\sigma_i} \varphi_i\omega = \sum_i \int_{\partial D} \varphi_i\omega = \int_{\partial D} \omega. \end{aligned} \tag{5}$$

Corollary *Let ω be a smooth $(n-1)$ form on a compact oriented n-dimensional manifold M. Then*

$$\int_M d\omega = 0.$$

4.10 Integration on a Riemannian Manifold Let M be a Riemannian manifold of dimension n. That is, M is an n-dimensional differentiable manifold with a positive definite inner product $\langle \ , \ \rangle_m$ on each tangent space M_m such that $m \mapsto \langle X, Y\rangle_m$ is a smooth function on M whenever X and Y are smooth vector fields. The existence of Riemannian metrics on differentiable manifolds was asserted in Exercise 23 of Chapter 1.

Given a point $m \in M$, one can find a neighborhood U of m and a collection $e_1, \ldots, e_n$ of C^∞ vector fields on U which are orthonormal in the sense that they form an orthonormal basis of the tangent space to M at each point of U. Start with a coordinate neighborhood $(U, x_1, \ldots, x_n)$, apply the usual Gram-Schmidt procedure to orthonormalize the vector fields $\partial/\partial x_1, \ldots, \partial/\partial x_n$, and do it simultaneously at all points of U. Such a collection $e_1, \ldots, e_n$ is called a *local orthonormal frame field.*

Since the inner product $\langle \ , \ \rangle_m$ is, in particular, a non-singular pairing of M_m with itself, it induces (see 2.7) a natural isomorphism of M_m with M_m^*, namely, $v \mapsto \varphi_v$ where

(1) $$\varphi_v(w) = \langle v, w\rangle_m .$$

Via this isomorphism, M_m^* inherits an inner product. Observe that the dual basis to an orthonormal basis of M_m is itself an orthonormal basis for M_m^*.

Now let $e_1, \ldots, e_n$ be a local orthonormal frame field on U, and let $\omega_1, \ldots, \omega_n$ be the dual 1-forms. That is,

(2) $$\omega_i(e_j) = \delta_{ij} \quad \text{on } U.$$

Then $\omega_1, \ldots, \omega_n$ form a *local orthonormal coframe field* on U. Consider now two local orthonormal coframe fields $\omega_1, \ldots, \omega_n$ on U and $\omega_1', \ldots, \omega_n'$ on U'. Then on $U \cap U'$

(3) $$\omega_1 \wedge \cdots \wedge \omega_n = \det(\sigma)\omega_1' \wedge \cdots \wedge \omega_n'$$

where σ is an orthogonal matrix whose entries are C^∞ functions on $U \cap U'$. Thus

(4) $$\omega_1 \wedge \cdots \wedge \omega_n = \pm\omega_1' \wedge \cdots \wedge \omega_n' .$$

Now assume that M is oriented. A local coframe field $\omega_1, \ldots, \omega_n$ on U will be called *oriented* if $\omega_1 \wedge \cdots \wedge \omega_n$ belongs to the orientation at each point of U. Choose a local oriented orthonormal coframe field about each point of M. Then the corresponding n-forms $\omega_1 \wedge \cdots \wedge \omega_n$ agree on overlaps, and therefore determine a globally defined nowhere-vanishing n-form ω on M. This form ω is called the *volume form* of the oriented Riemannian manifold M. Its integral over M is the *volume* of M.

In Exercise 13 of Chapter 2, the star operator $*$ was introduced on $\Lambda(V)$ for an oriented inner product space V. On an oriented Riemannian manifold M, we therefore have $*$ defined on $\Lambda(M_m^*)$ for each m. It is easy to see that $*$ takes smooth forms to smooth forms, so we have a linear operator

(5) $$*\colon E^p(M) \to E^{n-p}(M),$$

which, according to Exercise 13 of Chapter 2, satisfies

(6) $$** = (-1)^{p(n-p)}.$$

We already know from 4.8 how to integrate n-forms over an oriented n-dimensional manifold. Now in the case of an oriented Riemannian manifold M, we define the integral over M of a continuous function f with compact support to be the integral of the (continuous) n-form $*f = f\omega$. That is,

(7) $$\int_M f = \int_M *f = \int_M f\omega.$$

Actually, the orientation was convenient but not necessary in the definition of $\int_M f$. If M is a (not necessarily oriented) Riemannian manifold, we define the integral of continuous functions with compact support as follows. Let $\{U_\alpha\}$ be a cover of M by interiors of regular n-simplices σ_α, and let $\omega_1{}^\alpha, \ldots, \omega_n{}^\alpha$ be a local orthonormal coframe field defined on a neighborhood of $\sigma_\alpha(\Delta^n)$. Then there exist C^∞ functions h_α on neighborhoods of Δ^n such that

$$\delta\sigma_\alpha(\omega_1{}^\alpha \wedge \cdots \wedge \omega_n{}^\alpha) = h_\alpha\, dr_1 \wedge \cdots \wedge dr_n.$$

Let $\{\varphi_\alpha\}$ be a partition of unity subordinate to the cover $\{U_\alpha\}$, and let f be a continuous function with compact support on M. Then we define

(8) $$\int_M f = \sum_\alpha \int_{\Delta^n} (\varphi_\alpha f) \circ \sigma_\alpha\, |h_\alpha|\, dr_1 \wedge \cdots \wedge dr_n.$$

That this definition is independent of the cover and partition of unity chosen follows from an argument similar to the one at the end of 4.8. In the case of an oriented Riemannian manifold, (8) and (7) agree.

In classical vector analysis the gradient of a function f on $\mathbb{R}^n$ is defined to be the vector field $\sum_{i=1}^{n} (\partial f/\partial r_i)\partial/\partial r_i$, and the divergence of a vector field $V = \sum_{i=1}^{n} v_i\, \partial/\partial r_i$ is defined to be the function $\sum_{i=1}^{n} \partial v_i/\partial r_i$. We extend these notions to general Riemannian manifolds as follows. Recall that the metric gives us canonical isomorphisms $M_m \simeq M_m^*$. We shall for convenience denote such isomorphisms by a *tilde*, so that if $v \in M_m$, then $\tilde{v}$ will be the corresponding dual element in M_m^*; and if $\omega \in M_m^*$, then $\tilde{\omega}$ will be the corresponding vector in M_m. Then if f is a function on M, its gradient is the vector field

(9) $$\operatorname{grad} f = \widetilde{df}.$$

If V is a vector field on an oriented Riemannian manifold, then its divergence is the function

(10) $$\operatorname{div} V = *d*\tilde{V}.$$

Note that since div V involves $*$ twice, it is actually defined independent of orientability.

Stokes' theorem has an equivalent version on Riemannian manifolds known as the *divergence theorem.* This theorem says that *if V is a smooth vector field with compact support on a Riemannian manifold M, if D is a regular domain in M, and if $\vec{n}$ is the unit outer normal vector field on ∂D, then*

$$\int_D \operatorname{div} V = \int_{\partial D} \langle V, \vec{n} \rangle. \tag{11}$$

The proof is left to the reader as an exercise. (For a few additional remarks see Exercise 4.)

4.11 Integration on a Lie Group Let G be an n-dimensional Lie group. We observed in 4.3(a) that G is orientable. We now fix once and for all an orientation on G.

Consider the left invariant n-forms on G. Since such a form is uniquely determined by its value at one point, and since the nth exterior power of an n-dimensional vector space is one-dimensional, there is exactly a one-dimensional space of left invariant n-forms on G. Choose a non-zero left invariant n-form ω consistent with the fixed orientation on G.

Since G is oriented, the integral of compactly supported n-forms is defined on G as in 4.8. We now define, with respect to ω, the integral of a compactly supported continuous function f on G by setting

$$\int_G f = \int_G f\omega. \tag{1}$$

The integral (1) depends, of course, on the choice of the non-zero left invariant n-form ω consistent with the orientation on G. But since such forms are uniquely determined up to a positive constant multiple, so is the integral (1). In the case of a compact group G, we can and always will fix the choice of ω by requiring the normalization

$$\int_G \omega = 1. \tag{2}$$

Consider the diffeomorphism l_σ, which is left translation by the element σ of G. Then since $\delta l_\sigma(\omega) = \omega$, l_σ is orientation-preserving, so that, according to 4.8(5),

$$\int_G f = \int_G f\omega = \int_G \delta l_\sigma(f\omega) = \int_G (f \circ l_\sigma)\omega = \int_G f \circ l_\sigma. \tag{3}$$

In view of property (3)—that the integral of a function f on G is the same as the integral of any of its left translates $f \circ l_\sigma$—we call the integral (1) *left invariant.*

Now we ask to what extent the integral (1) is also right invariant. That is, when do we have

$$\int_G f = \int_G f \circ r_\sigma \tag{4}$$

for each $\sigma \in G$? The form $\delta r_\sigma \omega$ is still left invariant, since

$$\delta l_\tau \, \delta r_\sigma \omega = \delta r_\sigma \, \delta l_\tau \omega = \delta r_\sigma \omega.$$

Thus $\delta r_\sigma \omega$ is some constant multiple of ω. Thus there is defined a function $\tilde{\lambda}$ of G into the non-zero real numbers such that

$$\delta r_\sigma(\omega) = \tilde{\lambda}(\sigma)\omega. \tag{5}$$

It is easily checked that $\tilde{\lambda}$ is C^∞. We let

$$\lambda(\sigma) = |\tilde{\lambda}(\sigma)|. \tag{6}$$

Observe that

$$\lambda(\sigma\tau) = \lambda(\sigma)\lambda(\tau), \tag{7}$$

so that λ is a Lie group homomorphism of G into the multiplicative group of positive real numbers. λ is called the *modular function.* Now since, by 4.8(5), for each σ in G

$$\int_G f\omega = \int_G (f \circ r_\sigma)\lambda(\sigma)\omega,$$

it follows that the integral (1) is right invariant if and only if $\lambda \equiv 1$ on G. A Lie group G for which $\lambda \equiv 1$ is called *unimodular.* We observe that *each compact Lie group G is unimodular* since for each $\sigma \in G$

$$1 = \int_G \omega = \lambda(\sigma)\int_G \omega = \lambda(\sigma).$$

Thus the integral on a compact Lie group is both left and right invariant.

4.12 Application of 4.11 A typical application of the integral on a compact Lie group is the following.

Let G be a Lie group, and let $\alpha\colon G \to \operatorname{Aut}(V)$ be a representation into the automorphisms of a real or complex inner product space V. The representation α is called *unitary* (respectively *orthogonal*) in the case in which V is a complex (respectively real) inner product space if

$$\langle \alpha(\tau)v, \alpha(\tau)w \rangle = \langle v, w \rangle \tag{1}$$

for all v and w in V and for all $\tau \in G$.

Let G be compact and V complex (respectively real). Then there is an inner product on V with respect to which α is unitary (respectively orthogonal). The proofs in the real case and in the complex cases are similar. Let $\{\ ,\ \}$ be any inner product on V. We set

(2) $$\langle v,w\rangle = \int_G \{\alpha(\sigma)v,\alpha(\sigma)w\}\, d\sigma,$$

where we use $d\sigma$ to denote that we are considering the integrand as a function of σ in G. It is immediate that $\langle\ ,\ \rangle$ is again an inner product. That (1) holds follows from the right invariance of the integral on G:

$$\langle\alpha(\tau)v,\alpha(\tau)w\rangle = \int_G \{\alpha(\sigma)\alpha(\tau)v,\alpha(\sigma)\alpha(\tau)w\}\, d\sigma$$
$$= \int_G \{\alpha(\sigma\tau)v,\alpha(\sigma\tau)w\}\, d\sigma = \int_G \{\alpha(\sigma)v,\alpha(\sigma)w\}\, d\sigma = \langle v,w\rangle.$$

DE RHAM COHOMOLOGY

4.13 Definition A p-form α on a differentiable manifold M is called *closed* if $d\alpha = 0$. It is called *exact* if there is a $(p-1)$ form β such that $\alpha = d\beta$. Since $d^2 = 0$, every exact form is closed. The quotient space of the real vector space of closed p-forms modulo the subspace of exact p-forms is called the pth *de Rham cohomology group of M*.

(1) $$H^p_{\mathrm{de\,R}}(M) = \{\text{closed } p\text{-forms}\}/\{\text{exact } p\text{-forms}\}.$$

4.14 Example Consider the case of the unit circle S^1. Since there are no non-zero p-forms on S^1 for $p > 1$, all of the cohomology groups $H^p_{\mathrm{de\,R}}(S^1)$ are zero except possibly for $p = 0, 1$. There are no exact 0-forms, and a closed 0-form on a connected manifold is simply a constant function, so

(1) $$H^0_{\mathrm{de\,R}}(S^1) \cong \mathbb{R}.$$

The "polar coordinate function" θ on S^1 is not well-defined globally since it is defined only up to integral multiples of 2π. However, its differential $d\theta$ is a globally well-defined nowhere-vanishing 1-form on S^1. In fact, $d\theta$ is the volume form of the natural Riemannian metric which S^1 inherits from $\mathbb{R}^2$. Now, $d\theta$ is not exact, for if it were, its integral over S^1 would have to be 0 rather than 2π. All 1-forms on S^1 are closed. We claim that if α is a 1-form, then there is a constant c such that $\alpha - c\, d\theta$ is exact. For let $\alpha = f(\theta)\, d\theta$, let

$$c = \frac{1}{2\pi}\int_{S^1} \alpha,$$

and let

$$g(\theta) = \int_0^\theta (f(\theta) - c)\, d\theta.$$

Since $g(\theta + 2\pi n) = g(\theta)$ for every integer n, then g is a well-defined C^∞ function on S^1; and $dg = (f(\theta) - c)\, d\theta = \alpha - c\, d\theta$. Thus every 1-form on S^1 differs from a real multiple of $d\theta$ by an exact form. Consequently,

(2) $$H^1_{\mathrm{de\,R}}(S^1) \cong \mathbb{R}.$$

4.15 Effect of Mappings Let $f\colon M \to N$ be a C^∞ map. Then the algebra homomorphism $\delta f\colon E^*(N) \to E^*(M)$ commutes with d, according to 2.23, and hence maps closed forms to closed forms and exact forms to exact forms. Thus it induces a homomorphism

$$f^*\colon H^p_{\mathrm{de\,R}}(N) \to H^p_{\mathrm{de\,R}}(M) \tag{1}$$

for each integer $p \geq 0$. If, in addition, $g\colon N \to X$ is C^∞, then

$$(g \circ f)^* = f^* \circ g^*. \tag{2}$$

Clearly the identity map id: $M \to M$ induces the identity on de Rham cohomology:

$$(\mathrm{id})^* = \mathrm{id}. \tag{3}$$

It follows from (2) and (3) that a diffeomorphism $f\colon M \to N$ induces isomorphisms on de Rham cohomology. Thus the de Rham cohomology is a differentiable invariant of a differentiable manifold M. We shall prove in Chapter 5 that it is actually a topological invariant. That is, the de Rham cohomology groups depend only on the underlying topological structure of M and do not depend on the differentiable structure. A key part in the proof of this fact is the de Rham theorem, a version of which we shall formulate now. First, we need to define the real differentiable singular homology groups of M.

4.16 Real Differentiable Singular Homology For each integer $p \geq 0$ we let ${}_\infty S_p(M,\mathbb{R})$ denote the real vector space generated by the differentiable singular p-simplices in M. Hence the elements of ${}_\infty S_p(M,\mathbb{R})$ are precisely the differentiable singular p-chains in M with real coefficients. For $p < 0$, we let ${}_\infty S_p(M,\mathbb{R})$ be the zero vector space. The boundary operator ∂ induces linear transformations

$$\partial_p\colon {}_\infty S_p(M,\mathbb{R}) \to {}_\infty S_{p-1}(M,\mathbb{R}) \tag{1}$$

for each integer p, which for $p \leq 0$ are simply the zero transformation. According to 4.6(7), $\partial_p \circ \partial_{p+1} = 0$, so that the image of ∂_{p+1} lies in the kernel of ∂_p. The pth *differential singular homology group of M with real coefficients* is defined by

$${}_\infty H_p(M;\mathbb{R}) = \ker \partial_p/\mathrm{Im}\, \partial_{p+1}, \tag{2}$$

and is moreover a real vector space. Elements of $\ker \partial_p$ are called *differentiable p-cycles*, and elements of $\mathrm{Im}\, \partial_{p+1}$ are called *differentiable p-boundaries.*

4.17 The de Rham Theorem We shall define a linear mapping of the de Rham cohomology $H^p_{\mathrm{de\,R}}(M)$ into the dual space ${}_\infty H_p(M;\mathbb{R})^*$ of the real differentiable singular homology:

$$H^p_{\mathrm{de\,R}}(M) \to {}_\infty H_p(M;\mathbb{R})^*. \tag{1}$$

Let α be a closed p-form representing the de Rham cohomology class $\{\alpha\}$, and let z be a p-cycle representing the real differentiable singular homology class $\{z\}$. Then (1) is defined by

$$\{\alpha\}(\{z\}) = \int_z \alpha. \tag{2}$$

That (2) is independent of the representatives α and z chosen follows immediately from Stokes' theorem I (4.7).

The de Rham theorem asserts that (1) *is an isomorphism.* This will be proved in 5.36 and 5.37. The real numbers determined by the integrals of a differential form over differentiable cycles are called the *periods* of the differential form. Now, Stokes' theorem I says that the periods of an exact form are all zero. The injectiveness of the isomorphism (1) gives a converse, namely, if a closed form has all of its periods zero, then it is an exact form. The surjectivity of (1) says that if a real number per(z) is assigned to each cycle z in such a way that

$$\text{per}(az_1 + z_2) = a\,\text{per}(z_1) + \text{per}(z_2), \qquad \text{per}(\text{boundary}) = 0, \tag{3}$$

then there is a closed form α on M such that for all cycles z

$$\int_z \alpha = \text{per}(z). \tag{4}$$

A key ingredient in the proof of the de Rham theorem (see 5.28) is the

4.18 Poincaré Lemma *Let U be the open unit ball in Euclidean space $\mathbb{R}^n$, and let $E^k(U)$, as usual, be the space of differential k-forms on U. Then for each $k \geq 1$ there is a linear transformation $h_k\colon E^k(U) \to E^{k-1}(U)$ such that*

$$h_{k+1} \circ d + d \circ h_k = \text{id}. \tag{1}$$

PROOF We begin with formula 2.25(d), which expresses the Lie derivative in terms of exterior differentiation and interior multiplication:

$$L_X = i(X) \circ d + d \circ i(X). \tag{2}$$

We shall apply (2) to the radial vector field

$$X = \sum_{i=1}^{n} r_i \frac{\partial}{\partial r_i} \tag{3}$$

on U. We define a linear operator α_k on $E^k(U)$ by setting

$$\alpha_k(f\,dr_{i_1} \wedge \cdots \wedge dr_{i_k})(p) = \left(\int_0^1 t^{k-1} f(tp)\,dt\right) dr_{i_1} \wedge \cdots \wedge dr_{i_k}(p) \tag{4}$$

and extending linearly to all of $E^k(U)$. Now with X defined by (3), we show that

$$\alpha_k \circ L_X = \text{id} \quad \text{on} \quad E^k(U). \tag{5}$$

For, using the fact that L_X is a derivation which commutes with d, we have

$$\begin{aligned}
(6)\quad \alpha_k \circ L_X(f\,dr_{i_1} \wedge \cdots \wedge dr_{i_k})(p) \\
&= \alpha_k\left\{\left(kf + \sum r_i \frac{\partial f}{\partial r_i}\right) dr_{i_1} \wedge \cdots \wedge dr_{i_k}\right\}(p) \\
&= \left(\int_0^1 t^{k-1}\left(kf(tp) + \sum r_i(tp) \left.\frac{\partial f}{\partial r_i}\right|_{tp}\right) dt\right) dr_{i_1} \wedge \cdots \wedge dr_{i_k}(p) \\
&= \left(\int_0^1 \frac{d}{dt}\,(t^k f(tp))\,dt\right) dr_{i_1} \wedge \cdots \wedge dr_{i_k}(p) \\
&= f(p)\,dr_{i_1} \wedge \cdots \wedge dr_{i_k}(p).
\end{aligned}$$

From (5) and (2) we obtain

$$\text{(7)}\qquad \mathrm{id} = \alpha_k \circ i(X) \circ d + \alpha_k \circ d \circ i(X)$$

on $E^k(U)$ with X given by (3). Now α commutes with d; that is,

$$\text{(8)}\qquad \alpha_k \circ d = d \circ \alpha_{k-1}.$$

For

$$\begin{aligned}
\alpha_k \circ d(f\,dr_{i_1} \wedge \cdots \wedge dr_{i_{k-1}})(p) \\
&= \alpha_k\left(\sum \frac{\partial f}{\partial r_i}\,dr_i \wedge dr_{i_1} \wedge \cdots \wedge dr_{i_{k-1}}\right)(p) \\
&= \left(\int_0^1 t^{k-1} \sum \left.\frac{\partial f}{\partial r_i}\right|_{tp} dt\right) dr_i \wedge dr_{i_1} \wedge \cdots \wedge dr_{i_{k-1}}(p) \\
&= d\left(\int_0^1 t^{k-2} f(tp)\,dt\right) dr_{i_1} \wedge \cdots \wedge dr_{i_{k-1}}(p) \\
&= d \circ \alpha_{k-1}(f\,dr_{i_1} \wedge \cdots \wedge dr_{i_{k-1}})(p).
\end{aligned}$$

Thus from (8) and (7) we obtain

$$\text{(9)}\qquad \mathrm{id} = \alpha_k \circ i(X) \circ d + d \circ \alpha_{k-1} \circ i(X)$$

on $E^k(U)$. Thus the desired linear transformation h_k which yields (1) is obtained by setting

$$\text{(10)}\qquad h_k = \alpha_{k-1} \circ i\left(\sum r_i \frac{\partial}{\partial r_i}\right).$$

Corollary (a) *If ω is a k-form, $k \geq 1$, on the open unit ball in $\mathbb{R}^n$ and $d\omega = 0$, then there exists a $(k-1)$ form β (namely $h_k(\omega)$) such that $d\beta = \omega$.*

Corollary (b) *The de Rham cohomology groups of the open unit ball in $\mathbb{R}^n$ are all zero for $p \geq 1$.*

4.19 Remark Let f_1 and f_2 be C^∞ maps of M into N. Then we have induced maps

(1) $$\delta f_i \colon E^k(N) \to E^k(M) \quad \text{for each } k.$$

If we wish to prove that δf_1 and δf_2 both induce the same homomorphism on de Rham cohomology, then we need only find a collection of linear transformations

(2) $$h_k \colon E^k(N) \to E^{k-1}(M)$$

such that

(3) $$h_{k+1} \circ d + d \circ h_k = \delta f_1 - \delta f_2.$$

For then, if α is a closed form on N, $\delta f_1(\alpha)$ and $\delta f_2(\alpha)$ differ by an exact form, and thus lie in the same cohomology class. Such a collection of linear transformations $\{h_k\}$ is called a *homotopy operator* for f_1 and f_2. In the Poincaré lemma we found a homotopy operator between the identity map of the open unit ball U and any constant map (range 1 point) of U into itself.

EXERCISES

1 Prove the assertion of 4.3(c) that a d-dimensional manifold X for which there exists an immersion $f \colon X \to \mathbb{R}^{d+1}$ is orientable if and only if there is a smooth nowhere-vanishing normal vector field along (X,f).

2 Prove that the real projective space P^n is orientable if and only if n is odd. (*Hint:* Observe that the antipodal map on the n-sphere S^n is orientation-preserving if and only if n is odd.)

3 Carry out in detail the proof of the existence of local orthonormal frame fields on a Riemannian manifold.

4 Prove the divergence theorem 4.10(11). First assume M is oriented and use Stokes' theorem together with the identity

(1) $$\int_{\partial D} *\tilde{V} = \int_{\partial D} \langle V, \vec{n} \rangle.$$

The easiest way to see (1) is to choose a local oriented orthonormal frame field $e_1, \ldots, e_n$ on a neighborhood of a point of ∂D, such that at points of ∂D, e_1 is the outer unit normal vector and $e_2, \ldots, e_n$ form an oriented basis of the tangent space to ∂D. Then express $*\tilde{V}$ and $\langle V, \vec{n} \rangle$ in terms of this local frame field and its dual coframe field $\omega_1, \ldots, \omega_n$. Finally, show that the theorem holds for a regular domain D in a Riemannian manifold M which is not necessarily orientable.

5 Let M be an oriented Riemannian manifold, let f and g be C^∞ functions on M, and let D be a regular domain in M. The Laplacian of g, denoted Δg, is defined by

$$\Delta g = -*d*dg. \tag{1}$$

(For more on the Laplacian, see Chapter 6. Observe that our choice of sign for the Laplacian yields $\Delta g = -\sum_{i=1}^{n} \partial^2 g/\partial r_i^2$ for g a C^∞ function on Euclidean space $\mathbb{R}^n$ with its standard Riemannian structure in which $\{\partial/\partial r_i\}$ is an orthonormal basis of each tangent space.) If $\vec{n}$ is the unit outer normal vector field along ∂D, we let $\partial g/\partial n$ denote $\vec{n}(g)$. Prove the following two Green's identities:

Green's 1st: $$\int_{\partial D} f\frac{\partial g}{\partial n} = \int_D \langle \operatorname{grad} f, \operatorname{grad} g\rangle - \int_D f\,\Delta g.$$

Green's 2nd: $$\int_{\partial D}\left(f\frac{\partial g}{\partial n} - g\frac{\partial f}{\partial n}\right) = \int_D (g\,\Delta f - f\,\Delta g).$$

6 Let ω be the volume form of an oriented Riemannian manifold of dimension n. Let $X_1, \ldots, X_n$ and $Y_1, \ldots, Y_n$ be vector fields on M. Prove that

$$\omega(X_1, \ldots, X_n) \cdot \omega(Y_1, \ldots, Y_n) = \det\{\langle X_i, Y_j\rangle\}.$$

Prove also that

$$\omega(X_1, \ldots, X_n)\omega = \tilde{X}_1 \wedge \cdots \wedge \tilde{X}_n,$$

where $\tilde{X}_i$ is the 1-form dual (via the Riemannian structure) to the vector field X_i.

7 Prove the differentiability of the function $\tilde{\lambda}$ of 4.11(5).

8 Let G be a compact (oriented) Lie group, and let $\alpha(\sigma) = \sigma^{-1}$ for $\sigma \in G$. Prove that for every continuous function f on G

$$\int_G f = \int_G f \circ \alpha.$$

9 Prove that a Lie group G has a bi-invariant Riemannian metric (that is, a metric such that both dl_σ and dr_σ preserve the inner products for each $\sigma \in G$) if and only if the closure of $\operatorname{Ad}(G)$ in $\operatorname{Aut}(\mathfrak{g})$ is compact.

10 (a) Prove that $H^p_{\text{de R}}(\mathbb{R}^n) = 0$ for each $p \geq 1$ and each $n \geq 1$.

(b) Prove that $H^0_{\text{de R}}(M) \simeq \mathbb{R}$ for a connected manifold M.

11 Determine the de Rham cohomology of the annular region

$$1 < (r_1^2 + r_2^2)^{1/2} < 2 \text{ in } \mathbb{R}^2.$$

12 If α and β are closed differential forms, prove that $\alpha \wedge \beta$ is closed. If, in addition, β is exact, prove that $\alpha \wedge \beta$ is exact.

13 Consider the 1-form $\alpha = (x^2 + 7y)\,dx + (-x + y \sin y^2)\,dy$ on $\mathbb{R}^2$. Compute its integral over the following 1-cycle z.

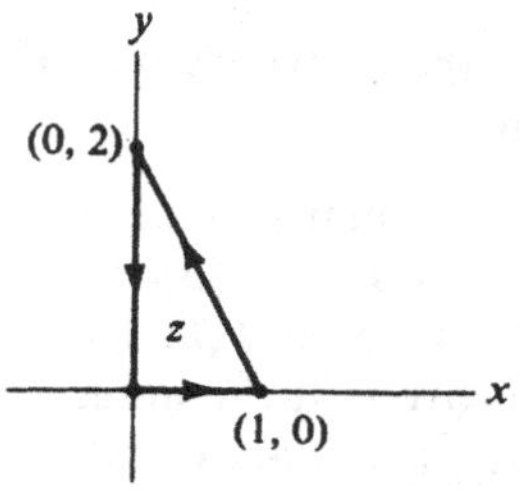

14 Let $\alpha = (2x + y \cos xy)\,dx + (x \cos xy)\,dy$ on $\mathbb{R}^2$. Show that α is closed. Show that α is exact by finding a function $f\colon \mathbb{R}^2 \to \mathbb{R}$ with $\alpha = df$. What would the integral of α over the cycle of Exercise 13 be?

15 Let

$$\alpha = \frac{1}{2\pi}\frac{x\,dy - y\,dx}{x^2 + y^2}.$$

Prove that α is a closed 1-form on $\mathbb{R}^2 - \{0\}$. Compute the integral of α over the unit circle S^1. How does this result show that α is not exact? How does this show that $\delta i(\alpha)$ is not exact, where $i\colon S^1 \to \mathbb{R}^2$ is the canonical imbedding?

16 (a) Prove that every closed 1-form on S^2 is exact.

(b) Let

$$\sigma = \frac{r_1\,dr_2 \wedge dr_3 - r_2\,dr_1 \wedge dr_3 + r_3\,dr_1 \wedge dr_2}{(r_1^2 + r_2^2 + r_3^2)^{3/2}}$$

in $\mathbb{R}^3 - \{0\}$. Prove that σ is closed.

(c) Evaluate $\int_{S^2}\sigma$. How does this show that σ is not exact?

(d) Let

$$\alpha = \frac{r_1\,dr_1 + r_2\,dr_2 + \cdots + r_n\,dr_n}{(r_1^2 + r_2^2 + \cdots + r_n^2)^{n/2}}$$

in $\mathbb{R}^n - \{0\}$. Find $*\alpha$, and prove that $*\alpha$ is closed.

(e) Evaluate $\int_{S^{n-1}} *\alpha$. Is $*\alpha$ exact?

17 Using de Rham cohomology, prove that the torus T^2 is not diffeomorphic with the 2-sphere S^2.

18 (a) Prove that every closed 1-form in the open shell

$$1 < \left(\sum_{i=1}^{3} r_i^2\right)^{1/2} < 2$$

in $\mathbb{R}^3$ is exact.

(b) Find a 2-form in the above shell that is closed but not exact.

(c) Prove that the above shell is not diffeomorphic with the open unit ball in $\mathbb{R}^3$.

19 Let f and g be C^∞ maps of M into N which are C^∞ homotopic; that is, there exists a C^∞ map F of $M \times (-\varepsilon, 1 + \varepsilon)$ into N, for some $\varepsilon > 0$, such that $F(m,0) = f(m)$ and $F(m,1) = g(m)$ for every $m \in M$. Prove that the induced homomorphisms f^* and g^* of $H^p_{\text{de R}}(N)$ into $H^p_{\text{de R}}(M)$ are equal for each integer p. (*Hint:* You will need to prove that the two injections $i_0(m) = (m,0)$ and $i_1(m) = (m,1)$ of M into $M \times (-\varepsilon, 1 + \varepsilon)$ induce the same homomorphisms on de Rham cohomology. To prove this, find suitable homotopy operators. The outline of the proof of the Poincaré lemma 4.18 should be helpful.)

20 (a) Let f: $M^n \to \mathbb{R}^{n+1}$ be an immersion, and let M^n be given the induced Riemannian structure; that is, for $m \in M$ and $u, v \in M_m$,

$$\langle u,v\rangle_m = \langle df(u), df(v)\rangle_{f(m)}.$$

Suppose that M is oriented, and that $\vec{n}$ is the oriented unit normal field along $f(M^n)$. (This means that $\vec{n}, df(v_1), \ldots, df(v_n)$ is to be an oriented orthonormal basis of the tangent space to Euclidean space at $f(m)$ whenever $v_1, \ldots, v_n$ is an oriented orthonormal basis of M_m.) Show that the volume form on M is given by

$$\omega = \delta f(i(\vec{n})(dr_1 \wedge \cdots \wedge dr_{n+1})).$$

(b) Let D be an open set in the xy plane, and let φ: $D \to \mathbb{R}^3$ be a smooth map of the form

$$\varphi(x,y) = (x,y,f(x,y)).$$

Thus φ determines an imbedded surface in $\mathbb{R}^3$. Give D and $\mathbb{R}^3$ the standard orientations. Use part (a) to prove that the induced volume form on D is given by

$$\omega = \left(\sqrt{\left(\frac{\partial f}{\partial x}\right)^2 + \left(\frac{\partial f}{\partial y}\right)^2 + 1}\right) dx \wedge dy.$$

5

SHEAVES, COHOMOLOGY, and the DE RHAM THEOREM

The principal objective in this chapter is a proof of the de Rham theorem, one version of which we have stated in 4.17. In its most complete form it asserts that the homomorphism from the de Rham cohomology ring to the differentiable singular cohomology ring given by integration of closed forms over differentiable singular cycles is a ring isomorphism. The approach will be to exhibit both the de Rham cohomology and the differentiable singular cohomology as special cases of sheaf cohomology and to use a basic uniqueness theorem for homomorphisms of sheaf cohomology theories to prove that the natural homomorphism between the de Rham and differentiable singular theories is an isomorphism. As an added dividend of this approach we shall also obtain the existence of canonical isomorphisms of the de Rham and differentiable singular cohomology theories with the continuous singular theory, the Alexander-Spanier theory, and the Čech cohomology theory for differentiable manifolds. From these isomorphisms we shall conclude that the de Rham cohomology theory is a topological invariant of a differentiable manifold.

No previous knowledge of sheaf theory nor of any of these cohomology theories is assumed. We shall develop those aspects of the theories necessary for our applications. We begin with an introduction to the theory of sheaves and sheaf cohomology. Our approach is based largely on the development given by Cartan in [4]. The interested reader will find more general treatments of sheaf theory in Cartan [4], Godement [7], and Bredon [3]. Sheaves are a very powerful tool in the study of complex manifolds. For an excellent account of their use in the theory of Riemann surfaces, see Gunning [8].

Let M be a differentiable manifold. Recall that M is assumed to be second countable and is therefore *paracompact*. It should be pointed out that nearly all of the sheaf theory presented in this chapter depends only on M being a paracompact Hausdorff space. The differentiable structure on M will be invoked only in a few examples and in the discussions of the de Rham cohomology (5.28) and the differentiable singular cohomology (5.31); and, in addition, the locally Euclidean structure will be invoked in the discussion of the singular cohomology (5.31). Otherwise, M need only be a paracompact Hausdorff space.

Throughout this chapter, K will be a fixed principal ideal domain. We shall treat sheaves of K-modules over M. The most important cases to keep in mind are these:

(i) K is the ring of integers Z, whence a K-module is simply an abelian group.

(ii) K is the field of real numbers $\mathbb{R}$, whence a K-module is simply a real vector space.

For the first few sections of this chapter, K could be any ring. The additional property that K is a principal ideal domain will be used first in 5.13.

SHEAVES AND PRESHEAVES

5.1 Definitions A *sheaf* $\mathcal{S}$ *of K-modules over M* consists of a topological space $\mathcal{S}$ together with a map $\pi: \mathcal{S} \to M$ satisfying:

(a) π is a local homeomorphism of $\mathcal{S}$ onto M.

(b) $\pi^{-1}(m)$ is a K-module for each $m \in M$.

(c) The composition laws are continuous in the topology on $\mathcal{S}$.

We elaborate on (c). Let $\mathcal{S} \circ \mathcal{S}$ be the subspace of $\mathcal{S} \times \mathcal{S}$ consisting of all pairs (s_1,s_2) such that $\pi(s_1) = \pi(s_2)$. Then (c) requires that the map $(s_1,s_2) \mapsto s_1 - s_2$ of $\mathcal{S} \circ \mathcal{S} \to \mathcal{S}$ be continuous, and that each of the maps $s \mapsto ks$ of $\mathcal{S} \to \mathcal{S}$ for $k \in K$ be continuous. It follows easily that the maps $s \mapsto (-s)$ and $(s_1,s_2) \mapsto s_1 + s_2$ also are continuous. Sheaves of K algebras are defined similarly, with the additional requirement in (c) that the map $(s_1,s_2) \mapsto s_1 \cdot s_2$ of $\mathcal{S} \circ \mathcal{S} \to \mathcal{S}$ be continuous.

The map π is called the *projection*; and the K-module $\mathcal{S}_m = \pi^{-1}(m)$ is called the *stalk over* $m \in M$. Let $U \subset M$ be open. A continuous map $f: U \to \mathcal{S}$ such that $\pi \circ f = \text{id}$ is called a *section of* $\mathcal{S}$ *over* U. The 0-*section* is the section which associates with $m \in U$ the zero element of $\mathcal{S}_m$. We let $\Gamma(\mathcal{S},U)$ denote the set of sections of $\mathcal{S}$ over U. We make $\Gamma(\mathcal{S},U)$ into a K-module as follows. Let f and g belong to $\Gamma(\mathcal{S},U)$, and let $k \in K$. We define the sum of f and g to be the section

$$(f + g)(m) = f(m) + g(m) \qquad (m \in U),$$

and we define a section kf by setting

$$(kf)(m) = k(f(m)) \qquad (m \in U).$$

With these operations, $\Gamma(\mathcal{S},U)$ becomes a K-module. The module of sections of $\mathcal{S}$ over M (*global sections*) will simply be denoted by $\Gamma(\mathcal{S})$. Observe that since π is a local homeomorphism, sections are open maps; and if sections f and g agree at $m \in M$, then they must agree on a neighborhood of m.

5.2 Examples The most elementary example of a sheaf over M is that of a so-called *constant sheaf* $\mathscr{G} = M \times G$, where G is a K-module with the discrete topology and $\mathscr{G}$ is given the product topology. Here the projection is simply $\pi(m,g) = m$.

A less trivial example is the following. Let $\tilde{F}_m$ denote the set of germs of C^∞ functions at $m \in M$ (as in 1.13), and let

$$\mathscr{C}^\infty(M) = \bigcup_{m \in M} \tilde{F}_m. \tag{1}$$

When it will introduce no confusion, we shall denote $\mathscr{C}^\infty(M)$ simply by $\mathscr{C}^\infty$. We define the projection $\pi\colon \mathscr{C}^\infty \to M$ in the obvious fashion so that $\mathbf{f} \in \tilde{F}_{\pi(\mathbf{f})}$. Associate with each open set U in M and each C^∞ function f on U the set

$$\bigcup_{m \in U} \mathbf{f}_m \subset \mathscr{C}^\infty,$$

where $\mathbf{f}_m$ is the germ of f at m. The collection of these sets forms a basis for a topology on $\mathscr{C}^\infty$ which makes $\mathscr{C}^\infty$ into a sheaf of real vector spaces (in fact, a sheaf of algebras since each $\tilde{F}_m$ has an algebra structure). $\mathscr{C}^\infty(M)$ is called the *sheaf of germs of C^∞ functions on M*. Similarly, one constructs the sheaf $\mathscr{C}^p(M)$ of germs of functions of class C^p on M for each integer $p \geq 0$.

5.3 Remark One should be cautioned that sheaves are generally not Hausdorff spaces. A simple example is provided by the sheaf $\mathscr{C}^0(\mathbb{R})$ of germs of continuous functions on the real line. Let $g(t) \equiv 0$, let $f(t) = 0$ for $t \leq 0$, and let $f(t) = t$ for $t > 0$. Then the germs of f and g at the origin are distinct elements of $\mathscr{C}^0(\mathbb{R})$; however, f and g have the same germ for each $t < 0$. It follows that the germs of f and g at the origin cannot be separated by disjoint open sets in $\mathscr{C}^0(\mathbb{R})$.

5.4 Definitions Let $\mathcal{S}$ and $\mathcal{S}'$ be sheaves on M with projections π and π' respectively. A continuous map $\varphi\colon \mathcal{S} \to \mathcal{S}'$ such that $\pi' \circ \varphi = \pi$ is called a *sheaf mapping*. Observe that sheaf mappings are necessarily local homeomorphisms, and they map stalks into stalks. A sheaf mapping φ which is a homomorphism (of K-modules) on each stalk is called a *sheaf homomorphism*. A *sheaf isomorphism* is a sheaf homomorphism with an inverse which is also a sheaf homomorphism.

An open set $\mathscr{R}$ in the sheaf $\mathcal{S}$ such that the subset $\mathscr{R}_m = \mathscr{R} \cap \mathcal{S}_m$ is a submodule of $\mathcal{S}_m$ for each $m \in M$ is called a *subsheaf* of $\mathcal{S}$. It is clear that a subsheaf of $\mathcal{S}$, with the natural topology and the projection map which it inherits from $\mathcal{S}$, is again a sheaf.

Let $\varphi\colon \mathcal{S} \to \mathscr{T}$ be a homomorphism of sheaves. The *kernel of* φ (ker φ) is the subset of $\mathcal{S}$ which maps under φ into the 0-section of $\mathscr{T}$, and is in fact a subsheaf of $\mathcal{S}$. The subsheaf $\varphi(\mathcal{S}) \subset \mathscr{T}$ is called the *image of* φ (im φ). In the case in which φ is one-to-one, we shall often identify $\mathcal{S}$ with $\varphi(\mathcal{S})$ and speak of the subsheaf $\mathcal{S}$ of $\mathscr{T}$.

Let $\mathscr{R}$ be a subsheaf of $\mathcal{S}$. For each $m \in M$, let $\mathscr{T}_m$ denote the quotient module $\mathcal{S}_m/\mathscr{R}_m$, and let

$$\mathscr{T} = \bigcup_{m \in M} \mathscr{T}_m. \tag{1}$$

Let $\tau\colon \mathcal{S} \to \mathscr{T}$ be the natural map which associates with each element of $\mathcal{S}_m$ its coset in $\mathscr{T}_m$, for each m, and give $\mathscr{T}$ the quotient topology. That is, a set U in $\mathscr{T}$ will be open if and only if $\tau^{-1}(U)$ is open in $\mathcal{S}$. Then, with this topology and with the natural projection which maps each element of $\mathscr{T}_m$ to m, $\mathscr{T}$ is a sheaf over M and $\tau\colon \mathcal{S} \to \mathscr{T}$ is a sheaf homomorphism. We leave the details as an exercise. $\mathscr{T}$ is called the *quotient sheaf* of $\mathcal{S}$ modulo $\mathscr{R}$.

If $\varphi\colon \mathcal{S} \to \mathscr{T}$ is a homomorphism of sheaves, then it is easily checked that the natural map $\mathcal{S}/\ker \varphi \to \operatorname{im} \varphi$ given by $(s + (\ker \varphi \mid \mathcal{S}_m)) \mapsto \varphi(s)$ for $s \in \mathcal{S}_m$ is a sheaf isomorphism.

A sequence of sheaves and homomorphisms

$$\cdots \to \mathcal{S}_i \to \mathcal{S}_{i+1} \to \mathcal{S}_{i+2} \to \cdots \tag{2}$$

is called *exact* if at each stage the image of a given homomorphism is the kernel of the next. Exact sequences of K-modules are defined similarly. Observe that (2) is exact if and only if for each $m \in M$ the induced sequence of homomorphisms of the stalks over m, namely,

$$\cdots \to (\mathcal{S}_i)_m \to (\mathcal{S}_{i+1})_m \to (\mathcal{S}_{i+2})_m \to \cdots, \tag{3}$$

is exact. If $\mathscr{R}$ is a subsheaf of $\mathcal{S}$, and $\mathscr{T}$ is the quotient sheaf $\mathcal{S}/\mathscr{R}$, then the natural sequence of homomorphisms

$$0 \to \mathscr{R} \to \mathcal{S} \to \mathscr{T} \to 0 \tag{4}$$

is exact, where 0 denotes the constant sheaf over M whose stalk over each point is the trivial K-module consisting of only one element. Exact sequences of the form (4) consisting of only five terms with the first and last being the 0 sheaf are called *short exact sequences.* Short exact sequences of K-modules are defined similarly.

5.5 Presheaves A *presheaf* $P = \{S_U; \rho_{U,V}\}$ *of K-modules on M* consists of a K-module S_U for each open set U in M and a homomorphism $\rho_{U,V}\colon S_V \to S_U$ for each inclusion $U \subset V$ of open sets in M, such that $\rho_{U,U} = \mathrm{id}$, and such that whenever $U \subset V \subset W$, the following diagram commutes:

$$\begin{array}{ccc} S_W & \xrightarrow{\rho_{V,W}} & S_V \\ & {\scriptstyle \rho_{U,W}} \searrow \quad \swarrow {\scriptstyle \rho_{U,V}} & \\ & S_U & \end{array} \tag{1}$$

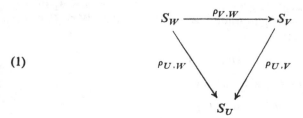

A typical example of a presheaf is the following. Associate with each open set U in M the algebra $C^\infty(U)$ of C^∞ functions on U, and let $\rho_{U,V}$ be the map which restricts a C^∞ function on V to a C^∞ function on U. This yields a presheaf $\{C^\infty(U);\rho_{U,V}\}$ of real vector spaces (in fact algebras) on M. Similarly, one could associate with U the vector space of p-forms on U for some fixed p and again let $\rho_{U,V}$ be the restriction mapping.

Let $P = \{S_U;\rho_{U,V}\}$ and $P' = \{S'_U;\rho'_{U,V}\}$ be presheaves on M. By a *presheaf homomorphism of P to P'* we mean a collection $\{\varphi_U\}$ of homomorphisms $\varphi_U: S_U \to S'_U$ such that

(2)
$$\rho'_{U,V} \circ \varphi_V = \varphi_U \circ \rho_{U,V}$$

whenever $U \subset V$. A *presheaf isomorphism* is a presheaf homomorphism $\{\varphi_U\}$ in which each φ_U is an isomorphism of K-modules.

5.6 The Relationship between Sheaves and Presheaves Each sheaf $\mathcal{S}$ gives rise canonically to a presheaf $\{\Gamma(\mathcal{S},U);\rho_{U,V}\}$. Here we associate with each open set U in M the K-module $\Gamma(\mathcal{S},U)$ of sections of $\mathcal{S}$ over U, and to each inclusion $U \subset V$ of open sets in M we associate the homomorphism

(1)
$$\rho_{U,V}: \Gamma(\mathcal{S},V) \to \Gamma(\mathcal{S},U)$$

which maps a section of $\mathcal{S}$ over V to its restriction to U. We shall denote this map of sheaves to presheaves by α. We call $\alpha(\mathcal{S})$ the *presheaf of sections of the sheaf* $\mathcal{S}$.

Conversely, we shall now show that each presheaf canonically determines a sheaf, which we call its *associated sheaf*. In practice, many sheaves will in this manner arise naturally from presheaves. The best example to keep in mind during the following construction is the presheaf $\{C^\infty(U);\rho_{U,V}\}$; its associated sheaf will be the sheaf of germs of C^∞ functions on M.

Let $P = \{S_U;\rho_{U,V}\}$ be a presheaf of K-modules on M. Let $m \in M$, and let S_m^* be the disjoint union of each of the modules S_U for which $m \in U$. If we set $f \in S_U$ equivalent to $g \in S_V$ if and only if there is a neighborhood W of m with $W \subset U \cap V$ such that $\rho_{W,U}f = \rho_{W,V}g$, we obtain an equivalence relation on S_m^*. The set $\mathcal{S}_m$ of equivalence classes of elements of S_m^* will be the stalk of the associated sheaf over m. If $m \in U$, let $\rho_{m,U}: S_U \to \mathcal{S}_m$ be the natural projection which assigns to each element of S_U its equivalence class. Let $f \in S_U$ and $g \in S_V$ be representatives of classes s_1 and s_2 in $\mathcal{S}_m$ respectively. There exists a neighborhood W of m such that $W \subset U \cap V$. Define addition in $\mathcal{S}_m$ by setting

(3)
$$s_1 + s_2 = \rho_{m,W}(\rho_{W,U}f + \rho_{W,V}g),$$

and define multiplication by $k \in K$ by setting

(4)
$$ks_1 = \rho_{m,U}(kf).$$

It is easy to check that the operations in (3) and (4) are well-defined and give $\mathcal{S}_m$ the structure of a K-module such that the maps $\rho_{m,U}$ are all homomorphisms. $\mathcal{S}_m$ is known as the *direct limit* of the modules S_U for U containing m. Now let

(5) $$\mathcal{S} = \bigcup_{m \in M} \mathcal{S}_m,$$

and let $\pi: \mathcal{S} \to M$ be the obvious projection such that $\pi(\mathcal{S}_m) = m$. We topologize $\mathcal{S}$ by taking for a basis of the topology the collection of subsets of $\mathcal{S}$ of the form

(6) $$O_f = \{\rho_{p,U}f : p \in U\}$$

for the various $f \in S_U$ and the various open sets U in M. This collection does indeed form a basis for a topology on $\mathcal{S}$, for if $s \in O_f \cap O_g$, say $s = \rho_{p,U}f = \rho_{p,V}g$, then there is a neighborhood W of p with $W \subset U \cap V$ for which $\rho_{W,U}f = \rho_{W,V}g$, from which it follows that $s \in O_{\rho_{W,U}f} \subset O_f \cap O_g$.

We claim that $\mathcal{S}$ is a sheaf over M with projection π. Now π is a local homeomorphism since it is a homeomorphism on each O_f. Each stalk $\mathcal{S}_m$ has been given the structure of a K-module, so finally we need only check that the composition operations are continuous. Consider first multiplication by $k \in K$. Let $s \in \mathcal{S}$, and let O be an open neighborhood of ks. Let $s = \rho_{p,U}f$. Then O_{kf} is also an open neighborhood of ks. Let V be a neighborhood of p contained in $\pi(O_{kf} \cap O)$. Then $O_{\rho_{V,U}f}$ is an open neighborhood of s which maps under multiplication by k into the open neighborhood $O_{\rho_{V,U}kf} \subset O$ of ks. Finally, we show that the map $(s_1, s_2) \mapsto s_1 - s_2$ of $\mathcal{S} \circ \mathcal{S} \to \mathcal{S}$ is continuous. For let O_f be an open neighborhood of $s_1 - s_2$, say $s_1 - s_2 = \rho_{p,U}f$, and let $s_1 = \rho_{p,V}g$ and $s_2 = \rho_{p,W}h$. Then there exists an open neighborhood Q of p with $Q \subset U \cap V \cap W$ such that

(7) $$\rho_{Q,U}f = \rho_{Q,V}g - \rho_{Q,W}h.$$

It follows that the open neighborhood

(8) $$O_{\rho_{Q,V}g} \times O_{\rho_{Q,W}h} \cap \mathcal{S} \circ \mathcal{S}$$

of (s_1, s_2) maps into O_f. Thus the composition operations are continuous, and $\mathcal{S}$ is a sheaf over M with projection π. $\mathcal{S}$ is called the sheaf *associated* with the presheaf P. We shall denote this map of presheaves to sheaves by β; thus $\mathcal{S} = \beta(P)$.

Clearly, each sheaf homomorphism $\varphi: \mathcal{S} \to \mathcal{T}$ gives rise to a presheaf homomorphism between the presheaves of sections $\alpha(\mathcal{S})$ and $\alpha(\mathcal{T})$ by composing elements of $\Gamma(\mathcal{S}, U)$ with φ. Conversely, a presheaf homomorphism $\{\varphi_U\}$ from P to P' canonically induces a homomorphism φ of the associated sheaves $\beta(P)$ and $\beta(P')$ such that

(9) $$\rho'_{p,U} \circ \varphi_U = \varphi \circ \rho_{p,U}$$

for each open set U in M and each $p \in U$. Moreover, in going in either direction, from homomorphisms of sheaves to homomorphisms of presheaves, or vice versa, the composition of two homomorphisms induces the composition of the corresponding homomorphisms.

Suppose now that we start with a sheaf $\mathcal{S}$, take its presheaf $\alpha(\mathcal{S})$ of sections, and then form the associated sheaf $\beta(\alpha(\mathcal{S}))$, thus obtaining the sheaf of germs of sections of $\mathcal{S}$. Then the sheaves $\mathcal{S}$ and $\beta(\alpha(\mathcal{S}))$ are canonically isomorphic. Indeed, let $\xi \in \beta(\alpha(\mathcal{S}))$ be the germ at p of some section f of $\mathcal{S}$ over an open set U containing p; that is, $\xi = \rho_{p,U}f$ for $f \in \Gamma(\mathcal{S},U)$. Then $\xi \mapsto f(p)$ determines a well-defined map $\beta(\alpha(\mathcal{S})) \to \mathcal{S}$ which is a sheaf isomorphism. We leave the details as an exercise.

Generally, however, if we start with a presheaf P and take the presheaf $\alpha(\beta(P))$ of sections of the sheaf associated with P, we may get a presheaf quite distinct from P. For example, let $P = \{S_U;\rho_{U,V}\}$ where S_U is the principal ideal domain K for each U, and all restrictions $\rho_{U,V}$ for $U \subsetneq V$ are identically 0. Then the presheaf $\alpha(\beta(P))$ assigns to each open set U in M the zero K-module. The conditions of the following definition are clearly necessary if P is to be isomorphic with $\alpha(\beta(P))$, and it will be shown in 5.8 that they are also sufficient.

5.7 Definition A presheaf $\{S_U;\rho_{U,V}\}$ on M is said to be *complete* if whenever the open set U is expressed as a union $\bigcup_\alpha U_\alpha$ of open sets in M, the following two conditions are satisfied:

(C_1) Whenever f and g in S_U are such that $\rho_{U_\alpha,U}f = \rho_{U_\alpha,U}g$ for all α, then $f = g$.

(C_2) Whenever there is an element $f_\alpha \in S_{U_\alpha}$ for each α such that $\rho_{U_\alpha \cap U_\beta,U_\alpha}f_\alpha = \rho_{U_\alpha \cap U_\beta,U_\beta}f_\beta$ for all α and β, then there exists $f \in S_U$ such that $f_\alpha = \rho_{U_\alpha,U}f$ for each α.

5.8 Proposition *If P is a complete presheaf, then $\alpha(\beta(P))$ is canonically isomorphic with P.*

PROOF Let $P = \{S_U;\rho_{U,V}\}$. We define a presheaf homomorphism from P to $\alpha(\beta(P))$ as follows. For each open set U in M we map

(1) $$S_U \to \Gamma(\beta(P),U)$$

by sending the element f of S_U to the section

(2) $$p \mapsto \rho_{p,U}f$$

of $\beta(P)$ over U. To prove that this presheaf homomorphism is an isomorphism, we need to prove that each of the homomorphisms (1) are isomorphisms. Injectivity follows from 5.7(C_1), for if f in S_U maps to the 0-section, then there exists a cover $\{U_\alpha\}$ of U such that $\rho_{U_\alpha,U}f = \rho_{U_\alpha,U}(0)$. Therefore $f = 0 \in S_U$ by (C_1). To prove surjectivity, let c

be a section of $\beta(P)$ over U. Then for each $p \in U$, there is a neighborhood U_p of p and an element $f_p \in S_{U_p}$ such that

$$\rho_{q,U_p} f_p = c(q) \tag{3}$$

for all $q \in U_p$. It follows from (3) and 5.7(C_1) that for each p and q in U,

$$\rho_{U_p \cap U_q, U_p} f_p = \rho_{U_p \cap U_q, U_q} f_q. \tag{4}$$

Thus, according to 5.7(C_2), there exists $f \in S_U$ such that

$$f_p = \rho_{U_p, U} f \tag{5}$$

for each $p \in U$. It follows from (5), (3), and (2) that f maps to c under the homomorphism (1).

5.9 Tensor Products Before we define the tensor products of sheaves and presheaves, we need a few preliminary remarks. The definition of the tensor product $S \otimes T$ of K-modules S and T is completely analogous to that for the special case 2.1 of tensor products of $\mathbb{R}$-modules. If $f\colon S \to S'$ and $h\colon T \to T'$ are homomorphisms of K-modules, then their tensor product $f \otimes h$ is the homomorphism of $S \otimes T$ into $S' \otimes T'$ uniquely associated by the universal mapping property 2.2(a) with the bilinear map

$$(s,t) \mapsto f(s) \otimes h(t) \tag{1}$$

of $S \times T$ into $S' \otimes T'$.

Observe that if S is a K-module, then there is a canonical isomorphism

$$S \otimes K \cong S. \tag{2}$$

Indeed, let $f\colon S \otimes K \to S$ be the homomorphism determined by the bilinear map $(s,k) \mapsto ks$ of $S \times K \to S$. Then f is clearly surjective, and is also injective since $\sum_i (s_i \otimes k_i) = \left(\sum_i k_i s_i\right) \otimes 1$, so if $f\left(\sum_i (s_i \otimes k_i)\right) = 0$, then $\sum_i k_i s_i = 0$, which implies $\sum_i (s_i \otimes k_i) = 0$.

Now let $P = \{S_U; \rho_{U,V}\}$ and $P' = \{S'_U; \rho'_{U,V}\}$ be presheaves on M. Then their tensor product is the presheaf

$$P \otimes P' = \{S_U \otimes S'_U;\ \rho_{U,V} \otimes \rho'_{U,V}\}. \tag{3}$$

If $\{\varphi_U\}\colon P \to Q$ and $\{\varphi'_U\}\colon P' \to Q'$ are homomorphisms of presheaves (5.5(2)), then their tensor product $\{\varphi_U\} \otimes \{\varphi'_U\}$ is by definition the homomorphism $\{\varphi_U \otimes \varphi'_U\}$ of $P \otimes P'$ into $Q \otimes Q'$.

If $\mathcal{S}$ and $\mathcal{T}$ are sheaves over M, we define their tensor product $\mathcal{S} \otimes \mathcal{T}$ to be the sheaf associated with the tensor product of the presheaves of sections of $\mathcal{S}$ and $\mathcal{T}$. That is,

$$\mathcal{S} \otimes \mathcal{T} = \beta(\alpha(\mathcal{S}) \otimes \alpha(\mathcal{T})). \tag{4}$$

Moreover, if $\varphi: \mathcal{S} \to \mathcal{T}$ and $\gamma: \mathcal{S}' \to \mathcal{T}'$ are sheaf homomorphisms, and if $\{\varphi_U\}: \alpha(\mathcal{S}) \to \alpha(\mathcal{T})$ and $\{\gamma_U\}: \alpha(\mathcal{S}') \to \alpha(\mathcal{T}')$ are the corresponding homomorphisms on the presheaves of sections, then we define the tensor product $\varphi \otimes \gamma$ to be the homomorphism $\mathcal{S} \otimes \mathcal{S}' \to \mathcal{T} \otimes \mathcal{T}'$ associated with the presheaf homomorphism

$$\{\varphi_U\} \otimes \{\gamma_U\}: \alpha(\mathcal{S}) \otimes \alpha(\mathcal{S}') \to \alpha(\mathcal{T}) \otimes \alpha(\mathcal{T}'). \tag{5}$$

The reader should check that there is a canonical isomorphism

$$(\mathcal{S} \otimes \mathcal{T})_m \cong \mathcal{S}_m \otimes \mathcal{T}_m, \tag{6}$$

so in particular if $\mathcal{K}$ is the constant sheaf $M \times K$, then in view of (2) there is a canonical isomorphism

$$\mathcal{S} \otimes \mathcal{K} \cong \mathcal{S}. \tag{7}$$

Moreover, if φ and γ are sheaf homomorphisms as above, then the reader should check that

$$(\varphi \otimes \gamma) \mid (\mathcal{S} \otimes \mathcal{S}')_m = \varphi \mid \mathcal{S}_m \otimes \gamma \mid \mathcal{S}'_m. \tag{8}$$

One might ask, "Why not define $\mathcal{S} \otimes \mathcal{T}$ and $\varphi \otimes \gamma$ by (6) and (8) rather than going to presheaves and back?" We could indeed proceed in this way, but it would involve a considerable duplication of previous work. The reason is that if we define the tensor products by (6) and (8), then we would also have to define the topology on $\mathcal{S} \otimes \mathcal{T}$ and check that the properties 5.1(a) and (c) are satisfied, and would also have to prove that $\varphi \otimes \gamma$ is continuous. In defining $\mathcal{S} \otimes \mathcal{T}$ and $\varphi \otimes \gamma$ by going to presheaves and back, the work of showing that $\mathcal{S} \otimes \mathcal{T}$ is really a sheaf and that $\varphi \otimes \gamma$ is a sheaf homomorphism has already been carried out in the analysis of the maps α and β of 5.6.

5.10 Fine Sheaves A sheaf $\mathcal{S}$ over M is said to be *fine* if for each locally finite cover $\{U_i\}$ of M by open sets there exists for each i an endomorphism l_i of $\mathcal{S}$ such that:

(a) $\operatorname{supp}(l_i) \subset U_i$.

(b) $\sum_i l_i = \mathrm{id}$.

Here, by $\operatorname{supp}(l_i)$ (the support of l_i) we mean the closure of the set of points in M for which $l_i \mid \mathcal{S}_m$ is not zero. We shall call $\{l_i\}$ a *partition of unity for* $\mathcal{S}$ *subordinate to the cover* $\{U_i\}$ *of* M.

An example of a fine sheaf is the sheaf $\mathcal{C}^\infty(M)$ of germs of C^∞ functions on M. If $\{U_i\}$ is a locally finite open cover of M, let $\{\varphi_i\}$ be a partition of unity on M subordinate to this cover (Theorem 1.11). We obtain endomorphisms $\tilde{l}_i$ of the presheaf $\{C^\infty(U); \rho_{U,V}\}$ by setting

$$\tilde{l}_i(f) = (\varphi_i \mid U) \cdot f \quad \text{for } f \in C^\infty(U). \tag{1}$$

The associated sheaf endomorphisms l_i of $\mathcal{C}^\infty(M)$ form a partition of unity subordinate to the cover $\{U_i\}$ of M.

Let $\mathcal{S}$ and $\mathcal{T}$ be sheaves over M, with $\mathcal{S}$ fine. Then $\mathcal{S} \otimes \mathcal{T}$ is itself a fine sheaf. Indeed, if $\{l_i\}$ is a partition of unity for $\mathcal{S}$ subordinate to the cover $\{U_i\}$ of M, then $\{l_i \otimes \mathrm{id}\}$ is a partition of unity for $\mathcal{S} \otimes \mathcal{T}$.

5.11 A sheaf homomorphism $\varphi: \mathcal{S} \to \mathcal{T}$ gives rise to a homomorphism $\Gamma(\mathcal{S}) \to \Gamma(\mathcal{T})$ of the modules of global sections by composing sections of $\mathcal{S}$ with φ. We leave as an exercise the fact that a short exact sequence

$$0 \to \mathcal{S}' \to \mathcal{S} \to \mathcal{S}'' \to 0 \tag{1}$$

gives rise to an exact sequence

$$0 \to \Gamma(\mathcal{S}') \to \Gamma(\mathcal{S}) \to \Gamma(\mathcal{S}''). \tag{2}$$

However, the homomorphism $\Gamma(\mathcal{S}) \to \Gamma(\mathcal{S}'')$ is generally not surjective. Consider the following example.

Let $\mathcal{S}_{p_1,p_2}$ be the "skyscraper" sheaf over a connected M whose stalk is the zero K-module over each point except over the distinct points p_1 and p_2 where the stalk is K. (Note that the topology on $\mathcal{S}_{p_1,p_2}$ is uniquely determined by the requirement that $\mathcal{S}_{p_1,p_2}$ be a sheaf.) Let $\mathcal{K}$ as usual be the constant sheaf with stalk K. There is an obvious homomorphism of $\mathcal{K}$ onto $\mathcal{S}_{p_1,p_2}$, namely, the homomorphism is zero on all stalks of $\mathcal{K}$ except on those over p_1 and p_2 where it is the identity map. However, the associated map $\Gamma(\mathcal{K}) \to \Gamma(\mathcal{S}_{p_1,p_2})$ cannot be surjective for $\Gamma(\mathcal{K}) \cong K$, whereas $\Gamma(\mathcal{S}_{p_1,p_2}) \cong K \oplus K$.

Exactness of (1) is a purely local property, whereas exactness of (2) is a global property. We will see that certain sheaf cohomology modules will provide an extension of the exact sequence (2) and thus will provide a measure of the extent to which $\Gamma(\mathcal{S}) \to \Gamma(\mathcal{S}'')$ fails to be surjective.

5.12 Theorem *Let $\varphi: \mathcal{S} \to \mathcal{T}$ be a surjective sheaf homomorphism with kernel $\mathcal{R}$. Suppose that $\mathcal{R}$ is a fine sheaf. Then the homomorphism $\Gamma(\mathcal{S}) \to \Gamma(\mathcal{T})$ is surjective.*

PROOF Let t be a global section of $\mathcal{T}$. We must construct a section s of $\mathcal{S}$ such that $\varphi \circ s = t$. By the continuity of φ and t and by the property 5.1(a) of $\mathcal{S}$ and $\mathcal{T}$ there is a covering $\{U_i\}$ of M by open sets and, for each i, a section s_i of $\mathcal{S}$ over U_i such that

$$\varphi \circ s_i = t \mid U_i. \tag{1}$$

Since M is paracompact, we can assume that the cover $\{U_i\}$ is locally finite. The difference

$$s_{ij} = s_i - s_j \tag{2}$$

is a section of the kernel $\mathcal{R}$ over $U_i \cap U_j$, and on $U_i \cap U_j \cap U_k$ the differences satisfy

$$s_{ij} + s_{jk} = s_{ik}. \tag{3}$$

Now, let $\{l_i\}$ be a partition of unity for $\mathscr{R}$ subordinate to the cover $\{U_i\}$ of M. Consider the section $l_j \circ s_{ij}$ of $\mathscr{R}$ over $U_i \cap U_j$. Since the support of l_j lies in U_j, we can extend the section $l_j \circ s_{ij}$ to be a continuous section of $\mathscr{R}$ over U_i by defining it to be zero on points of $U_i - U_j$. Let

$$s_i' = \sum_j l_j \circ s_{ij}. \tag{4}$$

Then s_i' is a section of $\mathscr{R}$ over U_i, and by (3) and (4) the difference

$$s_i' - s_j' = \sum_k l_k \circ s_{ik} - \sum_k l_k \circ s_{jk} = \sum_k l_k \circ s_{ij} = s_{ij} \tag{5}$$

over $U_i \cap U_j$. Thus

$$s_i - s_i' = s_j - s_j' \tag{6}$$

on $U_i \cap U_j$. It follows that if we set $s(m) = (s_i - s_i')(m)$ for $m \in U_i$, then s is a well-defined global section of $\mathcal{S}$ such that $\varphi \circ s = t$.

5.13 Definitions A K-module X is *torsionless* if there is no non-zero element $x \in X$ for which there exists a non-zero element $k \in K$ such that $kx = 0$. A sheaf of K-modules is said to be *torsionless* if each stalk is a torsionless K-module.

We shall need the following basic lemma concerning tensor products and exact sequences. This is the first of two fundamental algebraic propositions which we shall need in this chapter for which we shall not provide proofs. (The second proposition is the Kunneth formula, which we shall need in 5.42.) The proofs are rather long, are not of particular interest for our purposes, and would tend to obscure the main goals of the chapter. The interested reader can find a proof of the following lemma in Spanier [28, pp. 215, 221].

5.14 Lemma *Let*

$$0 \to A' \to A \to A'' \to 0 \tag{1}$$

be an exact sequence of K-modules, and let B be a K-module. Then the induced sequence

$$A' \otimes B \to A \otimes B \to A'' \otimes B \to 0 \tag{2}$$

(whose homomorphisms are the homomorphisms of (1) *tensored with the identity homomorphism of B) is exact, but $A' \otimes B \to A \otimes B$ is not necessarily injective. If, however, either A'' or B is torsionless, then the full sequence*

$$0 \to A' \otimes B \to A \otimes B \to A'' \otimes B \to 0 \tag{3}$$

is exact.

(We should point out that (3) is the first place in this chapter where the fact that the ring K is actually a principal ideal domain is used.)

5.15 Theorem *Let*

$$0 \to \mathcal{S}' \to \mathcal{S} \to \mathcal{S}'' \to 0 \tag{1}$$

be an exact sequence of sheaves over M, and let $\mathcal{T}$ be also a sheaf over M. Then if either $\mathcal{T}$ or $\mathcal{S}''$ is torsionless, then the sequence

$$0 \to \mathcal{S}' \otimes \mathcal{T} \to \mathcal{S} \otimes \mathcal{T} \to \mathcal{S}'' \otimes \mathcal{T} \to 0 \tag{2}$$

is exact. If, in addition, either $\mathcal{T}$ or $\mathcal{S}'$ is a fine sheaf, then the sequence

$$0 \to \Gamma(\mathcal{S}' \otimes \mathcal{T}) \to \Gamma(\mathcal{S} \otimes \mathcal{T}) \to \Gamma(\mathcal{S}'' \otimes \mathcal{T}) \to 0 \tag{3}$$

is exact.

PROOF The exactness of (2) follows from 5.14. If either $\mathcal{T}$ or $\mathcal{S}'$ is fine, then, according to 5.10, $\mathcal{S}' \otimes \mathcal{T}$ is fine; and this together with (2) above, 5.11(2), and 5.12 proves that (3) is exact.

COCHAIN COMPLEXES

5.16 Definitions A *cochain complex* C^* consists of a sequence of K-modules and homomorphisms

$$\cdots \to C^{q-1} \to C^q \to C^{q+1} \to \cdots \tag{1}$$

defined for all integers q such that at each stage the image of a given homomorphism is *contained* in the kernel of the next. The homomorphism $C^q \to C^{q+1}$ (which we shall refer to as d^q, or simply d if the index q is not needed for clarification) is called the qth *coboundary operator*. The kernel $Z^q(C^*)$ of d^q is the module of qth *degree cocycles* of the cochain complex C^*, and the image $B^q(C^*)$ of d^{q-1} is the module of qth *degree coboundaries*. The qth *cohomology module* $H^q(C^*)$ is defined to be the quotient module

$$H^q(C^*) = Z^q(C^*)/B^q(C^*). \tag{2}$$

Let C^* and D^* be cochain complexes. A *cochain map* $C^* \to D^*$ consists of a collection of homomorphisms $C^q \to D^q$ such that for each q, the diagram

$$\begin{array}{ccc} C^{q+1} & \longrightarrow & D^{q+1} \\ \uparrow & & \uparrow \\ C^q & \longrightarrow & D^q \end{array} \tag{3}$$

commutes. It follows from (3) that a cochain map sends the module of q-cocyles of C^* into the module of q-cocycles of D^* and maps the module of q-coboundaries of C^* into the module of q-coboundaries of D^*, and thus induces a homomorphism of the cohomology modules

$$H^q(C^*) \to H^q(D^*). \tag{4}$$

The composition $C^* \to E^*$ of two cochain maps $C^* \to D^*$ and $D^* \to E^*$ induces on the cohomology modules the homomorphism $H^q(C^*) \to H^q(E^*)$, which is the composition of $H^q(C^*) \to H^q(D^*)$ and $H^q(D^*) \to H^q(E^*)$.

A sequence of cochain maps

$$0 \longrightarrow C^* \longrightarrow D^* \longrightarrow E^* \longrightarrow 0 \tag{5}$$

forms a *short exact sequence* if for each q,

$$0 \longrightarrow C^q \longrightarrow D^q \longrightarrow E^q \longrightarrow 0 \tag{6}$$

is a short exact sequence of K-modules. A *homomorphism between short exact sequences* $0 \to C^* \to D^* \to E^* \to 0$ and $0 \to \tilde{C}^* \to \tilde{D}^* \to \tilde{E}^* \to 0$ *of cochain complexes* consists of cochain maps $C^* \to \tilde{C}^*$, $D^* \to \tilde{D}^*$, and $E^* \to \tilde{E}^*$ such that we have a commutative diagram:

$$\begin{array}{ccccccccc} 0 & \to & C^* & \to & D^* & \to & E^* & \to & 0 \\ & & \downarrow & & \downarrow & & \downarrow & & \\ 0 & \to & \tilde{C}^* & \to & \tilde{D}^* & \to & \tilde{E}^* & \to & 0. \end{array} \tag{7}$$

5.17 Proposition *Given the short exact sequence* 5.16(5) *of cochain maps, there are homomorphisms*

$$H^q(E^*) \xrightarrow{\partial} H^{q+1}(C^*) \tag{1}$$

for each q such that the sequence

(2)

$$\cdots \to H^{q-1}(E^*) \xrightarrow{\partial} H^q(C^*) \to H^q(D^*) \to H^q(E^*) \xrightarrow{\partial} H^{q+1}(C^*) \to \cdots$$

is exact, and such that given the homomorphism 5.16(7) *of short exact sequences of cochain complexes, the diagram*

$$\begin{array}{ccc} H^q(E^*) & \xrightarrow{\partial} & H^{q+1}(C^*) \\ \downarrow & & \downarrow \\ H^q(\tilde{E}^*) & \xrightarrow{\partial} & H^{q+1}(\tilde{C}^*) \end{array} \tag{3}$$

commutes.

PROOF To define ∂, we consider the commutative diagram (4), where for convenience we have labeled particular homomorphisms. Let σ be a cocycle in E^q. Since α is surjective, there is an element $\tilde{\sigma} \in D^q$ such that $\alpha(\tilde{\sigma}) = \sigma$. Since the square ① is commutative, and since σ is a cocycle, it follows that $d(\tilde{\sigma})$ maps to zero in E^{q+1}; therefore,

(4)

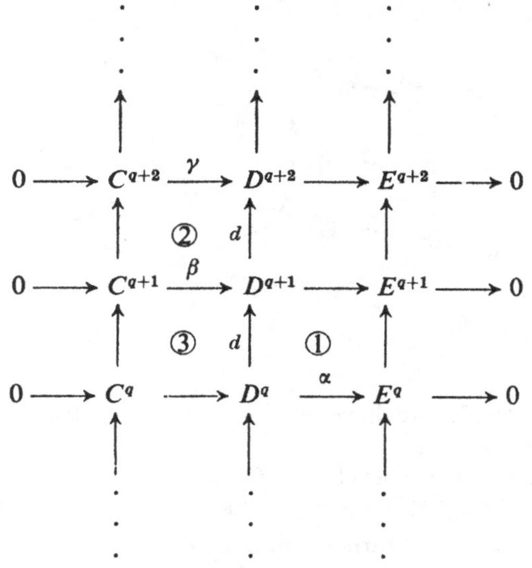

by the exactness of the horizontal sequences, $d(\tilde{\sigma})$ must be in the image of β. Thus there is an element $\sigma' \in C^{q+1}$ such that $\beta(\sigma') = d(\tilde{\sigma})$. It follows from the commutativity of the square ②, the injectivity of γ, and the fact that $d \circ d = 0$ that σ' is a cocycle. Now, σ' is not uniquely determined by σ because of the choice involved in the selection of $\tilde{\sigma} \in D^q$ such that $\alpha(\tilde{\sigma}) = \sigma$. However, it is easily seen by a quick chase around the square ③ that σ' is uniquely determined by σ up to a coboundary. Thus we have a well-defined map

(5) $$Z^q(E^*) \to H^{q+1}(C^*),$$

which is easily seen to be a homomorphism. It is also readily seen that $B^q(E^*)$ is in the kernel of (5); hence (5) yields a homomorphism defined on the quotient $Z^q(E^*)/B^q(E^*)$. This by definition is the desired homomorphism (1).

Checking the exactness of (2) involves checking exactness at $H^q(C^*)$, at $H^q(D^*)$, and at $H^q(E^*)$—and at each of these three stages we have to check two inclusion relations. These six steps in the proof of the exactness of (2) can be verified by simple chases around the diagram (4). We leave the details to the reader.

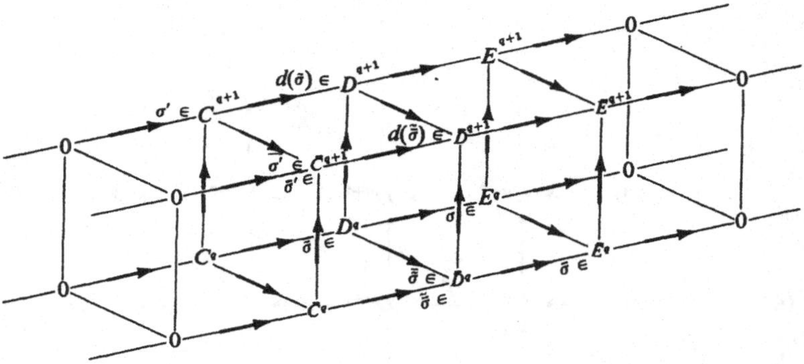

To prove (3), we have to follow some elements around the above commutative lattice.

We start with a cocycle $\sigma \in E^q$ and lift it back to $\tilde{\sigma} \in D^q$, which we map to $d(\tilde{\sigma}) \in D^{q+1}$, which we have seen is the image of an element $\sigma' \in C^{q+1}$, which in turn we map to $\overline{\sigma'} \in \bar{C}^{q+1}$. Then $\overline{\sigma'}$ is a cocycle in $\bar{C}^{q+1}$ and is a representative of the cohomology class in $H^{q+1}(\bar{C}^*)$ into which the cohomology class of σ in $H^q(E^*)$ is mapped under the composition $H^q(E^*) \xrightarrow{\partial} H^{q+1}(C^*) \to H^{q+1}(\bar{C}^*)$. On the other hand, we can first map σ to $\bar{\sigma} \in \bar{E}^q$, then lift back to $\tilde{\bar{\sigma}} \in \bar{D}^q$, map to $d(\tilde{\bar{\sigma}}) \in \bar{D}^{q+1}$, and lift back to $\bar{\sigma}' \in \bar{C}^{q+1}$. Then $\bar{\sigma}'$ is a cocycle in $\bar{C}^{q+1}$ which represents the cohomology class in $H^{q+1}(\bar{C}^*)$ into which the cohomology class of σ is mapped under the composition $H^q(E^*) \to H^q(\bar{E}^*) \xrightarrow{\partial} H^{q+1}(\bar{C}^*)$. To prove (3), we need only check that $\overline{\sigma'} - \bar{\sigma}'$ is a coboundary. Now $\tilde{\sigma} \in D^q$ maps to an element $\bar{\tilde{\sigma}} \in \bar{D}^q$, and elements $\bar{\tilde{\sigma}}$ and $\tilde{\bar{\sigma}}$ of $\bar{D}^q$ both map to $\bar{\sigma} \in \bar{E}^q$; hence there is an element $\gamma \in \bar{C}^q$ which maps to $(\bar{\tilde{\sigma}} - \tilde{\bar{\sigma}})$. One can easily check that $d\gamma = \overline{\sigma'} - \bar{\sigma}'$, which completes the proof of Proposition 5.17.

After a few trials the reader should find that these proofs by "diagram chase" become quite routine, albeit tedious. In the future we shall leave all such proofs as exercises.

AXIOMATIC SHEAF COHOMOLOGY

5.18 Definition A *sheaf cohomology theory $\mathscr{H}$ for M with coefficients in sheaves of K-modules over M* consists of

(I) a K-module $H^q(M,\mathcal{S})$ for each sheaf $\mathcal{S}$ and for each integer q,

(II) a homomorphism $H^q(M,\mathcal{S}) \to H^q(M,\mathcal{S}')$ for each homomorphism $\mathcal{S} \to \mathcal{S}'$ and for each integer q, and

(III) a homomorphism $H^q(M,\mathcal{S}'') \to H^{q+1}(M,\mathcal{S}')$ for each short exact sequence $0 \to \mathcal{S}' \to \mathcal{S} \to \mathcal{S}'' \to 0$ and for each integer q,

such that the properties (a)–(f) hold:

(a) $H^q(M,\mathcal{S}) = 0$ for $q < 0$, and there is an isomorphism $H^0(M,\mathcal{S}) \cong \Gamma(\mathcal{S})$ such that for each homomorphism $\mathcal{S} \to \mathcal{S}'$ the diagram

$$\begin{array}{ccc} H^0(M,\mathcal{S}) & \cong & \Gamma(\mathcal{S}) \\ \downarrow & & \downarrow \\ H^0(M,\mathcal{S}') & \cong & \Gamma(\mathcal{S}') \end{array}$$

commutes.

(b) $H^q(M,\mathcal{S}) = 0$ for all $q > 0$ if $\mathcal{S}$ is a fine sheaf.

(c) If $0 \to \mathcal{S}' \to \mathcal{S} \to \mathcal{S}'' \to 0$ is exact, then the following is exact:

$$\cdots \to H^q(M,\mathcal{S}') \to H^q(M,\mathcal{S}) \to H^q(M,\mathcal{S}'') \to H^{q+1}(M,\mathcal{S}') \to \cdots.$$

(d) The identity homomorphism id: $\mathcal{S} \to \mathcal{S}$ induces the identity homomorphism id: $H^q(M,\mathcal{S}) \to H^q(M,\mathcal{S})$.

(e) If the diagram

$$\begin{array}{ccc} \mathcal{S} & \longrightarrow & \mathcal{S}' \\ & \searrow & \downarrow \\ & & \mathcal{S}'' \end{array}$$

commutes, then for each q so does the diagram

$$\begin{array}{ccc} H^q(M,\mathcal{S}) & \longrightarrow & H^q(M,\mathcal{S}') \\ & \searrow & \downarrow \\ & & H^q(M,\mathcal{S}''). \end{array}$$

(f) For each homomorphism of short exact sequences of sheaves

$$\begin{array}{ccccccccc} 0 & \longrightarrow & \mathcal{S}' & \longrightarrow & \mathcal{S} & \longrightarrow & \mathcal{S}'' & \longrightarrow & 0 \\ & & \downarrow & & \downarrow & & \downarrow & & \\ 0 & \longrightarrow & \mathcal{T}' & \longrightarrow & \mathcal{T} & \longrightarrow & \mathcal{T}'' & \longrightarrow & 0, \end{array}$$

the following diagram commutes:

$$\begin{array}{ccc} H^q(M,\mathcal{S}'') & \to & H^{q+1}(M,\mathcal{S}') \\ \downarrow & & \downarrow \\ H^q(M,\mathcal{T}'') & \to & H^{q+1}(M,\mathcal{T}'). \end{array}$$

The module $H^q(M,\mathcal{S})$ is called the qth *cohomology module of M with coefficients in the sheaf* $\mathcal{S}$ relative to the cohomology theory $\mathcal{H}$.

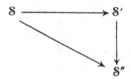

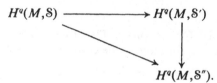

5.19 Definition An exact sheaf sequence

$$(1) \qquad 0 \to \mathscr{A} \to \mathscr{C}_0 \to \mathscr{C}_1 \to \mathscr{C}_2 \to \cdots$$

is called a *resolution* of the sheaf $\mathscr{A}$. The resolution (1) is called *fine* (respectively *torsionless*) if each of the sheaves $\mathscr{C}_i$ is fine (respectively torsionless).

With each resolution (1) of $\mathscr{A}$ and each sheaf $\mathcal{S}$ we associate a cochain complex

$$(2) \qquad \cdots \to 0 \to \Gamma(\mathscr{C}_0 \otimes \mathcal{S}) \to \Gamma(\mathscr{C}_1 \otimes \mathcal{S}) \to \Gamma(\mathscr{C}_2 \otimes \mathcal{S}) \to \cdots$$

which we shall denote by $\Gamma(\mathscr{C}^* \otimes \mathcal{S})$. For $q \geq 0$ the module of q-cochains is $\Gamma(\mathscr{C}_q \otimes \mathcal{S})$, whereas for $q < 0$ the module of q-cochains is the zero module. Note carefully that this cochain complex does not contain the module $\Gamma(\mathscr{A} \otimes \mathcal{S})$. The homomorphisms in (2) are those induced by the homomorphisms $\mathscr{C}_i \otimes \mathcal{S} \to \mathscr{C}_{i+1} \otimes \mathcal{S}$ which are the tensor product of the homomorphisms of (1) with id: $\mathcal{S} \to \mathcal{S}$. The exactness of (1) implies that in

$$(3) \qquad \cdots \to 0 \to \mathscr{C}_0 \otimes \mathcal{S} \to \mathscr{C}_1 \otimes \mathcal{S} \to \mathscr{C}_2 \otimes \mathcal{S} \to \cdots$$

the image of each homomorphism is contained in the kernel of the next, and this in turn implies that (2) is indeed a cochain complex.

A homomorphism $\mathcal{S} \to \mathcal{S}'$ when tensored with the identity homomorphisms of the sheaves $\mathscr{C}_i$ yields homomorphisms $\mathscr{C}_i \otimes \mathcal{S} \to \mathscr{C}_i \otimes \mathcal{S}'$. These, in turn, induce homomorphisms $\Gamma(\mathscr{C}_i \otimes \mathcal{S}) \to \Gamma(\mathscr{C}_i \otimes \mathcal{S}')$, which commute with the coboundary homomorphisms of the respective cochain complexes and thus determine a cochain map

$$(4) \qquad \Gamma(\mathscr{C}^* \otimes \mathcal{S}) \to \Gamma(\mathscr{C}^* \otimes \mathcal{S}').$$

5.20 Existence of Cohomology Theories We shall now show that each fine torsionless resolution of the constant sheaf $\mathscr{K} = M \times K$,

$$(1) \qquad 0 \to \mathscr{K} \to \mathscr{C}_0 \to \mathscr{C}_1 \to \mathscr{C}_2 \to \cdots,$$

canonically determines a cohomology theory for M with coefficients in sheaves of K-modules over M. The existence of such resolutions will be amply demonstrated in the later sections in which the classical cohomology theories are discussed. In particular, see 5.26(7) and (11). We therefore assume that we are given a fine torsionless resolution (1). Then we obtain a cohomology theory as follows:

(I) With a sheaf $\mathcal{S}$ and each integer q we associate the qth cohomology module of the cochain complex $\Gamma(\mathscr{C}^* \otimes \mathcal{S})$; that is, we set

$$H^q(M,\mathcal{S}) = H^q(\Gamma(\mathscr{C}^* \otimes \mathcal{S})).$$

(II) With each homomorphism $\mathcal{S} \to \mathcal{S}'$ and each integer q we associate the homomorphism $H^q(M,\mathcal{S}) \to H^q(M,\mathcal{S}')$ induced, according to 5.16(4), by the cochain map $\Gamma(\mathscr{C}^* \otimes \mathcal{S}) \to \Gamma(\mathscr{C}^* \otimes \mathcal{S}')$ of 5.19(4).

(III) Each short exact sheaf sequence

$$0 \to \mathcal{S}' \to \mathcal{S} \to \mathcal{S}'' \to 0$$

induces, in view of Theorem 5.15 and the fact that the $\mathscr{C}_i$ are fine torsionless sheaves, a short exact sequence of cochain maps

$$0 \to \Gamma(\mathscr{C}^* \otimes \mathcal{S}') \to \Gamma(\mathscr{C}^* \otimes \mathcal{S}) \to \Gamma(\mathscr{C}^* \otimes \mathcal{S}'') \to 0$$

with which, according to 5.17(1), there is associated a homomorphism $H^q(\Gamma(\mathscr{C}^* \otimes \mathcal{S}'')) \to H^{q+1}(\Gamma(\mathscr{C}^* \otimes \mathcal{S}'))$. This, by definition, is the homomorphism $H^q(M,\mathcal{S}'') \to H^{q+1}(M,\mathcal{S}')$ that we associate with the short exact sequence $0 \to \mathcal{S}' \to \mathcal{S} \to \mathcal{S}'' \to 0$ and the integer q.

The fact that axiom 5.18(d) for a cohomology theory is satisfied is immediately apparent. The remark immediately following 5.16(4) implies that axiom (e) is satisfied. Axioms (c) and (f) are consequences of 5.17.

Now let $\mathscr{Z}_q$ be the kernel of $\mathscr{C}_q \to \mathscr{C}_{q+1}$. Then it follows from the exactness of (1) that

(2) $$0 \to \mathscr{Z}_q \to \mathscr{C}_q \to \mathscr{Z}_{q+1} \to 0$$

is exact; and since $\mathscr{Z}_q$ is a subsheaf of the torsionless sheaf $\mathscr{C}_q$, then $\mathscr{Z}_q$ is also torsionless. It follows from Theorem 5.15 that for any sheaf $\mathcal{S}$ the sequence

(3) $$0 \to \mathscr{Z}_q \otimes \mathcal{S} \to \mathscr{C}_q \otimes \mathcal{S} \to \mathscr{Z}_{q+1} \otimes \mathcal{S} \to 0$$

is exact; and from (3) and 5.11(2) we obtain the exact sequence

(4) $$0 \to \Gamma(\mathscr{Z}_q \otimes \mathcal{S}) \to \Gamma(\mathscr{C}_q \otimes \mathcal{S}) \to \Gamma(\mathscr{Z}_{q+1} \otimes \mathcal{S}).$$

In particular, $\Gamma(\mathscr{Z}_q \otimes \mathcal{S}) \to \Gamma(\mathscr{C}_q \otimes \mathcal{S})$ is an injection, so since the homomorphism

(5) $$\Gamma(\mathscr{C}_q \otimes \mathcal{S}) \to \Gamma(\mathscr{C}_{q+1} \otimes \mathcal{S})$$

is the composition

(6) $$\Gamma(\mathscr{C}_q \otimes \mathcal{S}) \to \Gamma(\mathscr{Z}_{q+1} \otimes \mathcal{S}) \to \Gamma(\mathscr{C}_{q+1} \otimes \mathcal{S}),$$

it follows from (4) that the kernel of (5) is simply the submodule $\Gamma(\mathscr{Z}_q \otimes \mathcal{S})$ of $\Gamma(\mathscr{C}_q \otimes \mathcal{S})$. Thus for $q \geq 0$,

(7) $$H^q(M,\mathcal{S}) = H^q(\Gamma(\mathscr{C}^* \otimes \mathcal{S})) = \Gamma(\mathscr{Z}_q \otimes \mathcal{S})/\mathrm{Im}(\Gamma(\mathscr{C}_{q-1} \otimes \mathcal{S})).$$

Now, if $\mathcal{S}$ happens to be a fine sheaf, then according to Theorem 5.15 the full sequence

(8) $$0 \to \Gamma(\mathscr{Z}_{q-1} \otimes \mathcal{S}) \to \Gamma(\mathscr{C}_{q-1} \otimes \mathcal{S}) \to \Gamma(\mathscr{Z}_q \otimes \mathcal{S}) \to 0$$

is exact; and (8) together with (7) implies that

$$H^q(M,\mathcal{S}) = 0 \quad \text{for } q > 0.$$

Thus axiom (b) is satisfied.

Finally, it is evident that $H^q(M,\mathcal{S}) = 0$ for $q < 0$, and for $q = 0$ the isomorphism

$$\mathcal{K} \to \mathcal{Z}_0 \subset \mathcal{C}_0 \tag{9}$$

induces, for an arbitrary sheaf $\mathcal{S}$, an isomorphism

$$\mathcal{S} \cong \mathcal{K} \otimes \mathcal{S} \xrightarrow{\cong} \mathcal{Z}_0 \otimes \mathcal{S} \tag{10}$$

and thus an isomorphism

$$\Gamma(\mathcal{S}) \xrightarrow{\cong} \Gamma(\mathcal{Z}_0 \otimes \mathcal{S}) = H^0(M,\mathcal{S}),$$

which clearly satisfies

$$\begin{array}{ccc} \Gamma(\mathcal{S}) & \cong & H^0(M,\mathcal{S}) \\ \downarrow & & \downarrow \\ \Gamma(\mathcal{S}') & \cong & H^0(M,\mathcal{S}') \end{array}$$

for each homomorphism $\mathcal{S} \to \mathcal{S}'$. Thus axiom (a) is satisfied.

5.21 Definition Let $\mathcal{H}$ and $\tilde{\mathcal{H}}$ be two sheaf cohomology theories on M with coefficients in sheaves of K-modules over M. A *homomorphism of the cohomology theory* $\mathcal{H}$ *to the theory* $\tilde{\mathcal{H}}$ consists of a homomorphism

$$H^q(M,\mathcal{S}) \to \tilde{H}^q(M,\mathcal{S}) \tag{1}$$

for each sheaf $\mathcal{S}$ and for each integer q, such that the following conditions hold:

(a) For $q = 0$, the diagram

$$\begin{array}{ccc} H^0(M,\mathcal{S}) & \cong & \Gamma(\mathcal{S}) \\ \downarrow & & \downarrow \\ \tilde{H}^0(M,\mathcal{S}) & \cong & \Gamma(\mathcal{S}) \end{array}$$

commutes.

(b) For each homomorphism $\mathcal{S} \to \mathcal{T}$ and each integer q the diagram

$$\begin{array}{ccc} H^q(M,\mathcal{S}) & \to & H^q(M,\mathcal{T}) \\ \downarrow & & \downarrow \\ \tilde{H}^q(M,\mathcal{S}) & \to & \tilde{H}^q(M,\mathcal{T}) \end{array}$$

commutes.

(c) For each short exact sequence of sheaves

$$0 \to \mathcal{R} \to \mathcal{S} \to \mathcal{T} \to 0$$

the following diagram commutes for each integer q:

$$\begin{array}{ccc} H^q(M,\mathcal{T}) & \xrightarrow{\partial} & H^{q+1}(M,\mathcal{R}) \\ \downarrow & & \downarrow \\ \tilde{H}^q(M,\mathcal{T}) & \xrightarrow{\partial} & \tilde{H}^{q+1}(M,\mathcal{R}). \end{array}$$

An *isomorphism* $\mathscr{H} \to \tilde{\mathscr{H}}$ is a homomorphism in which each of the homomorphisms $H^q(M,\mathcal{S}) \to \tilde{H}^q(M,\mathcal{S})$ are isomorphisms. The *identity homomorphism of* $\mathscr{H}$ with itself is the homomorphism $\mathscr{H} \to \mathscr{H}$ which assigns the identity homomorphism $H^q(M,\mathcal{S}) \to H^q(M,\mathcal{S})$ to each sheaf $\mathcal{S}$ and each integer q.

We are now going to prove that there is a unique homomorphism between any two sheaf cohomology theories on M. However, first we need to introduce the notion of the sheaf of germs of discontinuous sections of a sheaf $\mathcal{S}$.

5.22 The Sheaf of Germs of Discontinuous Sections of $\mathcal{S}$ Let $\mathcal{S}$ be a sheaf on M. By a *discontinuous section of* $\mathcal{S}$ over the open set $U \subset M$ we mean any map $f: U \to \mathcal{S}$, continuous or not, such that $\pi \circ f = \text{id}$. The assignment to each open set $U \subset M$ of the module of all discontinuous sections of $\mathcal{S}$ over U yields a presheaf whose associated sheaf $\mathcal{S}_0$ is called the *sheaf of germs of discontinuous sections of* $\mathcal{S}$.

The important property of $\mathcal{S}_0$ that we shall need is that $\mathcal{S}_0$ is always a fine sheaf. For let $\{U_i\}$ be a locally finite open cover of M. Choose a refinement $\{V_i\}$ such that $\overline{V_i} \subset U_i$ for each i (Exercise 3, Chapter 1). (For such a refinement on a general paracompact Hausdorff space, see [13, Ch. 5, Problem V(a)].) Associate with each point of M a set V_i containing it, and then for each i define a function φ_i on M to have the value 1 at all points associated with V_i and to have the value 0 elsewhere. It follows that $\operatorname{supp} \varphi_i \subset U_i$ and $\sum \varphi_i \equiv 1$. We associate with φ_i an endomorphism $\tilde{l}_i$ of the presheaf of discontinuous sections of $\mathcal{S}$ by defining

$$\tilde{l}_i(s)(m) = \varphi_i(m)s(m) \tag{1}$$

for each discontinuous section s of $\mathcal{S}$ over an open set $U \subset M$ and each $m \in U$. The presheaf endomorphisms $\tilde{l}_i$ induce endomorphisms l_i of the sheaf $\mathcal{S}_0$ of germs of discontinuous sections of $\mathcal{S}$. It is immediate that the $\{l_i\}$ forms a partition of unity of $\mathcal{S}_0$ subordinate to the locally finite cover $\{U_i\}$ of M. Thus $\mathcal{S}_0$ is a fine sheaf, as claimed.

There is a natural injection of $\mathcal{S}$ into $\mathcal{S}_0$; namely, each element of the stalk $\mathcal{S}_m$ is the value at m of a (continuous) section s of $\mathcal{S}$ defined on some neighborhood of m, and we map $s(m)$ to the germ of s at m in $\mathcal{S}_0$. If we let $\overline{\mathcal{S}}$ denote the quotient sheaf $\mathcal{S}_0/\mathcal{S}$, then for each sheaf $\mathcal{S}$ we have constructed a short exact sequence

$$0 \to \mathcal{S} \to \mathcal{S}_0 \to \overline{\mathcal{S}} \to 0 \tag{2}$$

in which the middle sheaf is fine.

5.23 Theorem *Let $\mathscr{H}$ and $\tilde{\mathscr{H}}$ be cohomology theories on M with coefficients in sheaves of K-modules over M. Then there exists a unique homomorphism of $\mathscr{H}$ to $\tilde{\mathscr{H}}$.*

PROOF We first prove uniqueness. Suppose that we have a homomorphism of $\mathscr{H}$ to $\tilde{\mathscr{H}}$, and let a sheaf $\mathcal{S}$ be given. Then it follows from the axioms 5.18 and 5.21 applied to the exact sequence 5.22(2), in which, you recall, the sheaf $\mathcal{S}_0$ is fine, that we have a commutative diagram

$$
\text{(1)}\qquad
\begin{array}{ccccccc}
\Gamma(\mathcal{S}_0) & \rightarrow & \Gamma(\bar{\mathcal{S}}) & \rightarrow & H^1(M,\mathcal{S}) & \rightarrow & 0 \\
\downarrow\text{id} & & \downarrow\text{id} & & \downarrow & & \\
\Gamma(\mathcal{S}_0) & \rightarrow & \Gamma(\bar{\mathcal{S}}) & \rightarrow & \tilde{H}^1(M,\mathcal{S}) & \rightarrow & 0;
\end{array}
$$

and for each $q \geq 2$ we have a commutative diagram

$$
\text{(2)}\qquad
\begin{array}{ccccccc}
0 & \rightarrow & H^{q-1}(M,\bar{\mathcal{S}}) & \rightarrow & H^q(M,\mathcal{S}) & \rightarrow & 0 \\
 & & \downarrow & & \downarrow & & \\
0 & \rightarrow & \tilde{H}^{q-1}(M,\bar{\mathcal{S}}) & \rightarrow & \tilde{H}^q(M,\mathcal{S}) & \rightarrow & 0.
\end{array}
$$

Note that in both (1) and (2) the rows are exact. Now uniqueness of the homomorphisms 5.21(1) follows for $q = 0$ from 5.21(a), follows for $q = 1$ from (1), and follows inductively for $q > 1$ from (2).

We now show the existence of a homomorphism of the theory $\mathscr{H}$ to the theory $\tilde{\mathscr{H}}$. We define the homomorphisms for $q = 0$ by 5.21(a), and it follows immediately that 5.21(b) is satisfied for the case $q = 0$. We define the homomorphisms 5.21(1) for $q = 1$ by (1), and inductively for $q > 1$ by (2). To prove that this indeed defines a homomorphism of $\mathscr{H}$ to $\tilde{\mathscr{H}}$, it remains to show that 5.21(b) holds for $q > 0$ and that 5.21(c) holds for all q.

Now, 5.21(b) follows for $q = 1$ from the following lattice constructed from two copies of (1):

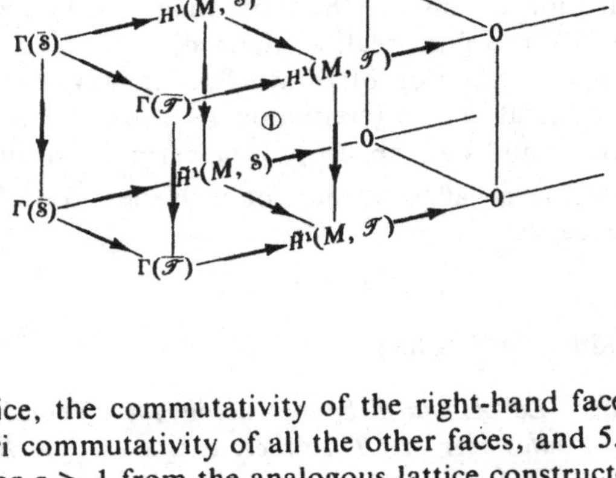

In the lattice, the commutativity of the right-hand face ① is forced by the apriori commutativity of all the other faces, and 5.21(b) follows inductively for $q > 1$ from the analogous lattice constructed from (2).

As for 5.21(c), suppose that a short exact sequence

$$(4) \qquad 0 \to \mathscr{R} \to \mathcal{S} \to \mathscr{T} \to 0$$

is given. Let $\mathscr{R}_0$ and $\mathcal{S}_0$ be the sheaves of germs of discontinuous sections of $\mathscr{R}$ and $\mathcal{S}$ respectively. The composition $\mathscr{R} \to \mathcal{S} \to \mathcal{S}_0$ is an injection of $\mathscr{R}$ into $\mathcal{S}_0$; let $\mathscr{G}$ be the quotient sheaf $\mathcal{S}_0/\mathscr{R}$. As in 5.22, let $\overline{\mathscr{R}} = \mathscr{R}_0/\mathscr{R}$. Then there are uniquely determined homomorphisms $\mathscr{T} \to \mathscr{G}$ and $\overline{\mathscr{R}} \to \mathscr{G}$ such that the following diagram, in which the rows are exact, commutes:

$$(5) \qquad \begin{array}{ccccccccc}
0 & \longrightarrow & \mathscr{R} & \longrightarrow & \mathcal{S} & \longrightarrow & \mathscr{T} & \longrightarrow & 0 \\
 & & \downarrow \text{id} & & \downarrow & & \downarrow & & \\
0 & \longrightarrow & \mathscr{R} & \longrightarrow & \mathcal{S}_0 & \longrightarrow & \mathscr{G} & \longrightarrow & 0 \\
 & & \uparrow \text{id} & & \uparrow & & \uparrow & & \\
0 & \longrightarrow & \mathscr{R} & \longrightarrow & \mathscr{R}_0 & \longrightarrow & \overline{\mathscr{R}} & \longrightarrow & 0.
\end{array}$$

It follows from the cohomology axioms 5.18 applied to (5) that there are commutative diagrams

$$(6) \qquad \begin{array}{ccccccccc}
\cdots & \longrightarrow & \Gamma(\mathcal{S}) & \longrightarrow & \Gamma(\mathscr{T}) & \longrightarrow & H^1(M,\mathscr{R}) & \longrightarrow & \cdots \\
 & & \downarrow & & \downarrow & & \downarrow \text{id} & & \\
\cdots & \longrightarrow & \Gamma(\mathcal{S}_0) & \longrightarrow & \Gamma(\mathscr{G}) & \longrightarrow & H^1(M,\mathscr{R}) & \longrightarrow & 0 \\
 & & \uparrow & & \uparrow & & \uparrow \text{id} & & \\
\cdots & \longrightarrow & \Gamma(\mathscr{R}_0) & \longrightarrow & \Gamma(\overline{\mathscr{R}}) & \longrightarrow & H^1(M,\mathscr{R}) & \longrightarrow & 0
\end{array}$$

and (for $q \geq 1$)

$$(7) \qquad \begin{array}{ccccccccc}
\cdots & \longrightarrow & H^q(M,\mathcal{S}) & \longrightarrow & H^q(M,\mathscr{T}) & \longrightarrow & H^{q+1}(M,\mathscr{R}) & \longrightarrow & \cdots \\
 & & & & \downarrow & & \downarrow \text{id} & & \\
 & & 0 & \longrightarrow & H^q(M,\mathscr{G}) & \xrightarrow{\cong} & H^{q+1}(M,\mathscr{R}) & \longrightarrow & 0 \\
 & & & & \uparrow \cong & & \uparrow \text{id} & & \\
 & & 0 & \longrightarrow & H^q(M,\overline{\mathscr{R}}) & \xrightarrow{\cong} & H^{q+1}(M,\mathscr{R}) & \longrightarrow & 0.
\end{array}$$

It follows from (6) that the homomorphism $H^0(M,\mathscr{T}) \to H^1(M,\mathscr{R})$ is the composition

$$(8) \quad H^0(M,\mathscr{T}) \xrightarrow{\cong} \Gamma(\mathscr{T}) \to \Gamma(\mathscr{G})/\operatorname{Im}\Gamma(\mathcal{S}_0) \xleftarrow{\cong} \Gamma(\overline{\mathscr{R}})/\operatorname{Im}\Gamma(\mathscr{R}_0) \xrightarrow{\cong} H^1(M,\mathscr{R}),$$

and it follows from (7) that the homomorphism $H^q(M,\mathscr{T}) \to H^{q+1}(M,\mathscr{R})$ for $q \geq 1$ is the composition

$$(9) \qquad H^q(M,\mathscr{T}) \to H^q(M,\mathscr{G}) \xleftarrow{\cong} H^q(M,\overline{\mathscr{R}}) \xrightarrow{\cong} H^{q+1}(M,\mathscr{R}).$$

From (8) and the corresponding sequence for the $\tilde{\mathscr{H}}$ theory we obtain the diagram

$$
(10)\qquad
\begin{array}{ccccccccc}
H^0(M,\mathscr{T}) & \xrightarrow{\cong} & \Gamma(\mathscr{T}) & \to & \Gamma(\mathscr{G})/\operatorname{Im}\Gamma(\mathcal{S}_0) & \xleftarrow{\cong} & \Gamma(\bar{\mathscr{R}})/\operatorname{Im}\Gamma(\mathscr{R}_0) & \xrightarrow{\cong} & H^1(M,\mathscr{R}) \\
\downarrow & & \downarrow \text{id} & & \downarrow \text{id} & & \downarrow \text{id} & & \downarrow \\
\tilde{H}^0(M,\mathscr{T}) & \xrightarrow{\cong} & \Gamma(\mathscr{T}) & \to & \Gamma(\mathscr{G})/\operatorname{Im}\Gamma(\mathcal{S}_0) & \xleftarrow{\cong} & \Gamma(\bar{\mathscr{R}})/\operatorname{Im}\Gamma(\mathscr{R}_0) & \xrightarrow{\cong} & \tilde{H}^1(M,\mathscr{R})
\end{array}
$$

in which the first and last squares commute by the definitions of the homomorphisms $H^0(M,\mathscr{T}) \to \tilde{H}^0(M,\mathscr{T})$ and $H^1(M,\mathscr{R}) \to \tilde{H}^1(M,\mathscr{R})$, and the middle two squares trivially commute. Thus 5.21(c) is proved for $q = 0$. From (9) and the corresponding sequence for the $\tilde{\mathscr{H}}$ theory we obtain the diagram

$$
(11)\qquad
\begin{array}{ccccccc}
H^q(M,\mathscr{T}) & \to & H^q(M,\mathscr{G}) & \xleftarrow{\cong} & H^q(M,\bar{\mathscr{R}}) & \xrightarrow{\cong} & H^{q+1}(M,\mathscr{R}) \\
\downarrow & & \downarrow & & \downarrow & & \downarrow \\
\tilde{H}^q(M,\mathscr{T}) & \to & \tilde{H}^q(M,\mathscr{G}) & \xleftarrow{\cong} & \tilde{H}^q(M,\bar{\mathscr{R}}) & \xrightarrow{\cong} & \tilde{H}^{q+1}(M,\mathscr{R})
\end{array}
$$

in which the last square commutes by definition of the homomorphism $H^{q+1}(M,\mathscr{R}) \to \tilde{H}^{q+1}(M,\mathscr{R})$, and the first two squares commute by 5.21(b). Thus 5.21(c) follows for $q \geq 1$, and the proof of Theorem 5.23 is complete.

Corollary *A homomorphism of the cohomology theory $\mathscr{H}$ to the theory $\tilde{\mathscr{H}}$ must necessarily be an isomorphism. Consequently, any two cohomology theories on M with coefficients in sheaves of K-modules over M are uniquely isomorphic.*

PROOF Assume that a homomorphism $\mathscr{H} \to \tilde{\mathscr{H}}$ is given. By Theorem 5.23 there must also exist a homomorphism $\tilde{\mathscr{H}} \to \mathscr{H}$. The composition $\mathscr{H} \to \tilde{\mathscr{H}} \to \mathscr{H}$ must necessarily, by uniqueness, be the identity homomorphism of $\mathscr{H}$, and similarly the composition $\tilde{\mathscr{H}} \to \mathscr{H} \to \tilde{\mathscr{H}}$ must be the identity homomorphism of $\tilde{\mathscr{H}}$. It follows that the homomorphisms $\mathscr{H} \to \tilde{\mathscr{H}}$ and $\tilde{\mathscr{H}} \to \mathscr{H}$ are both isomorphisms.

5.24 Assume that

$$
(1)\qquad
\begin{array}{ccccccccccc}
0 & \to & \mathscr{K} & \to & \mathscr{C}_0 & \to & \mathscr{C}_1 & \to & \mathscr{C}_2 & \to & \cdots \\
 & & \downarrow \text{id} & & \downarrow & & \downarrow & & \downarrow & & \\
0 & \to & \mathscr{K} & \to & \tilde{\mathscr{C}}_0 & \to & \tilde{\mathscr{C}}_1 & \to & \tilde{\mathscr{C}}_2 & \to & \cdots
\end{array}
$$

is a commutative diagram in which the rows are fine torsionless resolutions of the constant sheaf $\mathscr{K}$. We have seen in 5.20 that each of these resolutions determines a cohomology theory. Denote the cohomology theory obtained from the top resolution by $\mathscr{H}$ and that obtained from the second resolution by $\tilde{\mathscr{H}}$. We claim that the "homomorphism" (1) between these two fine torsionless resolutions canonically induces a homomorphism, and therefore, by the corollary of 5.23, induces an isomorphism, $\mathscr{H} \to \tilde{\mathscr{H}}$. For each sheaf $\mathcal{S}$, (1) induces a cochain map $\Gamma(\mathscr{C}^* \otimes \mathcal{S}) \to \Gamma(\tilde{\mathscr{C}}^* \otimes \mathcal{S})$ from which we obtain a homomorphism $H^q(M,\mathcal{S}) \to \tilde{H}^q(M,\mathcal{S})$ for each integer q. That the axioms 5.21(a), (b), and (c) for a homomorphism of cohomology theories are satisfied is easily demonstrated and will be left to the reader as an exercise. This procedure for obtaining a homomorphism of cohomology theories will be used in 5.36 to obtain an explicit isomorphism of the de Rham and differentiable singular cohomology theories for a differentiable manifold M.

5.25 Theorem *Assume that $\mathscr{H}$ is a cohomology theory for M with coefficients in sheaves of K-modules over M. Let*

(1) $$0 \to \mathcal{S} \to \mathscr{C}_0 \to \mathscr{C}_1 \to \mathscr{C}_2 \to \cdots$$

be a fine resolution of the sheaf $\mathcal{S}$. *Then there are canonical isomorphisms*

(2) $$H^q(M,\mathcal{S}) \cong H^q(\Gamma(\mathscr{C}^*)) \quad \textit{for all } q.$$

PROOF From the exactness of (1) it follows that $0 \to \Gamma(\mathcal{S}) \to \Gamma(\mathscr{C}_0) \to \Gamma(\mathscr{C}_1)$ is exact, whence $H^0(M,\mathcal{S}) \cong \Gamma(\mathcal{S}) \cong H^0(\Gamma(\mathscr{C}^*))$. Now let $\mathscr{K}_q$ be the kernel of $\mathscr{C}_q \to \mathscr{C}_{q+1}$ for $q \geq 1$. Then the exactness of (1) implies that

(3) $$0 \to \mathscr{K}_q \to \mathscr{C}_q \to \mathscr{K}_{q+1} \to 0$$

is exact for $q \geq 1$, and that

(4) $$0 \to \mathcal{S} \to \mathscr{C}_0 \to \mathscr{K}_1 \to 0$$

is exact. It follows from the long exact cohomology sequence for (4) and from the fact that $\mathscr{C}_0$ is a fine sheaf that

(5) $$H^q(M,\mathcal{S}) \cong H^{q-1}(M,\mathscr{K}_1) \quad \text{for } q > 1$$

and

(6) $$H^1(M,\mathcal{S}) \cong \Gamma(\mathscr{K}_1)/\text{Im }\Gamma(\mathscr{C}_0) \cong H^1(\Gamma(\mathscr{C}^*)).$$

By repeated application of the long exact sequence associated with (3) and of the fact that the sheaves $\mathscr{C}_i$ are fine, we obtain for $q > 1$

(7) $$H^q(M,\mathcal{S}) \cong H^{q-1}(M,\mathscr{K}_1) \cong H^{q-2}(M,\mathscr{K}_2) \cong \cdots \cong H^1(M,\mathscr{K}_{q-1}) \cong \Gamma(\mathscr{K}_q)/\text{Im }\Gamma(\mathscr{C}_{q-1}) \cong H^q(\Gamma(\mathscr{C}^*)).$$

This completes our development of axiomatic sheaf cohomology except for the multiplicative structure, which is discussed in 5.42. We shall now consider four sheaf cohomology theories for a differentiable manifold M—the Alexander-Spanier, de Rham, singular, and Čech. In view of the corollary of 5.23, these theories will be uniquely isomorphic. We shall also define the *classical* versions of these theories, and prove that they are canonically isomorphic with sheaf cohomology with coefficients in constant sheaves.

THE CLASSICAL COHOMOLOGY THEORIES

Alexander-Spanier Cohomology

5.26 For the Alexander-Spanier theory, M need only be a paracompact Hausdorff space. Let $U \subset M$ be open, and as usual let K be a fixed principal ideal domain. We use U^{p+1} to denote the $(p + 1)$ fold cartesian product of U with itself. Let $A^p(U,K)$ denote the K-module of functions $U^{p+1} \to K$ under pointwise addition. For each $p \geq 0$ we define a homomorphism

(1) $$d: A^p(U,K) \to A^{p+1}(U,K)$$

by setting

(2) $$df(m_0, \ldots, m_{p+1}) = \sum_{i=0}^{p+1} (-1)^i f(m_0, \ldots, \widehat{m_i}, \ldots, m_{p+1})$$

for each $f \in A^p(U,K)$ and $(m_0, \ldots, m_{p+1}) \in U^{p+2}$, where $\frown$ over an entry means that that entry is to be omitted. It is a straightforward exercise to check that $d \circ d = 0$, so for each $U \subset M$ we have a cochain complex

(3) $$\cdots \to 0 \to A^0(U,K) \xrightarrow{d} A^1(U,K) \xrightarrow{d} A^2(U,K) \xrightarrow{d} \cdots$$

which we shall denote by $A^*(U,K)$, and in which the modules of q-cochains for $q < 0$ are all assumed to be zero modules. If $V \subset U$, then $V^{p+1} \subset U^{p+1}$, and there is a corresponding restriction homomorphism

(4) $$\rho_{V,U}: A^p(U,K) \to A^p(V,K).$$

The collection

(5) $$\{A^p(U,K); \rho_{U,V}\}$$

forms a presheaf of K-modules on M called the *presheaf of Alexander-Spanier p-cochains.* Observe that these presheaves satisfy 5.7(C_2), but for $p \geq 1$ do not satisfy 5.7(C_1).

The associated sheaf of germs of Alexander-Spanier p-cochains we denote by $\mathscr{A}^p(M,K)$. The homomorphisms d give presheaf homomorphisms $\{A^p(U,K); \rho_{U,V}\} \to \{A^{p+1}(U,K); \rho_{U,V}\}$ for each $p \geq 0$, and induce sheaf homomorphisms (which we shall denote by the same symbol d):

(6) $$d: \mathscr{A}^p(M,K) \to \mathscr{A}^{p+1}(M,K) \qquad (p \geq 0).$$

These homomorphisms, together with the natural injection of the constant sheaf $\mathscr{K} \to \mathscr{A}^0(M,K)$ which sends $k \in \mathscr{K}_m$ to the germ of the constant function k at m, give us a sequence

(7) $$0 \to \mathscr{K} \to \mathscr{A}^0(M,K) \xrightarrow{d} \mathscr{A}^1(M,K) \xrightarrow{d} \mathscr{A}^2(M,K) \xrightarrow{d} \cdots.$$

We claim that (7) is a fine torsionless resolution of the constant sheaf $\mathscr{K}$. The elements of $\mathscr{A}^p(M,K)$ are equivalence classes of functions with values in K, and K is an integral domain; hence $\mathscr{A}^p(M,K)$ is torsionless. To prove that $\mathscr{A}^p(M,K)$ is a fine sheaf, let $\{U_i\}$ be a locally finite open cover of M, and take (as in 5.22) a partition of unity $\{\varphi_i\}$ subordinate to the cover $\{U_i\}$ in which the functions φ_i take values 0 or 1 only. For each i we define an endomorphism $\tilde{l}_i$ of $A^p(U,K)$ by setting

$$\tilde{l}_i(f)(m_0, \ldots, m_p) = \varphi_i(m_0)f(m_0, \ldots, m_p). \tag{8}$$

The endomorphisms $\tilde{l}_i$ commute with restrictions and thus determine presheaf endomorphisms of $\{A^p(U,K): \rho_{U,V}\}$. Let $l_i: \mathscr{A}^p(M,K) \to \mathscr{A}^p(M,K)$ be the sheaf endomorphism associated with $\tilde{l}_i$. Then it follows readily that $\operatorname{supp} l_i \subset U_i$ and that $\sum l_i \equiv 1$. Thus the sheaves $\mathscr{A}^p(M,K)$ are fine. The exactness of (7) follows directly from the fact that we have exactness on the presheaf level. That is, if $U \subset M$ is open, then

$$0 \to K \to A^0(U,K) \xrightarrow{d} A^1(U,K) \xrightarrow{d} A^2(U,K) \xrightarrow{d} \cdots \tag{9}$$

is exact. Here K is injected into $A^0(U,K)$ by sending $k \in K$ to the function that has the constant value k on U. That (9) is exact is apparent at K and at $A^0(U,K)$, and follows elsewhere from the fact that $d \circ d = 0$ and that if $f \in A^p(U,K)$ with $p \geq 1$ and $df = 0$, then $f = dg$, where $g \in A^{p-1}(U,K)$ is defined by

$$g(m_0, \ldots, m_{p-1}) = f(m, m_0, \ldots, m_{p-1}) \tag{10}$$

for some arbitrary but fixed $m \in U$. Thus (7) is a fine torsionless resolution of $\mathscr{K}$.

The explicit exhibition of the fine torsionless resolution (7) of the constant sheaf $\mathscr{K}$ completes the proof in 5.20 of the existence of a cohomology theory for M with coefficients in sheaves of K-modules. Thus by setting (as in 5.20(1))

$$H^q(M,\mathcal{S}) = H^q(\Gamma(\mathscr{A}^*(M,K) \otimes \mathcal{S})), \tag{11}$$

we obtain a cohomology theory with which, according to the corollary of 5.23, any other sheaf cohomology theory for M with coefficients in sheaves of K-modules is uniquely isomorphic.

We shall now define the classical Alexander-Spanier cohomology modules of M with coefficients in a K-module G, and show that they are canonically isomorphic with the cohomology modules $H^q(M,\mathscr{G})$ with coefficients in the constant sheaf $\mathscr{G} = M \times G$. We let $A^p(U,G)$ denote the K-module of functions $U^{p+1} \to G$, and similarly replace K by G and $\mathscr{K}$ by $\mathscr{G}$ in the constructions (1) through (7). Let

$$A_0^p(M,G) = \{f \in A^p(M,G): \rho_{m,M}(f) = 0 \text{ for all } m \in M\}. \tag{12}$$

Recall that $\rho_{m,M}$ is the homomorphism which assigns to each element of $A^p(M,G)$ its equivalence class in the stalk over m of the associated sheaf $\mathscr{A}^p(M,G)$. $A_0^p(M,G)$ is a submodule of $A^p(M,G)$, and the extent to which $A_0^p(M,G)$ is not zero measures the extent to which the presheaf $\{A^p(U,G);\rho_{U,V}\}$ fails to satisfy 5.7(C_1). The homomorphism (1) restricted to $A_0^p(M,G)$ has range in $A_0^{p+1}(M,G)$ and thus yields homomorphisms on quotients

$$A^p(M,G)/A_0^p(M,G) \to A^{p+1}(M,G)/A_0^{p+1}(M,G). \tag{13}$$

The sequence of modules and homomorphisms given by (13) for $p \geq 0$ form a cochain complex which we shall denote by $A^*(M,G)/A_0^*(M,G)$, and in which the modules of q cochains for $q < 0$ are as usual all assumed to be zero. The *classical Alexander-Spanier cohomology modules for M with coefficients in the K-module G* are given by definition by

$$H^q_{A-S}(M;G) = H^q\big(A^*(M,G)/A_0^*(M,G)\big). \tag{14}$$

Since the sequence (7) with K replaced by G and $\mathscr{K}$ by $\mathscr{G}$ is a fine resolution of $\mathscr{G}$, it follows from 5.25 that there are canonical isomorphisms

$$H^q(M,\mathscr{G}) \simeq H^q(\Gamma(\mathscr{A}^*(M,G))). \tag{15}$$

That there are canonical isomorphisms

$$H^q_{A-S}(M;G) \simeq H^q(M,\mathscr{G}) \tag{16}$$

follows from (14) and (15) and from the fact that the natural cochain map

$$A^*(M,G)/A_0^*(M,G) \to \Gamma\big(\mathscr{A}^*(M,G)\big) \tag{17}$$

is an isomorphism. This in turn is an immediate consequence of the following proposition.

5.27 Proposition *Let $\{S_U;\rho_{U,V}\}$ be a presheaf on M satisfying 5.7(C_2), and let $\mathcal{S}$ be the associated sheaf. As in 5.26(12), let*

$$(S_M)_0 = \{s \in S_M\colon \rho_{m,M}(s) = 0 \text{ for all } m \in M\}. \tag{1}$$

Then the sequence

$$0 \to (S_M)_0 \to S_M \xrightarrow{\gamma} \Gamma(\mathcal{S}) \to 0 \tag{2}$$

is exact.

PROOF Since γ is the homomorphism which sends $s \in S_M$ to the global section $m \mapsto \rho_{m,M}(s)$ of $\mathcal{S}$, the exactness of the sequence (2) at S_M is the result of the definition of $(S_M)_0$. Thus we need only prove that γ is surjective. Let $t \in \Gamma(\mathcal{S})$. Then there is a locally finite open cover $\{U_\alpha\}$ of M, and there are elements $s_\alpha \in S_{U_\alpha}$ such that

$$\gamma(s_\alpha) = t \,|\, U_\alpha. \tag{3}$$

Let $\{V_\alpha\}$ be a refinement such that $\overline{V_\alpha} \subset U_\alpha$. Let I_m be the (finite) collection of all those indices α for which $m \in \overline{V_\alpha}$. Choose a neighborhood W_m of m such that

(a) $W_m \cap \overline{V_\beta} = \varnothing$ if $\beta \notin I_m$,

(b) $W_m \subset \bigcap_{\alpha \in I_m} U_\alpha$,

(c) $\rho_{W_m, U_\alpha}(s_\alpha) = \rho_{W_m, U_{\alpha'}}(s_{\alpha'})$ if $\alpha, \alpha' \in I_m$.

Let $s_m \in S_{W_m}$ be the common image of the elements in (c). Then for all n and m in M,

(4) $$\rho_{W_m \cap W_n, W_m}(s_m) = \rho_{W_m \cap W_n, W_n}(s_n).$$

For let $p \in W_m \cap W_n$. Then it follows from (a) that $I_p \subset I_m \cap I_n$. So let $\alpha \in I_p$. Then according to (c),

(5) $$s_m = \rho_{W_m, U_\alpha}(s_\alpha) \quad \text{and} \quad s_n = \rho_{W_n, U_\alpha}(s_\alpha),$$

so that

(6) $$\rho_{W_m \cap W_n, W_m}(s_m) = \rho_{W_m \cap W_n, U_\alpha}(s_\alpha) = \rho_{W_m \cap W_n, W_n}(s_n),$$

which proves (4). It follows from (4) and the property 5.7(C_2) that there is an element $s \in S_M$ such that

(7) $$\rho_{W_m, M}(s) = s_m ;$$

and it follows from (7), (5), and (3) that $\gamma(s) = t$.

de Rham Cohomology

5.28 In this section we shall take the principal ideal domain K to be the real number field $\mathbb{R}$. Let $U \subset M$ be open. The set of differential p-forms on U forms a real vector space, which we have denoted by $E^p(U)$. These real vector spaces together with ordinary restriction homomorphisms $\rho_{U,V}$ form a presheaf

(1) $$\{E^p(U); \rho_{U,V}\}$$

which satisfies both 5.7(C_1) and (C_2) and thus is complete. From the exterior derivative operator d we obtain presheaf homomorphisms

(2) $$\{E^p(U); \rho_{U,V}\} \xrightarrow{d} \{E^{p+1}(U); \rho_{U,V}\} \qquad (p \geq 0).$$

We shall denote the associated sheaf of germs of differential p-forms by $\mathscr{E}^p(M)$, and we shall retain the symbol d for the sheaf homomorphisms induced by (2). The constant sheaf $\mathscr{R} = M \times \mathbb{R}$ can be naturally injected into $\mathscr{E}^0(M)$ by sending $a \in \mathscr{R}_m$ to the germ at m of the function with constant value a. Thus we have a sequence

$$(3) \qquad 0 \to \mathscr{R} \to \mathscr{E}^0(M) \xrightarrow{d} \mathscr{E}^1(M) \xrightarrow{d} \mathscr{E}^2(M) \xrightarrow{d} \cdots.$$

We claim that (3) is a fine torsionless resolution of the constant sheaf $\mathscr{R}$. Partitions of unity can be constructed for the sheaves $\mathscr{E}^p(M)$ exactly as they were constructed in 5.10 for $\mathscr{C}^\infty(M) = \mathscr{E}^0(M)$. Thus the sheaves $\mathscr{E}^p(M)$ are all fine. They are certainly torsionless since they are sheaves of real vector spaces. The sequence (3) is clearly exact at $\mathscr{R}$ and at $\mathscr{E}^0(M)$. Since the exterior derivative operator satisfies $d \circ d = 0$, the same is true of the sheaf homomorphisms in (3); hence the image of each homomorphism is contained in the kernel of the next. That the sequence is actually exact, and therefore a resolution, is a consequence of the corollaries of the Poincaré lemma 4.18. We shall return to the resolution (3) in a moment. First, we comment about the structure of the Poincaré lemma.

5.29 Remark The collection of operators h_k in the Poincaré lemma 4.18 is an example of a very useful tool in the cohomology theory of cochain complexes called a *homotopy operator*. We shall have further occasion to make use of such operators. They arise generally in the following situation. Suppose that we have two cochain maps, call them f and g, between cochain complexes C^* and D^*, and suppose that we want to show that f and g both induce the same homomorphisms of the cohomology modules. This will follow if we can find homotopy operators for f and g, namely, homomorphisms h_k: $C^k \to D^{k-1}$ such that

$$h_{k+1} \circ d + d \circ h_k = f_k - g_k$$

on C^k (where d indicates the coboundary operators in both C^* and D^*). For then if $\sigma \in C^k$ is a cocycle, then $f_k(\sigma) = g_k(\sigma)$ up to a coboundary, so $f_k(\sigma)$ and $g_k(\sigma)$ lie in the same cohomology class. In the Poincaré lemma, we found homotopy operators (for $k \geq 1$) between the identity cochain map of the cochain complex

$$\cdots \to 0 \to E^0(U) \xrightarrow{d} E^1(U) \xrightarrow{d} E^2(U) \xrightarrow{d} \cdots$$

and the zero cochain map, with U the open unit ball in Euclidean space. Consequently, all of the cohomology groups of this cochain complex vanish for $k \geq 1$.

5.30 Since 5.28(3) is exact, it is therefore a fine torsionless resolution of the constant sheaf $\mathscr{R}$. According to 5.20, the resolution 5.28(3) gives rise to a cohomology theory for M with coefficients in sheaves of real vector spaces by setting

$$(1) \qquad H^q(M,\mathscr{T}) = H^q(\Gamma(\mathscr{E}^*(M) \otimes \mathscr{T}))$$

for $q \geq 0$ and for $\mathscr{T}$ a sheaf of real vector spaces over M. In view of the corollary of 5.23, this theory is uniquely isomorphic with the theory we constructed in 5.26(11).

If we take $\mathscr{T}$ to be the constant sheaf $\mathscr{R}$, then

$$H^q(M,\mathscr{R}) = H^q(\Gamma(\mathscr{E}^*(M) \otimes \mathscr{R})) \simeq H^q(\Gamma(\mathscr{E}^*(M))). \tag{2}$$

Consider now the cochain complex $\Gamma(\mathscr{E}^*(M))$:

$$\cdots \to 0 \to \Gamma(\mathscr{E}^0(M)) \to \Gamma(\mathscr{E}^1(M)) \to \Gamma(\mathscr{E}^2(M)) \to \cdots \tag{3}$$

and the cochain complex $E^*(M)$:

$$\cdots \to 0 \to E^0(M) \xrightarrow{d} E^1(M) \xrightarrow{d} E^2(M) \xrightarrow{d} \cdots. \tag{4}$$

In view of the fact that the presheaves $\{E^p(U);\rho_{U,V}\}$ are complete, it follows from 5.8 (or is easily seen directly in this special case) that the natural homomorphisms

$$E^p(M) \to \Gamma(\mathscr{E}^p(M)) \tag{5}$$

are isomorphisms. Since the homomorphisms (5) commute with the coboundaries of the respective cochain complexes (3) and (4), they induce a cochain map $E^*(M) \to \Gamma(\mathscr{E}^*(M))$ which is an isomorphism of cochain complexes. Thus there are canonical isomorphisms

$$H^q(\Gamma(\mathscr{E}^*(M))) \simeq H^q(E^*(M)). \tag{6}$$

But $H^q(E^*(M))$ is the classical qth *de Rham cohomology group* $H^q_{\text{de R}}(M)$ for M which we introduced in 4.13(1), namely, the quotient of the vector space of *closed q-forms* (those annihilated by d) modulo the vector space of *exact q-forms* (those in the image of d). Thus from (2) and (6) we obtain canonical isomorphisms

$$H^q(M,\mathscr{R}) \simeq H^q_{\text{de R}}(M). \tag{7}$$

Singular Cohomology

5.31 In this section we again take K to be an arbitrary principal ideal domain, and we let $U \subset M$ be open. In 4.6 we introduced differentiable singular simplices for the theory of integration on manifolds. Now we need continuous singular simplices. We review the definitions. Recall that for each integer $p \geq 1$ we let

$$\Delta^p = \left\{(a_1, \ldots, a_p) \in \mathbb{R}^p: \sum_{i=1}^{p} a_i \leq 1 \quad \text{and} \quad \text{each} \quad a_i \geq 0\right\}. \tag{1}$$

Δ^p is called the *standard p-simplex* in $\mathbb{R}^p$. For $p = 0$ we set Δ^0 equal to the 1-point space $\{0\}$; and Δ^0 is the *standard* 0-*simplex*. A *(continuous) singular p-simplex* σ in U is a continuous map $\sigma: \Delta^p \to U$. If $p \geq 1$, we define a *differentiable singular p-simplex* in U to be a singular p-simplex σ which can be extended to be a differentiable (C^∞) map of a neighborhood of Δ^p in $\mathbb{R}^p$ into U.

We shall let $S_p(U)$ denote the free abelian group generated by the singular p-simplices in U. Elements of $S_p(U)$ are called *singular p-chains with integer coefficients*. If σ is a singular p-simplex in U, with $p \geq 1$, then its *boundary* is defined as in 4.6(4) to be the singular $(p - 1)$ chain

$$\partial\sigma = \sum_{i=0}^{p} (-1)^i \sigma^i \tag{2}$$

where σ^i is the ith face of σ (see 4.6(3)). The boundary operator extends to a homomorphism of $S_p(U)$ into $S_{p-1}(U)$ for each $p \geq 1$, and satisfies (as in 4.6(7))

$$\partial \circ \partial = 0. \tag{3}$$

Let $S^p(U,K)$ denote the K-module consisting of functions f which assign to each singular p-simplex in U an element of K. Such an f is called a *singular p-cochain* on U. Scalar multiplication and addition in the module $S^p(U,K)$ are defined by

$$\begin{aligned} (kf)(\sigma) &= k(f(\sigma)), \\ (f + g)(\sigma) &= f(\sigma) + g(\sigma). \end{aligned} \tag{4}$$

Each cochain in $S^p(U,K)$ canonically extends to a homomorphism of $S_p(U)$ into K. In fact, this determines an isomorphism of $S^p(U,K)$ with the K-module of homomorphisms of $S_p(U)$ into K. For convenience we shall at times regard elements of $S^p(U,K)$ as such homomorphisms. If $V \subset U$, we let

$$\rho_{V,U}\colon S^p(U,K) \to S^p(V,K) \tag{5}$$

be the homomorphism which assigns to each $f \in S^p(U,K)$ its restriction to singular p-simplices which lie in V. Then

$$\{S^p(U,K); \rho_{U,V}\} \tag{6}$$

forms a presheaf on M called the *presheaf of singular p-cochains*. Observe that these presheaves, for $p \geq 1$, satisfy 5.7(C_2) but not (C_1), and that the presheaf $\{S^0(U,K); \rho_{U,V}\}$ (which can be canonically identified with the presheaf $\{A^0(U,K); \rho_{U,V}\}$ of 5.26 since each singular 0-simplex in U can be identified with a point of U) is complete.

For $p \geq 1$, we let $S^p_\infty(U,K)$ denote the K-module consisting of functions f which assign to each differentiable singular p-simplex in U an element of K. Such an f is called a *differentiable singular p-cochain on U*. With restriction homomorphisms defined as in (5), we obtain the presheaf $\{S^p_\infty(U,K); \rho_{U,V}\}$ of differentiable singular p-cochains on M. Since the developments of the differentiable and the continuous singular cohomology theories are nearly word-for-word the same, we shall primarily discuss the continuous case. Just keep in mind that analogous constructions apply if $S^p_\infty(U,K)$, for $p \geq 1$, is substituted for $S^p(U,K)$. When substantial differences arise they will be discussed.

A coboundary homomorphism

(7) $$d\colon S^p(U,K) \to S^{p+1}(U,K)$$

is defined by setting

(8) $$df(\sigma) = f(\partial\sigma)$$

for $f \in S^p(U,K)$ and for σ a singular $(p + 1)$ simplex in U. It follows from item (3) that

(9) $$d \circ d = 0.$$

Since d commutes with restriction homomorphisms, d yields a presheaf homomorphism

(10) $$\{S^p(U,K);\rho_{U,V}\} \to \{S^{p+1}(U,K);\rho_{U,V}\};$$

and at the same time, in view of (9), d makes

(11) $$\cdots \to 0 \to S^0(U,K) \xrightarrow{d} S^1(U,K) \xrightarrow{d} S^2(U,K) \xrightarrow{d} \cdots$$

into a cochain complex (where the modules of q-cochains are zero for $q < 0$) which we denote by $S^*(U,K)$.

We denote the associated sheaf of germs of singular p-cochains by $\mathcal{S}^p(M,K)$ (and by $\mathcal{S}^p_\infty(M,K)$ in the differentiable case) and retain d to denote the induced sheaf homomorphisms

(12) $$\mathcal{S}^p(M,K) \xrightarrow{d} \mathcal{S}^{p+1}(M,K).$$

Observe that $\mathcal{S}^0(M,K)$ is simply the sheaf of germs of functions on M with values in K. The constant sheaf $\mathcal{K}$ can be canonically injected into $\mathcal{S}^0(M,K)$ by sending $k \in \mathcal{K}_m$ to the germ at m of the function on M with constant value k. Thus we have a sequence

(13) $$0 \to \mathcal{K} \to \mathcal{S}^0(M,K) \xrightarrow{d} \mathcal{S}^1(M,K) \xrightarrow{d} \mathcal{S}^2(M,K) \xrightarrow{d} \cdots$$

and an analogous sequence with $\mathcal{S}^p(M,K)$ replaced by $\mathcal{S}^p_\infty(M,K)$ for $p \geq 1$. We claim that in both the continuous and the differentiable cases, the sequence (13) is a fine torsionless resolution of the constant sheaf $\mathcal{K}$. That the sheaves $\mathcal{S}^p(M,K)$ (and $\mathcal{S}^p_\infty(M,K)$ for $p \geq 1$) are all torsionless is a consequence of the fact that they are all sheaves of germs of certain types of functions with values in K, and K is an integral domain. To see that the sheaves $\mathcal{S}^p(M,K)$ (and $\mathcal{S}^p_\infty(M,K)$ for $p \geq 1$) are all fine sheaves, let a locally finite open cover $\{U_i\}$ of M be given, and take (as in 5.22) a partition of unity $\{\varphi_i\}$ subordinate to the cover $\{U_i\}$ in which the functions φ_i take values 0 or 1 only. For each i, we define an endomorphism $\tilde{l}_i$ of $S^p(U,K)$ by setting

(14) $$\tilde{l}_i(f)(\sigma) = \varphi_i(\sigma(0))f(\sigma),$$

where if $p \geq 1$, then 0 denotes the origin in $\mathbb{R}^p$. The endomorphisms $\tilde{l}_i$ commute with restrictions, and thus determine presheaf endomorphisms of $\{S^p(U,K);\rho_{U,V}\}$. Let $l_i\colon \mathcal{S}^p(M,K)\to\mathcal{S}^p(M,K)$ be the sheaf endomorphism associated with $\tilde{l}_i$. Then it follows readily that $\operatorname{supp} l_i \subset U_i$ and that $\sum l_i \equiv 1$. Thus the sheaves $\mathcal{S}^p(M,K)$ (similarly $\mathcal{S}^p_\infty(M,K)$ for $p\geq 1$) are fine sheaves. The exactness of (13) is apparent at $\mathcal{K}$ and $\mathcal{S}^0(M,K)$, and we know that $d\circ d=0$. For the exactness of (13), it remains to be shown that at each stage $\mathcal{S}^p(M,K)$ for $p\geq 1$ the image of the preceding homomorphism contains the kernel of the next. This will follow if we can prove that for "sufficiently small" open sets U, if f is a singular p-cochain on U such that $df=0$ and if $p\geq 1$, then there is a singular $(p-1)$ cochain g on U such that $f=dg$. Here we shall make use of the fact that M is a manifold and is therefore locally Euclidean. It suffices to assume that U is the open unit ball in $\mathbb{R}^{\dim M}$. To prove that if $df=0$ then there is a g such that $dg=f$, we need only find a homotopy operator (see 5.29)

$$h_p\colon S^p(U,K)\to S^{p-1}(U,K)\qquad (p\geq 1) \tag{15}$$

such that

$$d\circ h_p + h_{p+1}\circ d = \mathrm{id}. \tag{16}$$

We define h_p as follows. Let $f\in S^p(U,K)$, where U is the open unit ball in $\mathbb{R}^{\dim M}$, and let σ be a singular $(p-1)$ simplex in U. Define

$$h_p(f)(\sigma) = f(\tilde{h}_p(\sigma)) \tag{17}$$

where $\tilde{h}_p(\sigma)$ is the singular p-simplex in U which maps the origin in Δ^p to the origin in U, and which is defined for $(a_1,\ldots,a_p)\neq 0$ by

$$\tilde{h}_p(\sigma)(a_1,\ldots,a_p) = \left(\sum_{i=1}^p a_i\right)\cdot\sigma\left(a_2\Big/\sum_{i=1}^p a_i,\ldots,a_p\Big/\sum_{i=1}^p a_i\right). \tag{18}$$

Geometrically, $\tilde{h}_p(\sigma)$ is the cone in U obtained by joining σ radially to the origin:

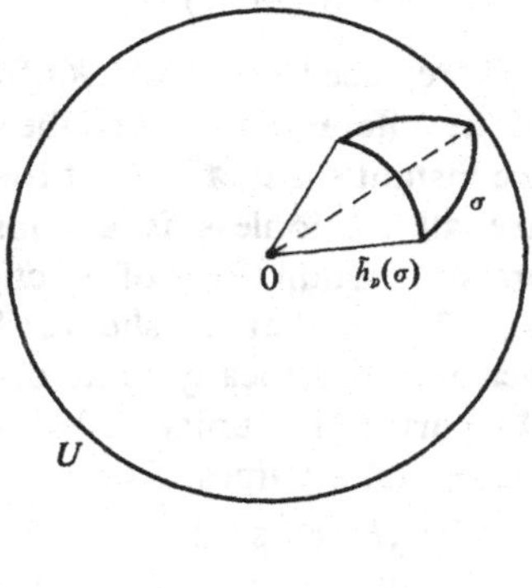

$\tilde{h}_p$ extends to a homomorphism $S_{p-1}(U)\to S_p(U)$. It follows from (18) and the definition 4.6(4) of the boundary operator ∂ that

$$\mathrm{id} = \partial\circ\tilde{h}_{p+1} + \tilde{h}_p\circ\partial \tag{19}$$

on $S_p(U)$ for $p \geq 1$. Before checking (19) in general, try the case in which $p = 1$, where it turns out that

$$(20) \qquad \sigma = (\tilde{h}_2(\sigma))^2 - (\tilde{h}_2(\sigma))^1 + (\tilde{h}_2(\sigma))^0 + \tilde{h}_1(\sigma^0) - \tilde{h}_1(\sigma^1),$$

and where the corresponding picture is

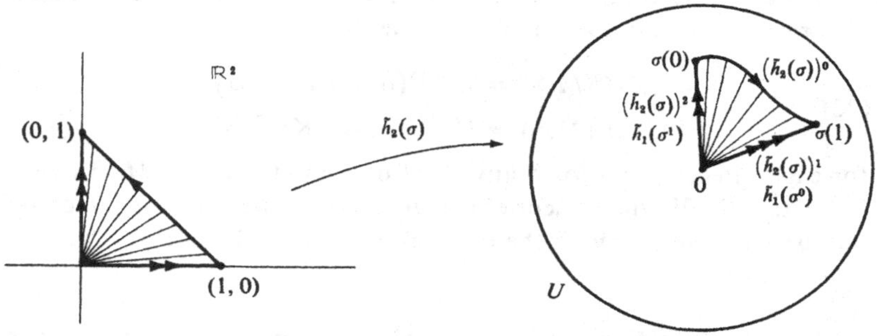

Now, (16) follows from (19) and (17) and from the definition (8) of the coboundary operator d. However, here is one point where the treatment of the continuous case does not quite suffice for the differentiable case. For if σ is a differentiable singular $(p - 1)$ simplex with $p \geq 2$, and if $\tilde{h}_p(\sigma)$ is defined as in (18), then $\tilde{h}_p(\sigma)$ will be a continuous but generally not a differentiable singular p-simplex—there are differentiability problems at the origin. For example, if σ were a differentiable singular 1-simplex in U, then for $\tilde{h}(\sigma)$ as defined in (18) to extend to be a smooth mapping on a neighborhood of the origin in $\mathbb{R}^2$, one would at least have to have the image of σ contained in a 2-dimensional plane passing through the origin in U. This, of course, may not be the case. The defect is easily remedied by means of a "smoothing" function. Let φ denote the real-valued C^∞ function on the real line defined by 1.10(3). Then $\varphi(t)$ takes values between 0 and 1, and has the value 1 for $t \geq 1$ and the value 0 for $t \leq 0$. Then if $p \geq 2$, and if σ is a differentiable $(p - 1)$ simplex, we define $\tilde{h}_p(\sigma)$ by

$$(21) \qquad \tilde{h}_p(\sigma)(a_1, \ldots, a_p) = \varphi\left(\sum_{i=1}^{p} a_i\right) \cdot \sigma\left(a_2\Big/\sum_{i=1}^{p} a_i, \ldots, a_p\Big/\sum_{i=1}^{p} a_i\right),$$

where, as in (18), we assume that $\tilde{h}_p(\sigma)$ maps the origin in Δ^p to the origin in U. Now extend σ to be differentiable on all of $\mathbb{R}^{p-1}$ so that σ and each of its derivatives is bounded. Then $\tilde{h}_p(\sigma)$is defined on all of $\mathbb{R}^p$, providing that we agree that $\tilde{h}_p(\sigma)$ maps any point where $\sum_{i=1}^{p} a_i = 0$ to the origin in U. Moreover, $\tilde{h}_p(\sigma)$ is differentiable of class C^∞ on all of $\mathbb{R}^p$. The only points where problems arise are those for which $\sum_{i=1}^{p} a_i = 0$; and since $\varphi(t)$ and all of its derivatives vanish faster than any polynomial in t as $t \to 0$, and since

σ and each of its derivatives is bounded on $\mathbb{R}^{p-1}$, it follows that all derivatives of $\tilde{h}_p(\sigma)$ of all orders exist, and are continuous, and are zero at points where $\sum_{i=1}^{p} a_i = 0$. This completes the proof that (13) is a fine torsionless resolution of the constant sheaf $\mathcal{K}$, both for the continuous and for the differentiable singular theories.

Thus both resolutions (13), for the continuous and the differentiable cases, give rise, as in 5.20, to cohomology theories if we set

$$H^q(M,\mathcal{S}) = H^q(\Gamma(\mathcal{S}^*(M,K) \otimes \mathcal{S})), \tag{22}$$
$$H^q(M,\mathcal{S}) = H^q(\Gamma(\mathcal{S}^*_\infty(M,K) \otimes \mathcal{S}))$$

for each integer q and for $\mathcal{S}$ any sheaf of K-modules over M. In view of the corollary of 5.23, these theories are uniquely isomorphic and, moreover, are uniquely isomorphic with the theories 5.26(11) and 5.30(1).

5.32 Let G be a K-module. We let $S^p(U,G)$ denote the K-module consisting of functions which assign to each singular p-simplex in U an element of G. Similarly, we may replace K by G and $\mathcal{K}$ by the constant sheaf $\mathcal{G}$ in the constructions 5.31(4)–(13).

The *classical singular cohomology groups of M* with coefficients in a K-module G are defined in the continuous and differentiable cases by:

$$H^q_\Delta(M;G) = H^q(S^*(M,G)), \tag{1}$$
$$H^q_{\Delta\infty}(M;G) = H^q(S^*_\infty(M,G)).$$

We shall now show that the classical cohomology groups (actually K-modules) are canonically isomorphic with the sheaf cohomology modules $H^q(M,\mathcal{G})$. It follows from Proposition 5.27 that

$$0 \to S^*_0(M,G) \to S^*(M,G) \to \Gamma(\mathcal{S}^*(M,G)) \to 0 \tag{2}$$

is a short exact sequence of cochain complexes (with a similar sequence in the differentiable case). Thus if we prove that

$$H^q(S^*_0(M,G)) = 0 \quad \text{for all} \quad q \tag{3}$$

(also in the differentiable case), then it follows from the long exact sequence 5.17(2) that there are canonical isomorphisms

$$H^q(S^*(M,G)) \cong H^q(\Gamma(\mathcal{S}^*(M,G))), \tag{4}$$
$$H^q(S^*_\infty(M,G)) \cong H^q(\Gamma(\mathcal{S}^*_\infty(M,G))).$$

Thus it follows from (1) and (4), and from 5.25 (applied to the fine resolutions of $\mathcal{G}$ obtained by replacing K by G and $\mathcal{K}$ by $\mathcal{G}$ in 5.31(13)) that we have canonical isomorphisms

$$H^q_\Delta(M;G) \cong H^q(M,\mathcal{G}) \cong H^q_{\Delta\infty}(M;G). \tag{5}$$

The proof of (3) for the continuous singular theory goes as follows. (The differentiable case is identical.) It is trivially satisfied for $q < 0$ since in this range the modules of the cochain complex $S_0^*(M,G)$ are all zero. The module $S_0^0(M,G)$ is also zero since the presheaf $\{S^0(U,G);\rho_{U,V}\}$ is complete. Thus $H^0(S_0^*(M,G)) = 0$. It remains to prove (3) for $q \geq 1$. Let $\mathfrak{U} = \{U_i\}$ be an arbitrary open cover of M. Let $S_{\mathfrak{U}}^*(M,G)$ be the cochain complex consisting of modules $S_{\mathfrak{U}}^p(M,G)$ of singular cochains f, with values in G, defined only on "$\mathfrak{U}$-small" singular p-simplices, that is, defined only on those singular p-simplices whose ranges lie in elements of $\mathfrak{U}$. Each element of $S^p(M,G)$ determines an element of $S_{\mathfrak{U}}^p(M,G)$ by restriction to $\mathfrak{U}$-small simplices. The restriction homomorphisms $j_{\mathfrak{U}}: S^p(M,G) \to S_{\mathfrak{U}}^p(M,G)$ yield a surjective cochain map

$$j_{\mathfrak{U}}: S^*(M,G) \to S_{\mathfrak{U}}^*(M,G). \tag{6}$$

The kernels of the homomorphisms $j_{\mathfrak{U}}$ form a cochain complex $K_{\mathfrak{U}}^*$ such that

$$0 \to K_{\mathfrak{U}}^* \to S^*(M,G) \to S_{\mathfrak{U}}^*(M,G) \to 0 \tag{7}$$

is an exact sequence of cochain complexes. The key ingredient in the proof of (3) is the fact that the cochain map $j_{\mathfrak{U}}$ induces isomorphisms in cohomology. Let us assume this for the moment. It follows from the long exact cohomology sequence for (7) that

$$H^q(K_{\mathfrak{U}}^*) = 0 \quad \text{for all } q. \tag{8}$$

So now let $q \geq 1$, and let f be a cocycle in $S_0^q(M,G)$; that is, $df = 0$. Then, by the definition of $S_0^q(M,G)$, there is an open cover $\mathfrak{U}$ of M consisting of sufficiently small open sets so that $f \in K_{\mathfrak{U}}^q$. It follows from (8) that there exists $g \in K_{\mathfrak{U}}^{q-1} \subset S_0^{q-1}(M,G)$ such that $dg = f$, which proves (3).

Now we return to the proof that the cochain map $j_{\mathfrak{U}}$ induces isomorphisms of the cohomology. Since we will work with a fixed cover $\mathfrak{U}$ of M, we drop the subscript $\mathfrak{U}$ from $j_{\mathfrak{U}}$. To prove that j: $S^*(M,G) \to S_{\mathfrak{U}}^*(M,G)$ induces isomorphisms of the cohomology modules, we shall construct a cochain map

$$k: S_{\mathfrak{U}}^*(M,G) \to S^*(M,G) \tag{9}$$

such that

$$j \circ k = \text{id}, \tag{10}$$

whence the cochain map j must induce surjections of the cohomology modules, and such that there exist homotopy operators h_p: $S^p(M,G) \to S^{p-1}(M,G)$ for all p such that

$$h_{p+1} \circ d + d \circ h_p = \text{id} - k_p \circ j_p, \tag{11}$$

whence $k \circ j$ induces the identity on cohomology, which implies that j must induce injections. Hence it will follow from (10) and (11) that j induces isomorphisms on the cohomology modules. The definitions of h and k require a few preliminary constructions. The details become a little technical, but the general idea is this. Let $f \in S^p_{\mathfrak{U}}(M,G)$. Thus f is defined only on $\mathfrak{U}$-small singular p-simplices. We want to define $k(f)$ to be an element of $S^p(M,G)$; that is, $k(f)$ is to be defined on all singular p-simplices. We will define an operation "subdivision" to break large singular p-simplices into chains of smaller ones. By subdividing any singular p-simplex sufficiently many times, we will obtain a chain of $\mathfrak{U}$-small singular p-simplices to which we can then apply f. So we will define $k(f)$ on the large singular simplex to be f on the subdivided ones. Simplices which are already $\mathfrak{U}$-small will not be subdivided at all. The technical difficulty which arises will be due to the fact that the number of times that a singular p-simplex σ needs to be subdivided in order to obtain a chain of $\mathfrak{U}$-small simplices depends on σ.

Let $q \geq 1$. A linear p-simplex in Δ^q is a singular p-simplex of the form

$$(12) \qquad (a_1, \ldots, a_p) \mapsto \left(1 - \sum_{i=1}^{p} a_i\right) v_0 + a_1 v_1 + \cdots + a_p v_p$$

($0 \mapsto v_0$ for $p = 0$), canonically determined by the ordered sequence of points $v_0, \ldots, v_p$ in Δ^q. We shall denote such a linear simplex by $(v_0, \ldots, v_p)$. The identity map of Δ^q onto itself is a linear q-simplex in Δ^q which we shall denote simply by Δ^q. The free abelian group generated by the linear p-simplices in Δ^q we shall denote by $L_p(\Delta^q)$. The *barycenter* of a linear p-simplex $\sigma = (v_0, \ldots, v_p)$ in Δ^q is the point

$$(13) \qquad b_\sigma = \frac{1}{p+1} v_0 + \cdots + \frac{1}{p+1} v_p .$$

Given a linear p-simplex $\sigma = (v_0, \ldots, v_p)$ in Δ^q and a point $v \in \Delta^q$, we define the *join* $v\sigma$ of v and σ to be the linear $(p + 1)$ simplex $(v, v_0, \ldots, v_p)$ in Δ^q. The join operation extends by linearity to $L_p(\Delta^q)$. A direct calculation shows that if $p \geq 1$, then

$$(14) \qquad \partial(v\sigma) = \sigma - v(\partial\sigma).$$

We define subdivision homomorphisms $\mathrm{Sd}\colon L_p(\Delta^q) \to L_p(\Delta^q)$ by setting $\mathrm{Sd} = \mathrm{id}$ for $p = 0$ and by setting

$$(15) \qquad \mathrm{Sd}(\sigma) = b_\sigma \mathrm{Sd}(\partial\sigma)$$

for σ a linear p-simplex in Δ^q with $p \geq 1$, and extending linearly to $L_p(\Delta^q)$. We define homomorphisms $R\colon L_p(\Delta^q) \to L_{p+1}(\Delta^q)$ by setting $R = 0$ for $p = 0$ and by setting

$$(16) \qquad R(\sigma) = b_\sigma(\sigma - \mathrm{Sd}(\sigma) - R(\partial\sigma))$$

for σ a linear p-simplex in Δ^q with $p \geq 1$, and extending linearly to $L_p(\Delta^q)$. It follows from (13), (14), (15), and (16), using an elementary induction argument, that

$$\partial \circ \mathrm{Sd} = \mathrm{Sd} \circ \partial,$$
(17)
$$\partial \circ R + R \circ \partial = \mathrm{id} - \mathrm{Sd}$$

on $L_p(\Delta^q)$ with $p \geq 1$. (The collection of modules $L_p(\Delta^q)$ and homomorphisms $\partial: L_p(\Delta^q) \to L_{p-1}(\Delta^q)$ form what is called a *chain complex* in contrast with a cochain complex 5.16(1) in which the homomorphisms go the other direction. The first formula in (17) says that subdivision Sd is a chain map of this chain complex with itself. The second formula in (17) shows that R is a homotopy operator for the chain maps id and Sd, from which it follows that Sd induces the identity map on the *homology* groups ($\ker \partial_p / \operatorname{Im} \partial_{p+1}$) of this chain complex.)

We define homomorphisms $\mathrm{Sd}: S_p(U) \to S_p(U)$ and $R: S_p(U) \to S_{p+1}(U)$ for $p \geq 0$ by setting

$$\mathrm{Sd}(\sigma) = \sigma \circ \mathrm{Sd}(\Delta^p),$$
(18)
$$R(\sigma) = \sigma \circ R(\Delta^p)$$

for σ a singular p-simplex in U, and then extending linearly to $S_p(U)$. It follows easily that formulas (17) hold on $S_p(U)$ for $p \geq 1$.

Now let $\sigma = (v_0, \ldots, v_p)$ be a linear p-simplex in Δ^q. Then the diameter of each simplex of $\mathrm{Sd}(\sigma)$ is at most $p/(p+1)$ times the diameter of σ, for the diameter of a linear simplex σ is the maximum distance between any two of the vertices v_i. Of any two vertices in a simplex in $\mathrm{Sd}(\sigma)$, at least one must be of the form $[1/(k+1)](v_{i_0} + \cdots + v_{i_k})$ for $1 \leq k \leq p$. The distance from this vertex to the other vertex is less than or equal to the distance to some v_j, and

$$\left| \frac{1}{k+1}(v_{i_0} + \cdots + v_{i_k}) - v_j \right|$$
$$= \frac{1}{k+1} \left| \sum_{l=0}^{k} (v_{i_l} - v_j) \right| \leq \frac{k}{k+1} \operatorname{diam} \sigma \leq \frac{p}{p+1} \operatorname{diam} \sigma.$$

We are now ready to construct the maps h_p and k_p. Let σ be a singular p-simplex in M. The open cover $\sigma^{-1}(\mathfrak{U})$ of Δ^p has a Lebesgue number δ [26, p. 122]. It follows that for s large enough, the diameter of each simplex in $(\mathrm{Sd})^s(\Delta^p)$ is less than δ where $(\mathrm{Sd})^s$ is the s-fold composition of Sd with itself. Thus each singular p-simplex in the chain $(\mathrm{Sd})^s(\sigma)$ lies in an element of the cover $\mathfrak{U}$. Now let $s(\sigma)$ be the smallest of those positive integers $s \geq 0$ for which each simplex of $(\mathrm{Sd})^s(\sigma)$ lies in an element of $\mathfrak{U}$. Then we can define homomorphisms $k_p: S^p_{\mathfrak{U}}(M,G) \to S^p(M,G)$ by setting $k_p = \mathrm{id}$ for $p \leq 0$, and for $p \geq 1$ by setting

$$(19) \quad k_p(f)(\sigma) = f\left((\mathrm{Sd})^{s(\sigma)}(\sigma) + \sum_{j=0}^{p}(-1)^j R\left(\sum_{s(\sigma^j) \leq i \leq s(\sigma)-1} (\mathrm{Sd})^i(\sigma^j)\right)\right).$$

Note that Sd raised to a negative power is to be interpreted as the zero homomorphism, and $(\mathrm{Sd})^0 = \mathrm{id}$. We also define homomorphisms $h_p : S^p(M,G) \to S^{p-1}(M,G)$ by setting $h_p = 0$ for $p \leq 1$ and

$$h_p(f)(\sigma) = f\left(R\left(\sum_{0 \leq i \leq s(\sigma)-1} (\mathrm{Sd})^i(\sigma)\right)\right) \tag{20}$$

for $p \geq 2$. That (10) holds is obvious, and (11) follows immediately from a straightforward calculation. Finally, the fact that the homomorphisms k_p yield a cochain map (9) follows from (11). For from (11) we obtain

$$d \circ k_p \circ j_p = k_{p+1} \circ j_{p+1} \circ d. \tag{21}$$

But j is a cochain map. Thus $d \circ k_p \circ j_p = k_{p+1} \circ d \circ j_p$. Since j is surjective, it follows that $d \circ k_p = k_{p+1} \circ d$. This completes the proof that j induces isomorphisms in cohomology.

Čech Cohomology

5.33 In the Alexander-Spanier, de Rham, and singular cases we obtained a sheaf cohomology theory by exhibiting explicit fine torsionless resolutions of a constant sheaf, and we then proved that there were canonical isomorphisms of the classical cohomology modules with the sheaf cohomology modules. The Čech theory, by contrast, arises from a direct construction that does not involve finding a fine torsionless resolution of $\mathscr{K}$. We will define modules $\check{H}^q(M,\mathcal{S})$ for $\mathcal{S}$ a sheaf of K-modules over M, and show that the axioms 5.18 for a cohomology theory are satisfied. We again direct your attention to the comment in the introduction to this chapter that for much of the chapter, M need not be a differentiable manifold. In particular, M in this section need only be a paracompact Hausdorff space.

Let $\mathfrak{U} = \{U_\alpha\}$ be an open cover of M. A collection $(U_0, \ldots, U_q)$ of members of the cover such that $U_0 \cap \cdots \cap U_q \neq \varnothing$ will be called a q-simplex. If $\sigma = (U_0, \ldots, U_q)$ is a q-simplex, its support $|\sigma|$ is by definition

$$|\sigma| = U_0 \cap \cdots \cap U_q. \tag{1}$$

The ith face of a q-simplex $\sigma = (U_0, \ldots, U_q)$ is the $(q-1)$ simplex $\sigma^i = (U_0, \ldots, U_{i-1}, U_{i+1}, \ldots, U_q)$. Let $C^q(\mathfrak{U},\mathcal{S})$, for $q \geq 0$, be the K-module consisting of functions which assign to each q-simplex σ an element of $\Gamma(\mathcal{S},|\sigma|)$, and let $C^q(\mathfrak{U},\mathcal{S}) = 0$ for $q < 0$. Elements of $C^q(\mathfrak{U},\mathcal{S})$ are called q-cochains. With a coboundary homomorphism

$$d\colon C^q(\mathfrak{U},\mathcal{S}) \to C^{q+1}(\mathfrak{U},\mathcal{S}) \tag{2}$$

defined by

$$df(\sigma) = \sum_{i=0}^{q} (-1)^i \rho_{|\sigma|,|\sigma^i|} f(\sigma^i), \tag{3}$$

one obtains a cochain complex $C^*(\mathfrak{U},\mathcal{S})$ whose qth cohomology module, which we denote by $\check{H}^q(\mathfrak{U},\mathcal{S})$, is called the qth *Čech cohomology module of* $(M,\mathfrak{U})$ *with coefficients in* $\mathcal{S}$. A homomorphism $\mathcal{S} \to \mathcal{S}'$ induces by composition a cochain map $C^*(\mathfrak{U},\mathcal{S}) \to C^*(\mathfrak{U},\mathcal{S}')$ and thus induces homomorphisms

$$\check{H}^q(\mathfrak{U},\mathcal{S}) \to \check{H}^q(\mathfrak{U},\mathcal{S}') \tag{4}$$

for each q.

A cochain f belongs to $C^0(\mathfrak{U},\mathcal{S})$ if and only if f assigns to each open set $U_\alpha \in \mathfrak{U}$ a section of $\mathcal{S}$ over U_α. f is a 0-cocycle, that is $df = 0$, if and only if

$$0 = df(U_{\alpha_0},U_{\alpha_1}) = \rho_{U_{\alpha_0}\cap U_{\alpha_1},U_{\alpha_1}} f(U_{\alpha_1}) - \rho_{U_{\alpha_0}\cap U_{\alpha_1},U_{\alpha_0}} f(U_{\alpha_0}) \tag{5}$$

for every 1-simplex $(U_{\alpha_0},U_{\alpha_1})$. Thus f is a 0-cocycle if and only if f defines a global section of $\mathcal{S}$ over M. Thus

$$\check{H}^0(\mathfrak{U},\mathcal{S}) = \Gamma(\mathcal{S}). \tag{6}$$

We now consider the effect of refining the cover $\mathfrak{U}$. If a cover $\mathfrak{V}$ is a refinement of the cover $\mathfrak{U}$, then there exists a map $\mu\colon \mathfrak{V} \to \mathfrak{U}$ such that $V \subseteq \mu(V)$ for each $V \in \mathfrak{V}$. If $\sigma = (V_0, \ldots, V_q)$ is a q-simplex of the cover $\mathfrak{V}$, then we let $\mu(\sigma)$ denote the q-simplex $(\mu(V_0), \ldots, \mu(V_q))$ of the cover $\mathfrak{U}$. Now, μ induces a cochain map $\mu\colon C^*(\mathfrak{U},\mathcal{S}) \to C^*(\mathfrak{V},\mathcal{S})$ if we set

$$\mu_q(f)(\sigma) = \rho_{|\sigma|,|\mu(\sigma)|} f(\mu(\sigma)) \tag{7}$$

for $f \in C^q(\mathfrak{U},\mathcal{S})$ and for σ a q-simplex of the cover $\mathfrak{V}$. This cochain map induces homomorphisms

$$\mu_q^*\colon \check{H}^q(\mathfrak{U},\mathcal{S}) \to \check{H}^q(\mathfrak{V},\mathcal{S}) \tag{8}$$

of the cohomology modules. We claim that if μ and τ are both refining maps of $\mathfrak{V}$ into $\mathfrak{U}$, then $\mu_q^* = \tau_q^*$ for each q. As usual, we prove this by finding a homotopy operator. If $\sigma = (V_0, \ldots, V_{q-1})$ is a $(q-1)$ simplex of the cover $\mathfrak{V}$, we let

$$\tilde{\sigma}_j = (\mu(V_0), \ldots, \mu(V_j), \tau(V_j), \ldots, \tau(V_{q-1})). \tag{9}$$

We then define homomorphisms $h_q\colon C^q(\mathfrak{U},\mathcal{S}) \to C^{q-1}(\mathfrak{V},\mathcal{S})$ by setting

$$h_q(f)(\sigma) = \sum_{j=0}^{q-1} (-1)^j \rho_{|\sigma|,|\tilde{\sigma}_j|} f(\tilde{\sigma}_j). \tag{10}$$

From a straightforward calculation, one obtains

$$h_{q+1} \circ d + d \circ h_q = \tau_q - \mu_q, \tag{11}$$

from which it follows that $\mu_q^* = \tau_q^*$ for each integer q. Thus if $\mathfrak{V}$ is a refinement of $\mathfrak{U}$, which we shall denote by $\mathfrak{V} < \mathfrak{U}$, then there are canonical homomorphisms $\check{H}^q(\mathfrak{U},\mathcal{S}) \to \check{H}^q(\mathfrak{V},\mathcal{S})$. Since the set of coverings of M forms a directed set under the relation $<$ of refinement, and since if $\mathfrak{S} < \mathfrak{V} < \mathfrak{U}$ then the homomorphism $\check{H}^q(\mathfrak{U},\mathcal{S}) \to \check{H}^q(\mathfrak{S},\mathcal{S})$ is the composition of the homomorphisms $\check{H}^q(\mathfrak{U},\mathcal{S}) \to \check{H}^q(\mathfrak{V},\mathcal{S})$ and $\check{H}^q(\mathfrak{V},\mathcal{S}) \to \check{H}^q(\mathfrak{S},\mathcal{S})$—then the collection of modules $\check{H}^q(\mathfrak{U},\mathcal{S})$ and refinement homomorphisms forms a direct system. Thus one can form the direct limit module

$$\check{H}^q(M,\mathcal{S}) = \operatorname*{dir\,lim}_{\mathfrak{U}} \check{H}^q(\mathfrak{U},\mathcal{S}). \tag{12}$$

(Construction is completely analogous to the construction of the module $\mathcal{S}_m$ from the modules S_U in 5.6.) $\check{H}^q(M,\mathcal{S})$ is the qth *Čech cohomology module for M with coefficients in the sheaf of K-modules* $\mathcal{S}$. *The classical qth Čech cohomology module* $\check{H}^q(M;G)$ *of M with coefficients in a K-module G* is by definition $\check{H}^q(M,\mathcal{G})$, where $\mathcal{G}$ is the constant sheaf $M \times G$.

We shall now show that the Čech cohomology modules (12) give a sheaf cohomology theory in the sense of 5.18. It is apparent that $\check{H}^q(M,\mathcal{S}) = 0$ for $q < 0$, and it follows from (6) that $\check{H}^0(M,\mathcal{S}) = \Gamma(\mathcal{S})$; thus 5.18(a) is satisfied. The homomorphisms (4) induced from a homomorphism $\mathcal{S} \to \mathcal{S}'$ commute with the refinement homomorphisms (8) and thus induce homomorphisms

$$\check{H}^q(M,\mathcal{S}) \to \check{H}^q(M,\mathcal{S}'). \tag{13}$$

It is immediate that these homomorphisms satisfy 5.18(d) and (e).

Consider now a fine sheaf $\mathcal{S}$ and an integer $q > 0$. Since every cover of M has a locally finite refinement, in order to prove that $\check{H}^q(M,\mathcal{S}) = 0$, it suffices to prove that $\check{H}^q(\mathfrak{U},\mathcal{S}) = 0$ for each locally finite open cover $\mathfrak{U}$. Let $\{l_\alpha\}$ be a partition of unity for $\mathcal{S}$ subordinate to the locally finite open cover $\mathfrak{U} = \{U_\alpha\}$ of M. We shall define homomorphisms $h_p\colon C^p(\mathfrak{U},\mathcal{S}) \to C^{p-1}(\mathfrak{U},\mathcal{S})$ for each $p \geq 1$. Let $f \in C^p(\mathfrak{U},\mathcal{S})$, and let $\sigma = (U_0, \ldots, U_{p-1})$ be a $(p-1)$ simplex of the cover $\mathfrak{U}$. Then $l_\alpha \circ \big(f(U_\alpha, U_0, \ldots, U_{p-1})\big)$ has support in $U_\alpha \cap U_0 \cap \cdots \cap U_{p-1}$, so we can extend $l_\alpha \circ \big(f(U_\alpha, U_0, \ldots, U_{p-1})\big)$ to a continuous section of $\mathcal{S}$ over $U_0 \cap \cdots \cap U_{p-1}$ by extending it to be zero outside $U_\alpha \cap U_0 \cap \cdots \cap U_{p-1}$. We consider $l_\alpha \circ \big(f(U_\alpha, U_0, \ldots, U_{p-1})\big)$ as this section over $U_0 \cap \cdots \cap U_{p-1}$. Now define

$$h_p(f)(\sigma) = \sum_\alpha l_\alpha \circ \big(f(U_\alpha, U_0, \ldots, U_{p-1})\big). \tag{14}$$

It follows that

$$d \circ h_p + h_{p+1} \circ d = \text{id} \qquad \text{for } p \geq 1. \tag{15}$$

So if f is a q-cocycle with $q > 0$, then there is a $(q-1)$ cochain g, namely $g = h_q(f)$, such that $dg = f$. It follows that $\check{H}^q(\mathfrak{U},\mathcal{S}) = 0$. Thus axiom 5.18(b) is satisfied.

A short exact sheaf sequence $0 \to \mathcal{S}' \to \mathcal{S} \to \mathcal{S}'' \to 0$ induces exact sequences

$$(16) \qquad 0 \to C^q(\mathfrak{U},\mathcal{S}') \to C^q(\mathfrak{U},\mathcal{S}) \to C^q(\mathfrak{U},\mathcal{S}'').$$

Let $\bar{C}^q(\mathfrak{U},\mathcal{S}'')$ be the image of $C^q(\mathfrak{U},\mathcal{S})$ in $C^q(\mathfrak{U},\mathcal{S}'')$. Then the short exact sequences

$$(17) \qquad 0 \to C^q(\mathfrak{U},\mathcal{S}') \to C^q(\mathfrak{U},\mathcal{S}) \to \bar{C}^q(\mathfrak{U},\mathcal{S}'') \to 0$$

yield a short exact sequence of cochain complexes

$$(18) \qquad 0 \to C^*(\mathfrak{U},\mathcal{S}') \to C^*(\mathfrak{U},\mathcal{S}) \to \bar{C}^*(\mathfrak{U},\mathcal{S}'') \to 0.$$

A refining map $\mu\colon \mathfrak{B} \to \mathfrak{U}$ induces a homomorphism of short exact sequences of cochain complexes

$$(19) \qquad \begin{array}{ccccccccc} 0 \to & C^*(\mathfrak{U},\mathcal{S}') & \to & C^*(\mathfrak{U},\mathcal{S}) & \to & \bar{C}^*(\mathfrak{U},\mathcal{S}'') & \to 0 \\ & \downarrow{\scriptstyle \mu} & & \downarrow{\scriptstyle \mu} & & \downarrow{\scriptstyle \mu} & \\ 0 \to & C^*(\mathfrak{B},\mathcal{S}') & \to & C^*(\mathfrak{B},\mathcal{S}) & \to & \bar{C}^*(\mathfrak{B},\mathcal{S}'') & \to 0 \end{array}$$

and thus by 5.17 induces a commutative diagram of the associated cohomology sequences

(20)

$$\begin{array}{ccccccccccc} \cdots \to & \bar{H}^{q-1}(\mathfrak{U},\mathcal{S}'') & \xrightarrow{\partial} & \check{H}^q(\mathfrak{U},\mathcal{S}') & \to & \check{H}^q(\mathfrak{U},\mathcal{S}) & \to & \bar{H}^q(\mathfrak{U},\mathcal{S}'') & \xrightarrow{\partial} & \check{H}^{q+1}(\mathfrak{U},\mathcal{S}') & \to \cdots \\ & \downarrow{\scriptstyle \mu^*_{q-1}} & & \downarrow{\scriptstyle \mu^*_q} & & \downarrow{\scriptstyle \mu^*_q} & & \downarrow{\scriptstyle \mu^*_q} & & \downarrow{\scriptstyle \mu^*_{q+1}} & \\ \cdots \to & \bar{H}^{q-1}(\mathfrak{B},\mathcal{S}'') & \xrightarrow{\partial} & \check{H}^q(\mathfrak{B},\mathcal{S}') & \to & \check{H}^q(\mathfrak{B},\mathcal{S}) & \to & \bar{H}^q(\mathfrak{B},\mathcal{S}'') & \xrightarrow{\partial} & \check{H}^{q+1}(\mathfrak{B},\mathcal{S}') & \to \cdots. \end{array}$$

On passing to the direct limit, we obtain a long exact sequence

$$(21) \qquad \cdots \longrightarrow \bar{H}^{q-1}(M,\mathcal{S}'') \xrightarrow{\partial} \check{H}^q(M,\mathcal{S}') \\ \longrightarrow \check{H}^q(M,\mathcal{S}) \longrightarrow \bar{H}^q(M,\mathcal{S}'') \xrightarrow{\partial} \check{H}^{q+1}(M,\mathcal{S}') \longrightarrow \cdots.$$

We shall now prove that the inclusion cochain map $\bar{C}^*(\mathfrak{U},\mathcal{S}'') \to C^*(\mathfrak{U},\mathcal{S}'')$ on passage to the direct limit in cohomology induces isomorphisms

$$(22) \qquad \bar{H}^q(M,\mathcal{S}'') \xrightarrow{\cong} \check{H}^q(M,\mathcal{S}'').$$

These isomorphisms, together with (21), will then yield a long exact sequence

$$(23) \qquad \cdots \longrightarrow \check{H}^{q-1}(M,\mathcal{S}'') \xrightarrow{\partial} \check{H}^q(M,\mathcal{S}') \\ \longrightarrow \check{H}^q(M,\mathcal{S}) \longrightarrow \check{H}^q(M,\mathcal{S}'') \xrightarrow{\partial} \check{H}^{q+1}(M,\mathcal{S}') \longrightarrow \cdots,$$

which proves axiom 5.18(c). The quotient modules

(24) $$\tilde{C}^q(\mathfrak{U}) = C^q(\mathfrak{U},\mathcal{S}'')/(\bar{C}^q(\mathfrak{U},\mathcal{S}'')$$

together with the induced coboundary homomorphisms form a cochain complex $\tilde{C}^*(\mathfrak{U})$ such that

(25) $$0 \to \bar{C}^*(\mathfrak{U},\mathcal{S}'') \to C^*(\mathfrak{U},\mathcal{S}'') \to \tilde{C}^*(\mathfrak{U}) \to 0$$

is exact. It follows from the long exact cohomology sequence for (25), upon passage to the direct limit, that (22) will follow if we prove that the direct limit of the modules $H^q(\tilde{C}^*(\mathfrak{U}))$ is 0 for all q. So let $\mathfrak{U}$ be a locally finite cover of M. That the direct limit of the modules $H^q(\tilde{C}^*(\mathfrak{U}))$ is 0 will certainly follow if we prove that for an arbitrary $f \in C^q(\mathfrak{U},\mathcal{S}'')$ there is a refinement $\mu: \mathfrak{V} \to \mathfrak{U}$ such that $\mu_q(f) \in \bar{C}^q(\mathfrak{V},\mathcal{S}'')$. Choose a refinement $\mathfrak{O} = \{O_\alpha\}$ of $\mathfrak{U} = \{U_\alpha\}$ such that $\overline{O_\alpha} \subset U_\alpha$ for each α. For each $p \in M$ choose a neighborhood V_p such that

(a) $V_p \subset O_\alpha$ for some α.

(b) If $V_p \cap O_\alpha \neq \varnothing$, then $V_p \subset U_\alpha$.

(c) V_p lies in the intersection of the U_α containing p.

(d) If σ is a q-simplex of the cover $\mathfrak{U}$, and $p \in |\sigma|$ (so $V_p \subset |\sigma|$), then $\rho_{V_p,|\sigma|}f(\sigma)$ is the image of a section of $\mathcal{S}$ over V_p.

It is possible to satisfy (d) since there are only finitely many q-simplices of the cover $\mathfrak{U}$ which contain p. Now let $\mathfrak{V}$ be the cover $\{V_p\}$, and for every p choose $O_p \in \mathfrak{O}$ and $U_p \in \mathfrak{U}$ such that $V_p \subset O_p \subset U_p$. Thus we have a refinement $\mu: \mathfrak{V} \to \mathfrak{U}$. Now let $\sigma = (V_{p_0}, \ldots, V_{p_q})$ be a q-simplex of the cover $\mathfrak{V}$. Then $V_{p_0} \cap O_{p_i} \neq \varnothing$, $0 \leq i \leq q$, so by (b), $V_{p_0} \subset U_{p_i}$. Thus $V_{p_0} \subset U_{p_0} \cap \cdots \cap U_{p_q} = |\mu(\sigma)|$. Therefore

$$\begin{aligned}\mu_q(f)(\sigma) &= \rho_{|\sigma|,|\mu(\sigma)|}f(U_{p_0}, \ldots, U_{p_q})\\ &= \rho_{|\sigma|,V_{p_0}} \circ \rho_{V_{p_0},|\mu(\sigma)|}f(U_{p_0}, \ldots, U_{p_q});\end{aligned}$$

hence by condition (d), $\mu_q(f) \in \bar{C}^q(\mathfrak{V},\mathcal{S}'')$.

Finally, axiom 5.18(f) follows readily from the above construction of (23) by making use of 5.17(3).

Thus the Čech cohomology satisfies the axioms 5.18 for a sheaf cohomology theory.

THE DE RHAM THEOREM

5.34 Convention Since according to the corollary of 5.23, any two sheaf cohomology theories on M are uniquely isomorphic, *we shall consider them as identified* via their unique isomorphisms, and shall henceforth use $H^p(M,\mathcal{S})$ to denote *the* pth cohomology module of M with coefficients in the sheaf $\mathcal{S}$. With this convention,

(1)
$$H^p(M,\mathcal{S}) = \check{H}^p(M,\mathcal{S});$$

and given any fine torsionless resolution

(2)
$$0 \to \mathscr{K} \to \mathscr{C}_0 \to \mathscr{C}_1 \to \mathscr{C}_2 \to \cdots,$$

we have

(3)
$$H^p(M,\mathcal{S}) = H^p(\Gamma(\mathscr{C}^* \otimes \mathcal{S})).$$

5.35 In the preceding sections we have obtained canonical isomorphisms

(1)
$$H^p_{A-S}(M;G) \cong H^p_\Delta(M;G) \cong H^p_{\Delta^\infty}(M;G) \cong \check{H}^p(M;G)$$

of the classical Alexander-Spanier, singular, differentiable singular, and Čech cohomology modules of a differentiable manifold M with coefficients in a K-module G over a principal ideal domain K. We have seen that each of these is canonically isomorphic with the sheaf cohomology module $H^p(M,\mathscr{G})$ with coefficients in the constant sheaf $\mathscr{G}$. If we take K to be the field of real numbers, we can add the de Rham cohomology group $H^p_{\text{de R}}(M)$ to the above list of isomorphisms. In particular, we have canonical isomorphisms

(2)
$$H^p_{\text{de R}}(M) \cong H^p(M,\mathscr{R}) \cong H^p_{\Delta^\infty}(M;\mathbb{R}).$$

We shall now prove, with the help of 5.24, that the explicit homomorphism from the de Rham to the differentiable singular cohomology theory obtained from integration of forms over differentiable singular simplices yields the canonical isomorphisms (2).

We define homomorphisms

(3)
$$k_p\colon E^p(M) \to S^p_\infty(M,\mathbb{R})$$

for each integer $p \geq 0$ by setting

(4)
$$k_p(\omega)(\sigma) = \int_\sigma \omega$$

for each differentiable p-form ω on M and differentiable singular p-simplex σ in M. It is an immediate consequence of Stokes' theorem 4.7 that the homomorphisms k_p induce a cochain map

(5)
$$k\colon E^*(M) \to S^*_\infty(M,\mathbb{R}).$$

Let

$$(6) \qquad k_p^*\colon H^p_{\mathrm{de\,R}}(M) \to H^p_{\Delta\infty}(M;\mathbb{R})$$

denote the induced homomorphism of the cohomology modules (real vector spaces). k_p^* is called the *de Rham homomorphism.*

5.36 The de Rham Theorem *The de Rham homomorphism k_p^* is the canonical isomorphism* 5.35(2) *for each integer p.*

PROOF The homomorphisms 5.35(3) can be defined for arbitrary open sets in M, and yield presheaf homomorphisms

$$\{E^p(U);\rho_{U,V}\} \xrightarrow{k_p} \{S^p_\infty(U,\mathbb{R});\rho_{U,V}\}$$

which commute, by Stokes' theorem, with the coboundary homomorphisms 5.28(2) and 5.31(10). Thus the induced homomorphisms of the associated sheaves form a commutative diagram:

$$(1) \qquad \begin{array}{ccccccccc} 0 \to \mathscr{R} & \longrightarrow & \mathscr{E}^0(M) & \longrightarrow & \mathscr{E}^1(M) & \longrightarrow & \mathscr{E}^2(M) & \to \cdots \\ \downarrow{\scriptstyle \mathrm{id}} & & \downarrow{\scriptstyle k_0} & & \downarrow{\scriptstyle k_1} & & \downarrow{\scriptstyle k_2} & \\ 0 \to \mathscr{R} & \to & \mathcal{S}^0_\infty(M,\mathbb{R}) & \to & \mathcal{S}^1_\infty(M,\mathbb{R}) & \to & \mathcal{S}^2_\infty(M,\mathbb{R}) & \to \cdots. \end{array}$$

Consider now the following commutative diagram of cochain complexes in which the rows, according to 5.30(5) and 5.32(2), are exact:

$$(2) \qquad \begin{array}{ccccccc} & 0 \longrightarrow & E^*(M) & \xrightarrow{①} & \Gamma(\mathscr{E}^*(M)) & \longrightarrow 0 \\ & & \downarrow{\scriptstyle k} & & \downarrow{\scriptstyle ③} & \\ 0 \to S^*_{\infty,0}(M,\mathbb{R}) & \to & S^*_\infty(M,\mathbb{R}) & \xrightarrow{②} & \Gamma(\mathcal{S}^*_\infty(M,\mathbb{R})) & \to 0. \end{array}$$

The cochain map ① induces the isomorphisms $H^p_{\mathrm{de\,R}}(M) \cong H^p(M,\mathscr{R})$. We proved in 5.32(4) that ② induces the isomorphisms $H^p_{\Delta\infty}(M;\mathbb{R}) \cong H^p(M,\mathscr{R})$. That ③ induces isomorphisms on cohomology follows from the corollary of 5.23 applied to the homomorphism of sheaf cohomology theories induced according to 5.24 by the homomorphism (1) of fine torsionless resolutions of $\mathscr{R}$. Thus from the uniqueness of the isomorphism between sheaf cohomology theories, ③ induces the identity isomorphism of $H^p(M,\mathscr{R})$. It follows that k_p^* is the canonical isomorphism 5.35(2) for each integer p.

5.37 Earlier, in 4.17, we stated a slightly different version of the de Rham theorem in terms of singular homology instead of cohomology. We shall now see that there is a natural isomorphism

$$(1) \qquad H^p_{\Delta\infty}(M;\mathbb{R}) \to {}_\infty H_p(M;\mathbb{R})^*$$

which composed with k_p^* yields the isomorphism 4.17(1). (For the definition of the real differentiable singular homology group ${}_\infty H_p(M;\mathbb{R})$ and other relevant notation, the reader should review section 4.16.) The map (1) is defined as follows. If f represents a cohomology class in $H^p_{\Delta\infty}(M;\mathbb{R})$, then f can be considered as a linear function on the vector space ${}_{\infty}S_p(M;\mathbb{R})$ of real differentiable singular p-chains in M. In particular, f is a linear function on the subspace of ${}_{\infty}S_p(M;\mathbb{R})$ consisting of the p-cycles, and f vanishes on the p-boundaries in ${}_{\infty}S_p(M;\mathbb{R})$ since $df = 0$; hence f determines a linear function on the homology group ${}_{\infty}H_p(M;\mathbb{R})$. Since a real differentiable singular coboundary necessarily vanishes on all p-cycles, each cohomology class in $H^p_{\Delta^\infty}(M;\mathbb{R})$ determines a well-defined element of ${}_{\infty}H_p(M;\mathbb{R})^*$ independent of the representative f chosen. This defines the map (1). We leave it to the reader as an exercise to establish that (1) is actually an isomorphism. Clearly the composition of (1) with k_p^* is exactly the homomorphism 4.17(1). Thus 4.17(1) is an isomorphism.

5.38 Remark The singular cohomology groups $H^p_\Delta(M;\mathbb{R})$ are topological invariants; that is, homeomorphic spaces have isomorphic real singular cohomology (see Exercise 19). As a consequence of this and the isomorphisms 5.35(1) and (2), the de Rham cohomology groups are also topological invariants of a differentiable manifold.

MULTIPLICATIVE STRUCTURE

In the case in which $\mathcal{S}$ is a sheaf of K-algebras over M, we shall make the direct sum of the cohomology modules $\sum_p H^p(M,\mathcal{S})$ into an associative algebra over K. But first we need a few preliminary constructions.

5.39 Definitions Let C^* and $'C^*$ be cochain complexes. Their tensor product $C^* \otimes {'C^*}$ is the cochain complex whose rth module is the direct sum

(1) $$\sum_{p+q=r} C_p \otimes {'C_q},$$

and whose rth coboundary homomorphism is the direct sum

(2) $$\sum_{p+q=r} (d_p \otimes {'(\mathrm{id})_q} + (-1)^p(\mathrm{id})_p \otimes {'d_q}).$$

If $\sigma \in Z^p(C^*)$ and $\tau \in Z^q('C^*)$, then $\sigma \otimes \tau \in Z^{p+q}(C^* \otimes {'C^*})$; whereas if $\sigma \in B^p(C^*)$ and $\tau \in Z^q('C^*)$ (or vice versa, if $\sigma \in Z^p(C^*)$ and $\tau \in B^q('C^*)$), then $\sigma \otimes \tau \in B^{p+q}(C^* \otimes {'C^*})$. Thus there is a well-defined homomorphism

(3) $$H^p(C^*) \otimes H^q('C^*) \to H^{p+q}(C^* \otimes {'C^*}).$$

5.40 Lemma *The tensor product of two torsionless K-modules, for K a principal ideal domain, is again a torsionless K-module.*

PROOF Let $\lambda \neq 0 \in K$, and let K/λ be the K-module whose elements are $\{k/\lambda : k \in K\}$, and in which addition and multiplication by K are defined by

$$(1)\qquad \begin{aligned} k_1/\lambda + k_2/\lambda &= (k_1 + k_2)/\lambda, \\ k(k_1/\lambda) &= kk_1/\lambda. \end{aligned}$$

Let A be a K-module. We define a sequence of homomorphisms

$$(2)\qquad A \to K/\lambda \otimes A \to K \otimes A \to A.$$

The first homomorphism is defined by $a \mapsto (\lambda/\lambda) \otimes a$, the second by $\sum (k_i/\lambda) \otimes a_i \mapsto \sum k_i \otimes a_i$, and the third by $\sum k_i \otimes a_i \mapsto \sum k_i a_i$. The second homomorphism obviously has an inverse, so is an isomorphism; and the third is an isomorphism according to 5.9(2). The composition (2) sends $a \in A$ to $\lambda a \in A$. By definition, A is torsionless if and only if the composition (2) has kernel zero for each $\lambda \neq 0 \in K$. Thus A is torsionless if and only if

$$(3)\qquad 0 \to A \to (K/\lambda) \otimes A$$

is exact for each $\lambda \neq 0 \in K$. Now let A and B be torsionless K-modules, and let $\lambda \neq 0 \in K$. Then (3) is exact, and so since B is torsionless it follows from 5.14 that

$$(4)\qquad 0 \to A \otimes B \to (K/\lambda) \otimes A \otimes B$$

is exact, whence $A \otimes B$ is torsionless.

5.41 Definition The definitions of the direct sum $\mathcal{S} \oplus \mathcal{T}$ of sheaves and the direct sum $\mathcal{S} \oplus \mathcal{T} \to \mathcal{S}' \oplus \mathcal{T}'$ of sheaf homomorphisms $\mathcal{S} \to \mathcal{S}'$ and $\mathcal{T} \to \mathcal{T}'$ are obtained by replacing $\otimes$ by $\oplus$ in 5.9(3)–(8), with the deletion of 5.9(7). Observe that the direct sum of fine sheaves is a fine sheaf.

5.42 The Multiplicative Structure Consider a resolution

$$(1)\qquad 0 \to \mathcal{K} \to \mathcal{C}_0 \xrightarrow{d_0} \mathcal{C}_1 \xrightarrow{d_1} \mathcal{C}_2 \xrightarrow{d_2} \cdots.$$

By the *tensor product of the resolution* (1) *with itself* we mean the sequence

$$(2)\qquad 0 \to \mathcal{K} \to \mathcal{C}_0 \otimes \mathcal{C}_0 \to (\mathcal{C}_0 \otimes \mathcal{C}_1) \oplus (\mathcal{C}_1 \otimes \mathcal{C}_0) \to \cdots$$
$$\cdots \to \sum_{p+q=r} \mathcal{C}_p \otimes \mathcal{C}_q \to \cdots$$

in which the homomorphism $\mathcal{K} \to \mathcal{C}_0 \otimes \mathcal{C}_0$ is the composition $\mathcal{K} \cong \mathcal{K} \otimes \mathcal{K} \to \mathcal{C}_0 \otimes \mathcal{C}_0$, and in which the homomorphism whose domain is $\sum_{p+q=r} \mathcal{C}_p \otimes \mathcal{C}_q$ is the direct sum

$$(3)\qquad \sum_{p+q=r} (d_p \otimes (\mathrm{id})_q + (-1)^p (\mathrm{id})_p \otimes d_q).$$

If (1) is a fine torsionless resolution of $\mathscr{K}$, we claim that (2) is also a fine torsionless resolution of $\mathscr{K}$. That the sheaves in (2) are fine is a consequence of the fact that tensor products and direct sums of fine sheaves are also fine sheaves. From 5.40 and the observation that direct sums of torsionless K-modules are torsionless, it follows that the sheaves in (2) are torsionless. That (2) is exact, therefore a resolution, is easily seen at $\mathscr{K}$ and at $\mathscr{C}_0 \otimes \mathscr{C}_0$, and is proved elsewhere by applying the Kunneth formula to the cochain complexes obtained for each $m \in M$ from

$$(4) \qquad \cdots \to 0 \to \mathscr{C}_0 \otimes \mathscr{C}_0 \to (\mathscr{C}_0 \otimes \mathscr{C}_1) \oplus (\mathscr{C}_1 \otimes \mathscr{C}_0) \to \cdots$$
$$\cdots \to \sum_{p+q=r} \mathscr{C}_p \otimes \mathscr{C}_q \to \cdots$$

by restricting to the stalks over m. The extensive algebra necessary for a proof of the Kunneth formula, which expresses the cohomology of the tensor product of cochain complexes in terms of the cohomology of the individual complexes, would not be particularly illuminating for our purposes, so we shall simply refer the interested reader to Spanier [28, Ch. 5, §4, THEOREM 2], where the Kunneth formula is proved in detail.

Let $\mathcal{S}$ be a sheaf over M. From the resolution (2) we obtain as in 5.19(2) a cochain complex which we shall denote by $\Gamma(\mathscr{C}^* \otimes \mathscr{C}^* \otimes \mathcal{S})$. If $\mathcal{S}$ is a sheaf of algebras over M, then the natural homomorphism $\mathcal{S} \otimes \mathcal{S} \to \mathcal{S}$ determined by the multiplicative structure of the stalks induces homomorphisms

$$(5) \quad \Gamma(\mathscr{C}_p \otimes \mathcal{S}) \otimes \Gamma(\mathscr{C}_q \otimes \mathcal{S}) \to \Gamma(\mathscr{C}_p \otimes \mathscr{C}_q \otimes \mathcal{S} \otimes \mathcal{S}) \to \Gamma(\mathscr{C}_p \otimes \mathscr{C}_q \otimes \mathcal{S})$$

which in turn induce a cochain map

$$(6) \qquad \Gamma(\mathscr{C}^* \otimes \mathcal{S}) \otimes \Gamma(\mathscr{C}^* \otimes \mathcal{S}) \to \Gamma(\mathscr{C}^* \otimes \mathscr{C}^* \otimes \mathcal{S}).$$

We have, according to 5.39(3), a well-defined homomorphism

$$(7) \qquad H^p(M,\mathcal{S}) \otimes H^q(M,\mathcal{S}) \to H^{p+q}(\Gamma(\mathscr{C}^* \otimes \mathcal{S}) \otimes \Gamma(\mathscr{C}^* \otimes \mathcal{S})).$$

In the case in which $\mathcal{S}$ is a sheaf of algebras over M, the cochain map (6) induces, together with (7), a homomorphism

$$(8) \qquad H^p(M,\mathcal{S}) \otimes H^q(M,\mathcal{S}) \to H^{p+q}(M,\mathcal{S}).$$

This will define the multiplicative structure in sheaf cohomology. But first we need to show that the multiplication defined by (8) is independent of the resolution (1) with which we began. Consider another fine torsionless resolution

$$(9) \qquad 0 \to \mathscr{K} \to \tilde{\mathscr{C}}_0 \to \tilde{\mathscr{C}}_1 \to \tilde{\mathscr{C}}_2 \to \cdots.$$

By tensoring the resolution (1) with the resolution (9) (construction completely analogous to the tensor product of (1) with itself given in (2) and (3)), we obtain another fine torsionless resolution

$$(10)\quad 0 \to \mathscr{K} \to \mathscr{C}_0 \otimes \tilde{\mathscr{C}}_0 \to (\mathscr{C}_0 \otimes \tilde{\mathscr{C}}_1) \oplus (\mathscr{C}_1 \otimes \tilde{\mathscr{C}}_0) \to \cdots \to \sum_{p+q=r} \mathscr{C}_p \otimes \tilde{\mathscr{C}}_q \to \cdots.$$

A homomorphism of the resolution (1) to the resolution (10), in the sense of 5.24, is generated by the homomorphisms

$$\mathscr{C}_p \cong \mathscr{C}_p \otimes \mathscr{K} \to \mathscr{C}_p \otimes \tilde{\mathscr{C}}_0 \to \sum_{r+s=p} \mathscr{C}_r \otimes \tilde{\mathscr{C}}_s,$$

where the last map is inclusion. This homomorphism of resolutions induces a cochain map $\Gamma(\mathscr{C}^* \otimes \mathcal{S}) \to \Gamma(\mathscr{C}^* \otimes \tilde{\mathscr{C}}^* \otimes \mathcal{S})$ which, according to 5.24 and the corollary of 5.23, induces the identity map $H^q(M,\mathcal{S}) \to H^q(M,\mathcal{S})$. By applying these constructions to various resolutions and then tensoring resulting cochain complexes, one obtains a commutative diagram of cochain complexes and cochain maps

(11)

$$\begin{array}{ccc}
\Gamma(\mathscr{C}^* \otimes \mathcal{S}) \otimes \Gamma(\mathscr{C}^* \otimes \mathcal{S}) & \longrightarrow & \Gamma(\mathscr{C}^* \otimes \mathscr{C}^* \otimes \mathcal{S}) \\
\downarrow & & \downarrow \\
\Gamma(\mathscr{C}^* \otimes \tilde{\mathscr{C}}^* \otimes \mathcal{S}) \otimes \Gamma(\mathscr{C}^* \otimes \tilde{\mathscr{C}}^* \otimes \mathcal{S}) & \longrightarrow & \Gamma(\mathscr{C}^* \otimes \tilde{\mathscr{C}}^* \otimes \mathscr{C}^* \otimes \tilde{\mathscr{C}}^* \otimes \mathcal{S}) \\
\uparrow & & \uparrow \\
\Gamma(\tilde{\mathscr{C}}^* \otimes \mathcal{S}) \otimes \Gamma(\tilde{\mathscr{C}}^* \otimes \mathcal{S}) & \longrightarrow & \Gamma(\tilde{\mathscr{C}}^* \otimes \tilde{\mathscr{C}}^* \otimes \mathcal{S})
\end{array}$$

from which is induced the following commutative diagram on cohomology:

(12)

$$\begin{array}{ccccc}
 & \nearrow & H^{p+q}(\Gamma(\mathscr{C}^* \otimes \mathcal{S}) \otimes \Gamma(\mathscr{C}^* \otimes \mathcal{S})) & \searrow & \\
 & & \downarrow & & \\
H^p(M,\mathcal{S}) \otimes H^q(M,\mathcal{S}) & \to & H^{p+q}(\Gamma(\mathscr{C}^* \otimes \tilde{\mathscr{C}}^* \otimes \mathcal{S}) \otimes \Gamma(\mathscr{C}^* \otimes \tilde{\mathscr{C}}^* \otimes \mathcal{S})) & \to & H^{p+q}(\\
 & & \uparrow & & \\
 & \searrow & H^{p+q}(\Gamma(\tilde{\mathscr{C}}^* \otimes \mathcal{S}) \otimes \Gamma(\tilde{\mathscr{C}}^* \otimes \mathcal{S})) & \nearrow &
\end{array}$$

From (12) it follows that the multiplicative structure (8) is independent of the choice of (1).

It follows from the construction of (8) and the associativity of tensor products that the multiplicative structure induced on $\sum_p H^p(M,\mathcal{S})$ by (8) is associative. Thus (8) makes $\sum_p H^p(M,\mathcal{S})$ into an associative algebra over K.

The homomorphisms $\mathscr{C}_p \otimes \mathscr{C}_q \to \mathscr{C}_q \otimes \mathscr{C}_p$ defined by $c_p \otimes c_q \to (-1)^{pq} c_q \otimes c_p$ induce a homomorphism of the resolution (2) with itself in the sense of 5.24. According to 5.24, this homomorphism of (2) induces a homomorphism of cohomology theories which by the corollary of 5.23 must be the identity isomorphism. But the induced homomorphism $H^{p+q}(M,\mathcal{S}) \to H^{p+q}(M,\mathcal{S})$ sends $u \cdot v$ into $(-1)^{pq}v \cdot u$ if $u \in H^p(M,\mathcal{S})$ and $v \in H^q(M,\mathcal{S})$. Thus the multiplicative structure of $\sum_p H^p(M,\mathcal{S})$ satisfies the anti-commutativity relation

$$u \cdot v = (-1)^{pq} v \cdot u \qquad (u \in H^p(M,\mathcal{S}); v \in H^q(M,\mathcal{S})). \tag{13}$$

5.43 The de Rham Cohomology Algebra Exterior multiplication of differential forms induces, by mapping $\sigma \otimes \alpha \mapsto \sigma \wedge \alpha$, a cochain map

$$E^*(M) \otimes E^*(M) \xrightarrow{\wedge} E^*(M) \tag{1}$$

which together with the natural homomorphisms

$$H^p_{\text{de R}}(M) \otimes H^q_{\text{de R}}(M) \to H^{p+q}(E^*(M) \otimes E^*(M)) \tag{2}$$

of 5.39(3) induces homomorphisms

$$H^p_{\text{de R}}(M) \otimes H^q_{\text{de R}}(M) \to H^{p+q}_{\text{de R}}(M), \tag{3}$$

which make the direct sum $\sum_p H^p_{\text{de R}}(M)$ into an associative algebra over $\mathbb{R}$. This is the classical multiplicative structure in de Rham cohomology. Now, $\sum_p H^p_{\text{de R}}(M)$ also inherits an associative algebra structure from the algebra structure 5.42(8) in sheaf cohomology via the canonical isomorphism

$$\sum_p H^p_{\text{de R}}(M) \cong \sum_p H^p(M,\mathcal{R}).$$

We shall now prove that these two algebra structures on $\sum_p H^p_{\text{de R}}(M)$ are identical. For each open set $U \subset M$, exterior multiplication induces homomorphisms

$$E^p(U) \otimes E^q(U) \to E^{p+q}(U) \tag{4}$$

which commute with appropriate restrictions and thus yield presheaf homomorphisms

$$\{E^p(U);\rho_{U,V}\} \otimes \{E^q(U);\rho_{U,V}\} \to \{E^{p+q}(U);\rho_{U,V}\} \tag{5}$$

which in turn induce sheaf homomorphisms

$$\mathscr{E}^p(M) \otimes \mathscr{E}^q(M) \to \mathscr{E}^{p+q}(M). \tag{6}$$

The homomorphisms (6) induce, as is easily checked, a homomorphism (in the sense of 5.24) of the tensor product of the resolution 5.28(3) with itself into itself:

$$\begin{array}{ccccccc}
0 \to \mathcal{R} & \to & \mathscr{E}^0(M) \otimes \mathscr{E}^0(M) & \to & (\mathscr{E}^0(M) \otimes \mathscr{E}^1(M)) \oplus (\mathscr{E}^1(M) \otimes \mathscr{E}^0(M)) & \to & \cdots \\
\downarrow \text{id} & & \downarrow & & \downarrow & & \\
0 \to \mathcal{R} & \to & \mathscr{E}^0(M) & \to & \mathscr{E}^1(M) & \to & \cdots
\end{array} \tag{7}$$

The homomorphism (7) induces a cochain map $\Gamma(\mathscr{E}^*(M) \otimes \mathscr{E}^*(M)) \to \Gamma(\mathscr{E}^*(M))$ which, according to 5.24 and the corollary of 5.23, induces the identity map $H^q(M,\mathscr{R}) \to H^q(M,\mathscr{R})$. Consider now the following commutative diagram of cochain complexes:

$$\begin{array}{ccc} \Gamma(\mathscr{E}^*(M)) \otimes \Gamma(\mathscr{E}^*(M)) \to \Gamma(\mathscr{E}^*(M) \otimes \mathscr{E}^*(M)) \to & & \Gamma(\mathscr{E}^*(M)) \\ \uparrow & & \uparrow \\ E^*(M) \otimes E^*(M) & \longrightarrow & E^*(M) \end{array} \tag{8}$$

The diagram (8), together with 5.39(3), induces the following commutative diagram on cohomology:

$$\begin{array}{ccccc} H^p(M,\mathscr{R}) \otimes H^q(M,\mathscr{R}) & \to H^{p+q}(\Gamma(\mathscr{E}^*(M)) \otimes \Gamma(\mathscr{E}^*(M))) & \to H^{p+q}(M,\mathscr{R}) & \xrightarrow{\text{id}} & H^{p+q}(M, \\ \uparrow & \uparrow & & & \uparrow \\ H^p_{\text{de R}}(M) \otimes H^q_{\text{de R}}(M) & \longrightarrow H^{p+q}(E^*(M) \otimes E^*(M)) & & \longrightarrow & H^{p+q}_{\text{de R}}(\end{array} \tag{9}$$

The composition of homomorphisms in the top row of (9) gives precisely the multiplicative structure 5.42(8) in sheaf cohomology; whereas the composition in the bottom row of (9) is precisely the classical multiplicative structure (2), and the first and last vertical arrows are the canonical isomorphisms. Thus the classical algebra structure on $\sum_p H^p_{\text{de R}}(M)$ is identical with that induced from the algebra structure on sheaf cohomology.

5.44 The Singular Cohomology Algebra This section will be written in terms of the continuous singular cohomology. All considerations, however, apply in exactly the same manner to the differentiable singular theory. Let $f \in S^p(M,K)$ be a singular p-cochain, and let $g \in S^q(M,K)$ be a singular q-cochain. We shall define a singular $(p + q)$ cochain $f \smile g$ called the *cup product* of f and g. Let σ be a singular $(p + q)$ simplex in M. Define

$$(1) \quad (f \smile g)(\sigma) = f(\sigma \circ k_{p+q}^{p+q-1} \circ k_{p+q-1}^{p+q-2} \circ \cdots \circ k_{p+1}^{p})g(\sigma \circ k_0^{p+q-1} \circ k_0^{p+q-2} \circ \cdots \circ k_0{}^q),$$

where if $q = 0$, then the first factor on the right-hand side of (1) is $f(\sigma)$; and if $p = 0$, then the second factor on the right-hand side of (1) is $g(\sigma)$. The k's are the mappings defined in 4.6(2). In other words, (1) says that we start with σ and take the top face q times to get a p-simplex to which we apply f, and we start with σ and take the 0-face p times to get a q-simplex to which we apply g. The multiplication on the right-hand side of (1) takes place in the principal ideal domain K. Associativity, namely,

$$(2) \qquad f \smile (g \smile h) = (f \smile g) \smile h,$$

follows from (1) and from the identity

(3) $$k_0^{p+q+r-1}\circ\cdots\circ k_0^{q+r}\circ k_{q+r}^{q+r-1}\circ\cdots\circ k_{q+1}^{q}$$
$$= k_{p+q+r}^{p+q+r-1}\circ\cdots\circ k_{p+q+1}^{p+q}\circ k_0^{p+q-1}\circ\cdots\circ k_0{}^q$$

which follows from repeated applications of 4.6(5). The bilinear map $(f,g)\mapsto f\smile g$ induces a homomorphism

(4) $$S^p(M,K)\otimes S^q(M,K)\to S^{p+q}(M,K).$$

We claim that

(5) $$(df)\smile g+(-1)^p f\smile(dg)=d(f\smile g).$$

We suggest that the reader check (5) first for the case in which f and g are 1-cochains and σ is a 3-simplex. In this case some simple pictures will aid in following the computation. In general, let τ be a singular $(p+q+1)$ simplex. Then from (1) and 5.31(8), and from repeated applications of 4.6(5), we obtain

$$d(f\smile g)(\tau)=\sum_{l=0}^{p+q+1}(-1)^l(f\smile g)(\tau\circ k_l^{p+q})$$

$$=\sum_{l=0}^{p+q+1}(-1)^l f(\tau\circ k_l^{p+q}\circ k_{p+q}^{p+q-1}\circ\cdots\circ k_{p+1}^{p})\,g(\tau\circ k_l^{p+q}\circ k_0^{p+q-1}\circ\cdots\circ k_0{}^q)$$

$$=\sum_{l=0}^{p}(-1)^l f(\tau\circ k_l^{p+q}\circ k_{p+q}^{p+q-1}\circ\cdots\circ k_{p+1}^{p})\,g(\tau\circ k_0^{p+q}\circ\cdots\circ k_0{}^q)$$

$$+\sum_{l=p+1}^{p+q+1}(-1)^l f(\tau\circ k_{p+q+1}^{p+q}\circ\cdots\circ k_{p+1}^{p})\,g(\tau\circ k_0^{p+q}\circ\cdots\circ k_0^{q+1}\circ k_{l-p}^{q})$$

$$=\sum_{l=0}^{p}(-1)^l f(\tau\circ k_{p+q+1}^{p+q}\circ k_{p+q}^{p+q-1}\circ\cdots\circ k_{p+2}^{p+1}\circ k_l{}^p)\,g(\tau\circ k_0^{p+q}\circ\cdots\circ k_0{}^q)$$

$$+(-1)^p f(\tau\circ k_{p+q+1}^{p+q}\circ\cdots\circ k_{p+1}^{p})\sum_{j=1}^{q+1}(-1)^j g(\tau\circ k_0^{p+q}\circ\cdots\circ k_0^{q+1}\circ k_j{}^q)$$

$$=\sum_{l=0}^{p+1}(-1)^l f(\tau\circ k_{p+q+1}^{p+q}\circ k_{p+q}^{p+q-1}\circ\cdots\circ k_{p+2}^{p+1}\circ k_l{}^p)\,g(\tau\circ k_0^{p+q}\circ\cdots\circ k_0{}^q)$$

$$+(-1)^p f(\tau\circ k_{p+q+1}^{p+q}\circ\cdots\circ k_{p+1}^{p})\sum_{j=0}^{q+1}(-1)^j g(\tau\circ k_0^{p+q}\circ\cdots\circ k_0^{q+1}\circ k_j{}^q)$$

$$=(df\smile g)(\tau)+(-1)^p(f\smile dg)(\tau).$$

It follows from (5) that the homomorphisms (4) determine a cochain map

(6) $$S^*(M,K) \otimes S^*(M,K) \to S^*(M,K).$$

The cochain map (6) together with the natural homomorphisms

(7) $$H^p_\Delta(M;K) \otimes H^q_\Delta(M;K) \to H^{p+q}(S^*(M,K) \otimes S^*(M,K))$$

of 5.39(3) induces homomorphisms

(8) $$H^p_\Delta(M;K) \otimes H^q_\Delta(M;K) \to H^{p+q}_\Delta(M;K)$$

which give the direct sum $\sum_p H^p_\Delta(M;K)$ the structure of an associative algebra over K. This is the classical multiplicative structure of singular cohomology. Now, $\sum_p H^p_\Delta(M;K)$ also inherits an associative algebra structure from the algebra structure 5.42(8) on sheaf cohomology via the canonical isomorphism $\sum_p H^p_\Delta(M;K) \simeq \sum_p H^p_\Delta(M,\mathscr{K})$. The proof that these two algebra structures on $\sum_p H^p_\Delta(M;K)$ are identical is exactly the same as the proof of the corresponding statement for the de Rham cohomology as given in 5.43(4) through (9), but with the replacement of $\mathscr{E}^*(M)$ by $\mathcal{S}^*(M,K)$ and $E^*(M)$ by $S^*(M,K)$, and so on.

The results of 5.43 and 5.44 provide the following more complete version of the de Rham theorem 5.36.

5.45 The de Rham Theorem *The de Rham homomorphism*

(1) $$k^*\colon \sum_p H^p_{\mathrm{de\,R}}(M) \to \sum_p H^p_{\Delta\infty}(M,\mathbb{R})$$

is an algebra isomorphism.

SUPPORTS

5.46 Definition A *family of supports on M* is a family Φ of closed subsets of M satisfying:

(a) The union of any two members of Φ is again in Φ.

(b) Any closed subset of a member of Φ is also a member of Φ.

(c) Each member of Φ has a neighborhood whose closure is in Φ.

Let Φ be a family of supports on M. We recall that M is assumed to be at least a paracompact Hausdorff space and where necessary is actually a differentiable manifold. If $\mathcal{S}$ is a sheaf over M, we let $\Gamma_\Phi(\mathcal{S})$ be the set of sections of $\mathcal{S}$ whose supports are members of Φ. It follows from condition (a) that $\Gamma_\Phi(\mathcal{S})$ is a submodule of $\Gamma(\mathcal{S})$. It follows from (b) that a sheaf homomorphism $\mathcal{S} \to \mathcal{S}'$ induces a homomorphism $\Gamma_\Phi(\mathcal{S}) \to \Gamma_\Phi(\mathcal{S}')$.

Theorem 5.12 has an exact analog with supports. One needs only to make a slight modification in the proof by using condition (c) to help choose a cover which will guarantee that the resulting section of $\mathcal{S}$ actually lies in $\Gamma_\Phi(\mathcal{S})$. The details are left to the reader as an exercise.

A *cohomology theory* $\mathscr{H}_\Phi$ *for* M *with supports* Φ *and with coefficients in sheaves of K-modules over* M is defined exactly as in 5.18, with the exceptions that $\Gamma(\mathcal{S})$ is replaced by $\Gamma_\Phi(\mathcal{S})$ in 5.18(a) and the cohomology modules are now denoted by ${}_\Phi H^p(M,\mathcal{S})$. The entire development of this chapter can be carried through for cohomology with supports with very few modifications. We shall briefly sketch those modifications.

The proof of existence and uniqueness of a cohomology theory with supports (which will make use of the support version of Theorem 5.12) proceeds as in 5.18 through 5.25 with the exception that when given a fine torsionless resolution $0 \to \mathscr{K} \to \mathscr{C}_0 \to \mathscr{C}_1 \to \mathscr{C}_2 \to \cdots$, we now define the sheaf cohomology modules by setting

$$(1) \qquad {}_\Phi H^q(M,\mathcal{S}) = H^q(\Gamma_\Phi(\mathscr{C}^* \otimes \mathcal{S})).$$

Let $\{S_U;\rho_{U,V}\}$ be a presheaf with associated sheaf $\mathcal{S}$. We let ${}_\Phi S_M$ be the submodule of S_M consisting of those elements mapping into $\Gamma_\Phi(\mathcal{S})$ under the canonical homomorphism $S_M \to \Gamma(\mathcal{S})$. It then follows from Proposition 5.27 and the definition of ${}_\Phi S_M$ that 5.27 holds if we replace 5.27(2) by

$$(2) \qquad 0 \to (S_M)_0 \to {}_\Phi S_M \to \Gamma_\Phi(\mathcal{S}) \to 0.$$

The classical cohomology theories with supports are defined by:

$$(3) \qquad \begin{aligned} {}_\Phi H^q_{A-S}(M;G) &= H^q({}_\Phi A^*(M,G)/A^*_0(M,G)),\\ {}_\Phi H^q_{\mathrm{de\,R}}(M) &= H^q({}_\Phi E^*(M)),\\ {}_\Phi H^q_\Delta(M;G) &= H^q({}_\Phi S^*(M,G)),\\ {}_\Phi H^q_{\Delta\infty}(M;G) &= H^q({}_\Phi S^*_\infty(M,G)). \end{aligned}$$

In the Čech theory, the support of a q-cochain $f \in C^q(\mathfrak{U},\mathcal{S})$ is by definition the union of the supports of the sections $f(\sigma)$ as σ runs over the q-simplices of the cover $\mathfrak{U}$. If we set ${}_\Phi C^q(\mathfrak{U},\mathcal{S})$ equal to the module of q-cochains with supports in Φ, then the development of the Čech theory proceeds as before and yields the Čech cohomology modules ${}_\Phi \check{H}^q(M,\mathcal{S})$ with supports.

The resultant cohomology theories will depend on the family of supports chosen. Ordinary cohomology is the same as the special case of cohomology with supports in which Φ is taken to be the family of all closed sets.

The isomorphism theorems and the development of the multiplicative structure all proceed for supports exactly as before. In particular, we have the de Rham theorem with supports:

$$(4) \qquad k^*: \sum_p {}_\Phi H^p_{\mathrm{de\,R}}(M) \cong \sum_p {}_\Phi H^p_{\Delta\infty}(M;\mathbb{R})$$

is an algebra isomorphism.

EXERCISES

1 Let $\mathcal{S}$ be a sheaf over M. Prove that the mapping $m \mapsto 0 \in \mathcal{S}_m$ (the "0-section") is continuous.

2 Prove that sections of sheaves are local homeomorphisms and therefore are open maps.

3 Prove that if two sections of a sheaf agree at one point, then they agree on a neighborhood of that point. Conclude that the set of points on which a section is not zero is a closed set.

4 A C^∞ function on M determines a cross section of the sheaf of germs of C^∞ functions $\mathscr{C}^\infty(M)$. The set of points in M on which a C^∞ function is not zero is an open set; whereas, according to Exercise 3, the set of points in M where a section of $\mathscr{C}^\infty(M)$ is not zero is a closed set. Can you reconcile these two facts? Consider examples.

5 Prove that sheaf mappings are local homeomorphisms, and therefore are open maps.

6 Prove that if two sheaf mappings agree at a point then they agree on a neighborhood of that point. Conclude that the set of points where a sheaf mapping is not zero is a closed set.

7 Complete the details of the construction in 5.4 of quotient sheaves.

8 Complete the details of the proof begun in 5.6 that $\beta(\alpha(\mathcal{S}))$ is canonically isomorphic with $\mathcal{S}$.

9 Prove that there is a canonical isomorphism $(\mathcal{S} \otimes \mathscr{T})_m \cong \mathcal{S}_m \otimes \mathscr{T}_m$.

10 Prove that the tensor product of two complete presheaves need not be a complete presheaf.

11 Let φ be a C^∞ function on M. Define a mapping $\mathscr{C}^\infty(M) \to \mathscr{C}^\infty(M)$ by sending $\mathbf{f}_m \to \varphi(m) \cdot \mathbf{f}_m$. Here, as usual, $\mathbf{f}_m$ denotes the germ at m of a C^∞ function f defined on a neighborhood of m. Prove that this mapping is *not* in general continuous and therefore cannot be a sheaf homomorphism. (*Caution:* Errors are often made at this point in the construction of partitions of unity on sheaves—review the correct procedure for the construction of a partition of unity on $\mathscr{C}^\infty(M)$ given in 5.10.)

12 Make the necessary modifications in the proof of Theorem 5.12 to prove that if Φ is a system of supports on M and $\mathcal{S} \to \mathscr{T}$ is a surjective sheaf homomorphism with its kernel a fine sheaf, then $\Gamma_\Phi(\mathcal{S}) \to \Gamma_\Phi(\mathscr{T})$ is a surjection.

13 Carry out the details in the proof of the exactness of 5.17(2).

14 Complete the details of 5.24.

15 Prove that the presheaves 5.26(5) satisfy 5.7(C_2), but for $p \geq 1$ do not satisfy 5.7(C_1).

16 Give an example of an exact sequence of modules

$$0 \to A \to B \to C \to 0$$

and a module D such that

$$0 \to A \otimes D \to B \otimes D \to C \otimes D \to 0$$

is not exact.

17 Find an example of a fine sheaf which has a subsheaf which is not fine.

18 Prove that if you tensor a long exact sequence of torsionless sheaves with a sheaf $\mathcal{S}$, then the resulting long sequence is still exact.

19 Prove that a continuous map $f: M \to N$ induces in a natural way a homomorphism

$$f_*: H^p_\Delta(N;G) \to H^p_\Delta(M;G)$$

such that if $g: N \to X$, then

$$(g \circ f)_* = f_* \circ g_*,$$

and such that

$$(\text{id})_* = \text{id}.$$

Conclude that homeomorphic spaces have isomorphic singular cohomology.

20 Prove that 5.37(1) is an isomorphism. Keep in mind that these vector spaces are generally infinite dimensional.

21 Prove that if σ and τ are closed differential forms all of whose periods are integer-valued (see 4.17), then $\sigma \wedge \tau$ has integer periods.

6

THE HODGE THEOREM

Throughout this chapter, M will be a compact oriented Riemannian manifold of dimension n unless otherwise indicated. We will see that the ordinary Laplacian $(-1)\sum_i \partial^2/\partial x_i^2$ has a generalization to an operator Δ on differential forms, known as the Laplace-Beltrami operator. Our main objective in this chapter is a proof of the Hodge decomposition theorem, which says that the equation $\Delta\omega = \alpha$ has a solution ω in the smooth p-forms on M if and only if the p-form α is orthogonal (in a suitable inner product on $E^p(M)$) to the space of harmonic p-forms (those for which $\Delta\varphi = 0$). From the Hodge decomposition theorem we will conclude that there exists a unique harmonic form in each de Rham cohomology class. As another simple application we will obtain the Poincaré duality theorem for de Rham cohomology and, from it, the Poincaré duality theorem for real singular cohomology. To prove the Hodge theorem, we shall give a complete self-contained exposition of the local theory of elliptic operators, using Fourier series as our basic tool. The eigenfunctions of the Laplace-Beltrami operator and their use in a proof of the Peter-Weyl theorem are discussed in the exercises at the end of this chapter.

THE LAPLACE-BELTRAMI OPERATOR

6.1 Definitions Recall (from 4.10(6) and Exercise 13 of Chapter 2) that there is a linear operator $*$ which assigns to each p-form on M an $(n-p)$ form and which satisfies

(1) $$** = (-1)^{p(n-p)}.$$

We define an operator δ from p-forms to $(p-1)$ forms by setting

(2) $$\delta = (-1)^{n(p+1)+1} * d *.$$

On 0-forms, δ is simply the zero linear functional. The *Laplace-Beltrami operator* Δ (*Laplacian* for short) is defined by

(3) $$\Delta = \delta d + d\delta,$$

and is a linear operator on $E^p(M)$ for each p with $0 \leq p \leq n$. We leave it to the reader as an exercise to check that on $E^0(\mathbb{R}^n)$, that is, on C^∞ functions on Euclidean space $\mathbb{R}^n$, the Laplacian is simply the operator $(-1)\sum_{i=1}^{n} \partial^2/\partial x_i^2$. Also, it is a straightforward exercise to check that the Laplacian commutes with $*$, that is,

$$(4) \qquad *\Delta = \Delta *.$$

We define an inner product on the vector space $E^p(M)$ of p-forms on M by setting

$$(5) \qquad \langle\alpha,\beta\rangle = \int_M \alpha \wedge *\beta \quad \text{for } \alpha, \beta \in E^p(M),$$

and we denote the corresponding norm by $\|\alpha\|$. It follows from (6) of Exercise 13 in Chapter 2 that the bilinear form defined in (5) is actually symmetric and positive definite. We extend the inner products (5) for $0 \leq p \leq n$ to an inner product on the direct sum $\sum_{p=0}^{n} E^p(M)$ simply by requiring the various $E^p(M)$ to be orthogonal.

6.2 Proposition *δ is the adjoint of d on $\sum_{p=0}^{n} E^p(M)$; that is,*

$$(1) \qquad \langle d\alpha,\beta\rangle = \langle\alpha,\delta\beta\rangle.$$

PROOF By linearity, and the orthogonality of the $E^p(M)$, the proof reduces to consideration of the case in which α is a $(p-1)$ form and β is a p-form. In this case,

$$(2) \qquad \begin{aligned} d(\alpha \wedge *\beta) &= d\alpha \wedge *\beta + (-1)^{p-1}\alpha \wedge d{*\beta} \\ &= d\alpha \wedge *\beta - \alpha \wedge *\delta\beta. \end{aligned}$$

By integrating both sides over M and applying the special case of Stokes' theorem contained in the corollary of 4.9 to the left-hand side, we obtain

$$(3) \qquad 0 = \int_M (d\alpha \wedge *\beta - \alpha \wedge *\delta\beta) = \langle d\alpha,\beta\rangle - \langle\alpha,\delta\beta\rangle.$$

Thus

$$(4) \qquad \langle d\alpha,\beta\rangle = \langle\alpha,\delta\beta\rangle.$$

Corollary *Δ is self-adjoint, that is,*

$$(5) \qquad \langle\Delta\alpha,\beta\rangle = \langle\alpha,\Delta\beta\rangle \qquad (\alpha, \beta \in E^p(M); 0 \leq p \leq n).$$

6.3 Proposition $\Delta\alpha = 0$ *if and only if* $d\alpha = 0$ *and* $\delta\alpha = 0$.

PROOF Clearly $\Delta\alpha = 0$ if $d\alpha = 0$ and $\delta\alpha = 0$. Now,

$$(1) \qquad \langle\Delta\alpha,\alpha\rangle = \langle(d\delta + \delta d)\alpha, \alpha\rangle = \langle\delta\alpha,\delta\alpha\rangle + \langle d\alpha,d\alpha\rangle.$$

Thus if $\Delta\alpha = 0$, it follows that $d\alpha = 0$ and $\delta\alpha = 0$.

Corollary *The only harmonic functions ($\Delta f = 0$) on a compact, connected, oriented, Riemannian manifold are the constant functions.*

THE HODGE THEOREM

6.4 Definition We shall let Δ^* denote the adjoint of the Laplacian on $E^p(M)$. This operator is, of course, precisely Δ itself since the Laplacian is self-adjoint on $E^p(M)$, and usually we make no distinction between Δ and Δ^*. However, this distinction will be important for the form of the following definition.

We shall be interested in finding necessary and sufficient conditions for there to exist a solution ω of the equation $\Delta\omega = \alpha$. Suppose that ω is a solution of $\Delta\omega = \alpha$. Then

$$\text{(1)} \qquad \langle \Delta\omega, \varphi \rangle = \langle \alpha, \varphi \rangle \quad \text{for all } \varphi \in E^p(M),$$

from which it follows that

$$\text{(2)} \qquad \langle \omega, \Delta^* \varphi \rangle = \langle \alpha, \varphi \rangle \quad \text{for all } \varphi \in E^p(M).$$

Now, (2) suggests that we can view a solution of $\Delta\omega = \alpha$ as a certain type of linear functional on $E^p(M)$, namely, ω determines a bounded linear functional l on $E^p(M)$ by

$$\text{(3)} \qquad l(\beta) = \langle \omega, \beta \rangle;$$

and in view of (2), the functional l satisfies

$$\text{(4)} \qquad l(\Delta^* \varphi) = \langle \alpha, \varphi \rangle \quad \text{for all } \varphi \in E^p(M).$$

This view of a solution turns out to be extremely useful, for it will allow us to bring various techniques of functional analysis to bear on the problem of solving $\Delta\omega = \alpha$. We shall call such a linear functional a *weak solution* of $\Delta\omega = \alpha$. That is, a *weak solution* of $\Delta\omega = \alpha$ is a bounded linear functional $l: E^p(M) \to \mathbb{R}$ such that

$$\text{(5)} \qquad l(\Delta^* \varphi) = \langle \alpha, \varphi \rangle \quad \text{for all } \varphi \in E^p(M).$$

Later we will deal with weak solutions of a general partial differential operator L defined on an open set in Euclidean space, and in place of Δ^* in (5), there will be the formal adjoint L^* of L (see 6.24(3) and 6.31).

We have seen that each ordinary solution $\omega \in E^p(M)$ of $\Delta\omega = \alpha$ determines a weak solution by (3). It turns out that the major effort of this chapter will be to prove a regularity theorem which says that the converse of this is true; that is, each weak solution determines an ordinary solution. The main step in proving this converse is to show that if l is a weak solution of $\Delta\omega = \alpha$, then l is represented by a smooth form ω in the sense that there is a form $\omega \in E^p(M)$ such that (3) holds. That the form ω is then an ordinary solution follows from the fact that

(6) $$\langle \Delta\omega,\beta\rangle = \langle \omega,\Delta^*\beta\rangle = l(\Delta^*\beta) = \langle \alpha,\beta\rangle$$

for all $\beta \in E^p(M)$, which implies that $\Delta\omega = \alpha$.

6.5 Regularity Theorem *Let $\alpha \in E^p(M)$, and let l be a weak solution of $\Delta\omega = \alpha$. Then there exists $\omega \in E^p(M)$ such that*

$$l(\beta) = \langle \omega,\beta\rangle$$

for every $\beta \in E^p(M)$. Consequently, $\Delta\omega = \alpha$.

We shall assume Theorem 6.5 for the moment as well as the following.

6.6 Theorem *Let $\{\alpha_n\}$ be a sequence of smooth p-forms on M such that $\|\alpha_n\| \leq c$ and $\|\Delta\alpha_n\| \leq c$ for all n and for some constant $c > 0$. Then a subsequence of $\{\alpha_n\}$ is a Cauchy sequence in $E^p(M)$.*

We will begin the machinery necessary for the proofs of Theorems 6.5 and 6.6 in the unit beginning with 6.15. We eventually return to the proofs of Theorems 6.5 and 6.6 in 6.32 and 6.33 respectively. Meanwhile, we shall assume the two theorems as proved and shall proceed to the Hodge theorem.

6.7 Definition We let

(1) $$H^p = \{\omega \in E^p(M): \Delta\omega = 0\}.$$

The elements of H^p are called *harmonic p-forms.*

6.8 The Hodge Decomposition Theorem *For each integer p with $0 \leq p \leq n$, H^p is finite dimensional, and we have the following orthogonal direct sum decompositions of the space $E^p(M)$ of smooth p-forms on M:*

(1) $$\begin{aligned} E^p(M) &= \Delta(E^p) \oplus H^p \\ &= d\delta(E^p) \oplus \delta d(E^p) \oplus H^p \\ &= d(E^{p-1}) \oplus \delta(E^{p+1}) \oplus H^p. \end{aligned}$$

Consequently, the equation $\Delta\omega = \alpha$ has a solution $\omega \in E^p(M)$ if and only if the p-form α is orthogonal to the space of harmonic p-forms.

PROOF If H^p were not finite dimensional, then H^p would contain an infinite orthonormal sequence. But by Theorem 6.6, this orthonormal sequence would contain a Cauchy subsequence, which is impossible. Thus H^p is finite dimensional.

It is sufficient to prove the decomposition in the first line of (1), for the other two lines of (1) then follow from 6.1(3), 6.2, and 6.3.

Let $\omega_1, \ldots, \omega_l$ be an orthonormal basis of H^p. Then an arbitrary form $\alpha \in E^p(M)$ can uniquely be written

$$\alpha = \beta + \sum_{i=1}^{l} \langle \alpha, \omega_i \rangle \omega_i \tag{2}$$

where β lies in $(H^p)^\perp$, the subspace of $E^p(M)$ consisting of all elements orthogonal to H^p. Thus we have an orthogonal direct sum decomposition

$$E^p(M) = (H^p)^\perp \oplus H^p. \tag{3}$$

The theorem will be proved by showing that $(H^p)^\perp = \Delta(E^p)$. We let H denote the projection operator of $E^p(M)$ onto H^p so that $H(\alpha)$ is the harmonic part of α.

Now $\Delta(E^p) \subset (H^p)^\perp$. For if $\omega \in E^p$ and $\alpha \in H^p$, then

$$\langle \Delta\omega, \alpha \rangle = \langle \omega, \Delta\alpha \rangle = 0.$$

Conversely, we claim that $(H^p)^\perp \subset \Delta(E^p)$. In order to prove this, we first need the following inequality.

We claim that there is a constant $c > 0$ such that

$$\|\beta\| \leq c\,\|\Delta\beta\| \quad \text{for all} \quad \beta \in (H^p)^\perp. \tag{4}$$

Suppose the contrary. Then there exists a sequence $\beta_j \in (H^p)^\perp$ with $\|\beta_j\| = 1$ and $\|\Delta\beta_j\| \to 0$. By Theorem 6.6, a subsequence of the β_j, which for convenience we can assume to be $\{\beta_j\}$ itself, is Cauchy. Thus $\lim_{j\to\infty} \langle \beta_j, \psi \rangle$ exists for each $\psi \in E^p(M)$. We define a linear functional l on $E^p(M)$ by setting

$$l(\psi) = \lim_{j\to\infty} \langle \beta_j, \psi \rangle \quad \text{for} \quad \psi \in E^p(M). \tag{5}$$

Now l is clearly bounded, and

$$l(\Delta\varphi) = \lim_{j\to\infty} \langle \beta_j, \Delta\varphi \rangle = \lim_{j\to\infty} \langle \Delta\beta_j, \varphi \rangle = 0, \tag{6}$$

so l is a weak solution of $\Delta\beta = 0$. By Theorem 6.5, there exists $\beta \in E^p(M)$ such that $l(\psi) = \langle \beta, \psi \rangle$. Consequently, $\beta_j \to \beta$. Since $\|\beta_j\| = 1$ and $\beta_j \in (H^p)^\perp$, it follows that $\|\beta\| = 1$ and $\beta \in (H^p)^\perp$. But by Theorem 6.5, $\Delta\beta = 0$, so $\beta \in H^p$, which is a contradiction. Thus (4) is proved.

Now we shall use (4) to prove that $(H^p)^\perp \subset \Delta(E^p)$. Let $\alpha \in (H^p)^\perp$. We define a linear functional l on $\Delta(E^p)$ by setting

$$l(\Delta\varphi) = \langle \alpha, \varphi \rangle \quad \text{for all } \varphi \in E^p(M). \tag{7}$$

Now l is well-defined; for if $\Delta\varphi_1 = \Delta\varphi_2$, then $\varphi_1 - \varphi_2 \in H^p$, so that $\langle \alpha, \varphi_1 - \varphi_2 \rangle = 0$. Also l is a bounded linear functional on $\Delta(E^p)$, for let $\varphi \in E^p(M)$ and let $\psi = \varphi - H(\varphi)$. Then using (4), we obtain

$$(8) \qquad |l(\Delta\varphi)| = |l(\Delta\psi)| = |\langle \alpha,\psi\rangle| \leq \|\alpha\| \, \|\psi\| \leq c\,\|\alpha\|\,\|\Delta\psi\| = c\,\|\alpha\|\,\|\Delta\varphi\|.$$

By the Hahn-Banach theorem [26, p. 228], l extends to a bounded linear functional on $E^p(M)$. Thus l is a weak solution of $\Delta\omega = \alpha$. By Theorem 6.5, there exists $\omega \in E^p(M)$ such that $\Delta\omega = \alpha$. Hence

$$(9) \qquad (H^p)^\perp = \Delta(E^p),$$

and the Hodge decomposition theorem is proved.

6.9 Definition We define the *Green's operator* $G\colon E^p(M) \to (H^p)^\perp$ by setting $G(\alpha)$ equal to the unique solution of $\Delta\omega = \alpha - H(\alpha)$ in $(H^p)^\perp$. We leave it to the reader as an exercise to prove that G is a bounded self-adjoint linear operator which takes bounded sequences into sequences with Cauchy subsequences.

6.10 Proposition *G commutes with d, δ, and Δ. In fact, G commutes with any linear operator which commutes with the Laplacian Δ.*

PROOF Suppose that $T\Delta = \Delta T$ with, say, $T\colon E^p(M) \to E^q(M)$. Let $\pi_{(H^p)^\perp}$ denote the projection mapping of $E^p(M)$ onto $(H^p)^\perp$. By definition, $G = (\Delta \mid (H^p)^\perp)^{-1} \circ \pi_{(H^p)^\perp}$. Now, the fact that $T\Delta = \Delta T$ implies that $T(H^p) \subset H^q$; and since $(H^p)^\perp = \Delta(E^p)$, it implies also that $T((H^p)^\perp) \subset (H^q)^\perp$. It follows that

$$(1) \qquad T \circ \pi_{(H^p)^\perp} = \pi_{(H^q)^\perp} \circ T,$$

and on $(H^p)^\perp$,

$$(2) \qquad T \circ (\Delta \mid (H^p)^\perp) = (\Delta \mid (H^q)^\perp) \circ T,$$

and hence on $(H^p)^\perp$,

$$(3) \qquad T \circ (\Delta \mid (H^p)^\perp)^{-1} = (\Delta \mid (H^q)^\perp)^{-1} \circ T.$$

It follows from (1) and (3) that G commutes with T.

6.11 Theorem *Each de Rham cohomology class on a compact oriented Riemannian manifold M contains a unique harmonic representative.*

PROOF Let α be an arbitrary p-form on M. From the Hodge decomposition theorem and from the definition of the Green's operator G, we have

$$(1) \qquad \alpha = d\delta G\alpha + \delta d G\alpha + H\alpha.$$

Since G, by 6.10, commutes with d, we have

$$(2) \qquad \alpha = d\delta G\alpha + \delta G d\alpha + H\alpha.$$

Thus if α is a closed p-form,

$$(3) \qquad \alpha = d\delta G\alpha + H\alpha,$$

so $H\alpha$ is a harmonic p-form in the same de Rham cohomology class as is α. If two harmonic forms α_1 and α_2 differ by an exact form $d\beta$, then we have

$$0 = d\beta + (\alpha_1 - \alpha_2).$$

But $d\beta$ and $(\alpha_1 - \alpha_2)$ are orthogonal since

$$\langle d\beta, \alpha_1 - \alpha_2 \rangle = \langle \beta, \delta\alpha_1 - \delta\alpha_2 \rangle = \langle \beta, 0 \rangle = 0.$$

Thus $d\beta = 0$ and $\alpha_1 = \alpha_2$. Thus there is a unique harmonic form in each de Rham cohomology class.

Corollary *The de Rham cohomology groups for a compact, orientable, differentiable manifold are all finite dimensional.*

PROOF Any differentiable manifold can be equipped with a Riemannian metric (Exercise 23 of Chapter 1), and so the corollary follows immediately from Theorem 6.11 and from the finite dimensionality (6.8) of the spaces H^p of harmonic forms.

6.12 Let M be a compact, oriented, differentiable manifold of dimension n. We define a bilinear function

(1) $$H^p_{\text{de R}}(M) \times H^{n-p}_{\text{de R}}(M) \to \mathbb{R}$$

by sending

(2) $$(\{\varphi\},\{\psi\}) \mapsto \int_M \varphi \wedge \psi,$$

where φ and ψ are closed forms representing the cohomology classes $\{\varphi\}$ in $H^p_{\text{de R}}(M)$ and $\{\psi\}$ in $H^{n-p}_{\text{de R}}(M)$. Observe that the bilinear map (2) is well-defined. For example, if φ_1 is another representative of the de Rham class $\{\varphi\}$, then $\varphi_1 = \varphi + d\xi$ for some form ξ, and by the special case of Stokes' theorem which is contained in the corollary of 4.9,

(3) $$\int_M \varphi_1 \wedge \psi = \int_M \varphi \wedge \psi + \int_M d\xi \wedge \psi$$
$$= \int_M \varphi \wedge \psi + \int_M d(\xi \wedge \psi) = \int_M \varphi \wedge \psi.$$

Observe also from its definition that the bilinear function (2) depends on the orientation on M.

6.13 Theorem (Poincaré duality for the de Rham cohomology of a compact oriented n-dimensional manifold M) *The bilinear function 6.12(2) is a non-singular pairing and consequently determines isomorphisms of $H^{n-p}_{\text{de R}}(M)$ with the dual space of $H^p_{\text{de R}}(M)$:*

(1) $$H^{n-p}_{\text{de R}}(M) \cong (H^p_{\text{de R}}(M))^*.$$

PROOF Given a non-zero cohomology class $\{\varphi\} \in H^p_{\text{de R}}(M)$, we must find a non-zero cohomology class $\{\psi\} \in H^{n-p}_{\text{de R}}(M)$ such that $(\{\varphi\},\{\psi\}) \neq 0$. Choose a Riemannian structure on M. We can assume, according to 6.11, that φ is the harmonic representative of $\{\varphi\}$. Since the cohomology class $\{\varphi\}$ is not zero, φ is not identically zero. Since $*\Delta = \Delta *$, it follows that $*\varphi$ is also harmonic, and therefore closed by 6.3, and so $*\varphi$ represents a cohomology class $\{*\varphi\} \in H^{n-p}_{\text{de R}}(M)$. Now

(2) $$(\{\varphi\},\{*\varphi\}) = \int_M \varphi \wedge *\varphi = \|\varphi\|^2 \neq 0.$$

Thus the pairing 6.12(2) is non-singular, and the isomorphism (1) follows from 2.7.

Corollary *If M is a compact, connected, orientable, differentiable manifold of dimension n, then $H^n_{\text{de R}}(M) \cong \mathbb{R}$.*

6.14 Remark The *real continuous singular homology groups* $H_p(M;\mathbb{R})$ are defined just as in 4.16, with the exception that all continuous simplices in M are allowed, not just differentiable simplices. The isomorphism 5.37(1) holds for the continuous case exactly as for the differentiable theory. By combining 6.13 with the de Rham theorem 5.36, with the canonical isomorphism 5.32(5) of the differentiable with the continuous real singular cohomology of M, and with the isomorphism 5.37(1) for the continuous real cohomology and homology, we obtain the Poincaré duality between the real singular cohomology and the real singular homology of M:

(1) $$H^p_\Delta(M;\mathbb{R}) \cong H_{n-p}(M;\mathbb{R}).$$

SOME CALCULUS

We now begin to develop the machinery necessary for the proofs of Theorems 6.5 and 6.6.

6.15 Notation We shall be using multi-index notation $\alpha = (\alpha_1, \ldots, \alpha_n)$ where the α_i are integers. We let $|\alpha|$ denote the ordinary Euclidean norm of α; that is,

(1) $$|\alpha| = (\alpha_1^2 + \cdots + \alpha_n^2)^{1/2},$$

and if the α_i are all non-negative, then we let

(2) $$[\alpha] = \alpha_1 + \cdots + \alpha_n.$$

If α and η are both n-tuples of integers, then

(3) $$\eta^\alpha = \eta_1^{\alpha_1} \cdots \eta_n^{\alpha_n}, \quad \text{where we set } 0^0 = 1.$$

We shall use x_i for the ith canonical coordinate function on $\mathbb{R}^n$, and let

(4) $$x = (x_1, \ldots, x_n).$$

The αth derivative operator D^α is defined for each n-tuple α of non-negative integers by

(5) $$D^\alpha u = \left(\frac{1}{i}\right)^{[\alpha]} \frac{\partial^{[\alpha]} u}{\partial x_1^{\alpha_1} \cdots \partial x_n^{\alpha_n}},$$

where $i = \sqrt{-1}$. (The addition of the factor of i will be convenient later.)

We let $\mathscr{P}$ denote the complex vector space consisting of C^∞ functions defined on $\mathbb{R}^n$ which have values in complex m space $\mathbb{C}^m$ and are periodic of period 2π in each variable. I would suggest that the reader assume m to be 1 for the first reading of this section since that case contains all of the essential ideas and the computations are somewhat simpler. We will arrange the notation so that it is essentially the same for general m. One note of caution concerning the notation is this. If $\gamma, \beta \in \mathbb{C}^m$, then $\gamma \cdot \beta$ means the Hermitian product $\gamma_1\overline{\beta_1} + \cdots + \gamma_m\overline{\beta_m}$. If m happens to be 1, then $\gamma \cdot \beta$ means $\gamma\bar{\beta}$. The associated norm is denoted by $|\gamma|$.

If $\varphi, \psi \in \mathscr{P}$, then $\varphi \cdot \psi$ is the complex-valued function which is the Hermitian product of φ and ψ; that is,

(6) $$\varphi \cdot \psi = \varphi_1\overline{\psi_1} + \cdots + \varphi_m\overline{\psi_m}.$$

We let $|\psi|$ denote the real-valued function

(7) $$|\psi| = (\psi \cdot \psi)^{1/2}.$$

If $\varphi \in \mathscr{P}$ and if f is a periodic complex-valued C^∞ function on $\mathbb{R}^n$ (all periods always 2π), then φf shall denote the element of $\mathscr{P}$ whose ith component function is $\varphi_i f$; that is,

(8) $$\varphi f = (\varphi_1 f, \ldots, \varphi_m f).$$

Let $Q \subset \mathbb{R}^n$ be the open cube

(9) $$Q = \{p \in \mathbb{R}^n \colon 0 < x_i(p) < 2\pi,\ i = 1, \ldots, n\}.$$

We shall be introducing a number of different norms on $\mathscr{P}$. By $\|\psi\|$ we shall mean the ordinary L_2 norm of ψ over Q,

(10) $$\|\psi\| = \frac{1}{(2\pi)^{n/2}}\left(\int_Q \psi \cdot \psi\right)^{1/2},$$

and $\langle\psi,\varphi\rangle$ shall denote the L_2 inner product,

(11) $$\langle\psi,\varphi\rangle = \frac{1}{(2\pi)^n}\int_Q \psi \cdot \varphi.$$

The norm $\|\psi\|_\infty$ shall denote the uniform norm of ψ,

(12) $$\|\psi\|_\infty = \sup_Q |\psi|.$$

6.16 Some Facts about Fourier Series If $\varphi \in \mathscr{P}$ and if $\xi = (\xi_1, \ldots, \xi_n)$ where the ξ_i are integers, then the ξth Fourier coefficient $\varphi_\xi \in \mathbb{C}^m$ is defined by

(1) $$\varphi_\xi = \frac{1}{(2\pi)^n} \int_Q \varphi(x) e^{-ix\cdot\xi}\, dx,$$

where $x \cdot \xi = x_1\xi_1 + \cdots + x_n\xi_n$.

First we are going to show that the Fourier series $\sum_\xi \varphi_\xi e^{ix\cdot\xi}$ of φ converges uniformly to φ. Let an integer $k > 0$ be given. By integrating (1) repeatedly by parts, differentiating φ and integrating $e^{-ix\cdot\xi}$, and observing that the boundary terms drop out since the integrand is periodic, one sees that there is a constant c_k' depending on φ and its derivatives up to order at most $2nk$ such that

(2) $$|\varphi_\xi| \leq \frac{c_k'}{(\prod \xi_i)^{2k}} \quad \text{for all} \quad \xi \neq 0,$$

where $\prod \xi_i$ denotes the product of all the non-zero ξ_i. It follows that there is a constant c_k such that

(3) $$|\varphi_\xi| \leq \frac{c_k}{(1 + |\xi|^2)^k} \quad \text{for all } \xi.$$

Consider now the question of convergence of the series $\sum_\xi (1 + |\xi|^2)^{-k}$. If we let

(4) $$S_j = \left\{\xi = (\xi_1, \ldots, \xi_n) \colon \max_{1\leq i\leq n} |\xi_i| = j\right\},$$

then the number of elements of S_j is at most $2n(2j+1)^{n-1}$, and for each $\xi \in S_j$, $|\xi|^2 \geq j^2$, so that

(5) $$s_j = \sum_{\xi\in S_j} \frac{1}{(1 + |\xi|^2)^k} \leq \frac{2n(2j+1)^{n-1}}{(1 + j^2)^k} \leq cj^{n-1-2k}$$

for $j \geq 1$, where c is a constant depending only on n. Consequently,

(6) $$\sum_\xi \frac{1}{(1 + |\xi|^2)^k} = 1 + \sum_{j=1}^\infty s_j \leq 1 + c\sum_{j=1}^\infty \frac{1}{j^{1+2k-n}},$$

and so the series $\sum_\xi (1 + |\xi|^2)^{-k}$ converges for $1 + 2k - n > 1$, or in other words, for

(7) $$k \geq \left[\frac{n}{2}\right] + 1,$$

where $[n/2]$ denotes the greatest integer less than or equal to $n/2$. It would be an interesting exercise for the reader to reach the same conclusion by using an integral test.

It follows from (3) that if we take $k \geq [n/2] + 1$ as in (7), then the Fourier series

$$\sum_{\xi} \varphi_\xi e^{ix\cdot\xi} \tag{8}$$

converges uniformly to some continuous function Φ. We claim that $\Phi = \varphi$. This is of course due to the completeness of the trigonometric system and may be deduced as follows from the Stone-Weierstrass theorem [13], [26]. Let $\psi = \varphi - \Phi$, and let $t \in \mathscr{P}$ be a trigonometric polynomial; that is, t is a finite linear combination of terms of the form $a_\xi e^{ix\cdot\xi}$ for $a_\xi \in \mathbb{C}^m$. Then since φ and Φ have the same Fourier coefficients,

$$\int_Q \psi \cdot t = 0. \tag{9}$$

Now if $\varepsilon > 0$ is given, then by the Stone-Weierstrass theorem there is a trigonometric polynomial $t \in \mathscr{P}$ such that $\|\psi - t\|_\infty < \varepsilon$. Thus, using (9), we see that

$$\left|\int_Q \psi \cdot \psi\right| = \left|\int_Q \psi \cdot (\psi - t)\right| \leq \varepsilon(2\pi)^n \,\|\psi\|. \tag{10}$$

Consequently $\|\psi\| \leq \varepsilon$. But $\varepsilon > 0$ was arbitrary and ψ is continuous, so $\psi = 0$. Thus the Fourier series of a periodic C^∞ function φ converges uniformly to φ,

$$\varphi(x) = \sum_{\xi} \varphi_\xi e^{ix\cdot\xi}. \tag{11}$$

It follows from integration by parts that the ξth Fourier coefficient of $D^\alpha\varphi$ is $\xi_1^{\alpha_1} \cdots \xi_n^{\alpha_n}\varphi_\xi$. Thus

$$D^\alpha\varphi(x) = \sum_{\xi} \xi^\alpha \varphi_\xi e^{ix\cdot\xi}. \tag{12}$$

From (11) and the orthogonality of the trigonometric system, we obtain the Parseval identity:

$$\int_Q |\varphi|^2 = (2\pi)^n \sum_{\xi} |\varphi_\xi|^2. \tag{13}$$

Applying (13) to $D^\alpha\varphi$, we obtain

$$\|D^\alpha\varphi\|^2 = \sum_{\xi} \xi^{2\alpha} |\varphi_\xi|^2. \tag{14}$$

It follows from (14) that given a non-negative integer t, there is a constant c greater than zero and depending only on t and n such that

$$c \sum_{\xi} (1 + |\xi|^2)^t \,|\varphi_\xi|^2 \;\leq\; \sum_{[\alpha]=0}^{t} \|D^\alpha\varphi\|^2 \;\leq\; \sum_{\xi} (1 + |\xi|^2)^t \,|\varphi_\xi|^2. \tag{15}$$

6.17 The Sobolev spaces H_s Let $\mathcal{S}$ denote the complex vector space consisting of all sequences of complex vectors in $\mathbb{C}^m$ indexed by n-tuples of integers $\xi = (\xi_1, \ldots, \xi_n)$. Thus if $u \in \mathcal{S}$, $u = \{u_\xi\}$ where ξ runs over all n-tuples of integers and where each $u_\xi \in \mathbb{C}^m$. For each integer s (positive, negative, or zero) the *Sobolev space* H_s is the subspace of $\mathcal{S}$ defined by

$$(1) \qquad H_s = \left\{u \in \mathcal{S} : \sum_\xi (1 + |\xi|^2)^s \, |u_\xi|^2 < \infty\right\}.$$

It follows from the Schwarz inequality that

$$(2) \quad \left|\sum_\xi (1 + |\xi|^2)^{(s+t)/2} u_\xi \cdot v_\xi\right|^2 \leq \left(\sum_\xi (1 + |\xi|^2)^s \, |u_\xi|^2\right)\left(\sum_\xi (1 + |\xi|^2)^t \, |v_\xi|^2\right);$$

hence the left-hand side of (2) is finite whenever each member of the right-hand side is finite. We can therefore define an inner product on H_s by

$$(3) \qquad \langle u,v\rangle_s = \sum_\xi (1 + |\xi|^2)^s u_\xi \cdot v_\xi .$$

The associated norm is

$$(4) \qquad \|u\|_s = \langle u,u\rangle_s^{1/2}.$$

It also follows from (2) that $\langle u,v\rangle_s$ exists if $u \in H_t$ and $v \in H_{t'}$, where $(t + t')/2 = s$.

Since H_s is simply an l_2-space in which the measure space is the set of all n-tuples of integers ξ, and the measure is the counting measure weighted by $(1 + |\xi|^2)^s$, it follows that H_s is a Hilbert space.

We define a linear transformation K^t on $\mathcal{S}$ for each integer t by setting

$$(5) \qquad (K^t u)_\xi = (1 + |\xi|^2)^t u_\xi.$$

Finally, we identify $\mathcal{P}$ with a subspace of $\mathcal{S}$ by associating with each $\varphi \in \mathcal{P}$ its sequence of Fourier coefficients $\{\varphi_\xi\}$. In view of 6.16(12) we can extend the derivative operator D^α from $\mathcal{P}$ to all of $\mathcal{S}$ by setting

$$(6) \qquad (D^\alpha u)_\xi = \xi^\alpha u_\xi .$$

The inequality 6.16(3) together with 6.16(7) shows that the more differentiable a function is, the greater the order s of the Sobolev space H_s in which its sequence of Fourier coefficients lies. In particular, $\mathcal{P} \subset H_s$ for each s. Moreover, $\mathcal{P}$ is dense in each H_s since each $u \in H_s$ for which all but a finite number of the u_ξ are zero belongs to $\mathcal{P}$. We consider elements of the Sobolev spaces H_s as formal Fourier series or "generalized functions." A fundamental lemma due to Sobolev says that if $u \in H_s$ for sufficiently large s, then the formal Fourier series determined by u actually converges to a function with a certain number of derivatives depending on s. This lemma will be one of the key steps in the proof of the regularity theorem, for it will allow us to conclude that a generalized solution of a partial differential equation which belongs to a sufficiently high Sobolev space is an actual solution.

Some salient features of the H_s spaces are collected in the following theorem, which is divided into parts (a)–(j).

6.18 Theorem

(a) *Let s be a non-negative integer. Then there are constants c and c′, depending at most on s and n, such that*

(1) $$c\,\|\varphi\|_s \le \sum_{[\alpha]=0}^{s} \|D^\alpha\varphi\| \le c'\,\|\varphi\|_s \quad \textit{for all } \varphi \in \mathscr{P}.$$

Moreover, for the case $s = 0$, we actually have the equality

(2) $$\|\varphi\| = \|\varphi\|_0 \quad \textit{for all } \varphi \in \mathscr{P}.$$

(b) *If $t < s$, then $\|u\|_t \le \|u\|_s$, so $H_s \subset H_t$. Thus the union of the H_s spaces over all integers s is a subspace of $\mathcal{S}$, which we denote by $H_{-\infty}$.*

(c) *$\mathscr{P}$ is a dense subspace of H_s for each s.*

(d) *K^t is an isometry of H_s onto H_{s-2t} with inverse K^{-t},*

(3) $$\|u\|_s = \|K^t u\|_{s-2t};$$

and K^t maps $\mathscr{P}$ into $\mathscr{P}$. If $\varphi \in \mathscr{P}$ and $t \ge 0$, then

(4) $$K^t\varphi = \left(1 - \sum_{i=1}^{n} \frac{\partial^2}{\partial x_i^{\,2}}\right)^t \varphi.$$

Moreover, for all s and t

(5) $$\langle u,v\rangle_s = \langle u,K^t v\rangle_{s-t} = \langle K^t u,v\rangle_{s-t} \quad \textit{for } u, v \in H_s.$$

(e) **Schwartz Inequality** *If $u \in H_{s+t}$ and $v \in H_{s-t}$, then*

$$|\langle u,v\rangle_s| \le \|u\|_{s+t}\,\|v\|_{s-t}.$$

(f) *If $u \in H_{s+t}$, then*

$$\|u\|_{s+t} = \sup_{\substack{v\in H_{s-t}\\ v\neq 0}} \frac{|\langle u,v\rangle_s|}{\|v\|_{s-t}}.$$

(g) **"Peter-Paul" Inequality** *Given integers $t' < t < t''$ and $\varepsilon > 0$, there is a constant $c(\varepsilon) > 0$ such that*

$$\|u\|_t^2 \le \varepsilon\,\|u\|_{t''}^2 + c(\varepsilon)\,\|u\|_{t'}^2$$

for all $u \in H_{t''}$. (We refer to this inequality as the Peter-Paul inequality in view of the comparison with part (b), although the morality has been twisted around. Rather than robbing Peter to pay Paul, here we are paying Paul ($\|u\|_{t'}$) in order to rob Peter ($\|u\|_{t''}$).)

(h) *D^α is a bounded operator from $H_{s+[\alpha]}$ to H_s for each s; indeed,*

$$\|D^\alpha u\|_s \le \|u\|_{s+[\alpha]} \quad \textit{for all} \quad u \in H_{s+[\alpha]}.$$

(i) *Let ω be a C^{∞} complex-valued periodic function on $\mathbb{R}^n$. Then given an integer s, there are positive integers c and c′, with c depending only on s and n, and c′ depending on s, n, and on ω and its derivatives, such that*

(6) $$\|\omega\varphi\|_s \leq c\,\|\omega\|_{\infty}\,\|\varphi\|_s + c'\,\|\varphi\|_{s-1}, \quad \textit{for } \varphi \in \mathscr{P}.$$

In particular, there is a constant c″ depending on ω, s, and n such that

(7) $$\|\omega\varphi\|_s \leq c''\,\|\varphi\|_s,$$

so multiplication by ω extends by continuity to a bounded operator on H_s.

(j) *Let ω be a C^{∞} complex-valued periodic function on $\mathbb{R}^n$. Then given an integer s, there is a positive constant c such that*

(8) $$|\langle\omega u,v\rangle_s - \langle u,\bar{\omega}v\rangle_s| \leq c(\|u\|_s\,\|v\|_{s-1} + \|u\|_{s-1}\,\|v\|_s)$$

for each $u, v \in H_s$. For the case $s = 0$, we have

(9) $$\langle\omega u,v\rangle_0 = \langle u,\bar{\omega}v\rangle_0.$$

PROOF The inequality (1) follows from 6.16(15) and the fact that whenever $a_1, \ldots, a_n$ are positive numbers, then

(10) $$\frac{1}{n}\left(\sum_{i=1}^{n} a_i\right)^2 \leq \sum_{i=1}^{n} a_i^2 \leq \left(\sum_{i=1}^{n} a_i\right)^2.$$

The equality (2) is simply the Parseval identity 6.16(13).

Part (b) is obvious. The fact that 6.16(3) holds for each integer $k > 0$ implies that $\{\varphi_\xi\} \in H_s$ whenever $\varphi \in \mathscr{P}$, and as we have already indicated in the remarks immediately preceding the theorem, $\mathscr{P}$ is dense in H_s since each $u \in H_s$ for which all but a finite number of the u_ξ are zero belongs to $\mathscr{P}$.

The identities (3) and (5) in (d) are obvious; that the inner products are all well-defined follows from the remark immediately following 6.17(4). Let $\varphi \in \mathscr{P}$. We have $K^t\varphi = \{(1 + |\xi|^2)^t\varphi_\xi\}$. It follows from 6.16(3) and (7) and from the fact that there exists the inequality

(11) $$\xi^{2\alpha} \leq (1 + |\xi|^2)^{[\alpha]}$$

that the series

(12) $$\sum_{\xi} (1 + |\xi|^2)^t\varphi_\xi e^{ix\cdot\xi}$$

and all its formal derivatives

(13) $$\sum_{\xi} D^{\alpha}(1 + |\xi|^2)^t\varphi_\xi e^{ix\cdot\xi} = \sum_{\xi} \xi^{\alpha}(1 + |\xi|^2)^t\varphi_\xi e^{ix\cdot\xi}$$

converge uniformly. Thus (12) converges to a periodic C^{∞} function and therefore is an element of $\mathscr{P}$. Hence K^t maps $\mathscr{P}$ into $\mathscr{P}$. For (4), simply observe that the ξth Fourier coefficient of the right-hand side equals $(1 + |\xi|^2)^t\varphi_\xi$.

The Schwartz inequality (e) is none other than 6.17(2). It follows from (e) that

$$\|u\|_{s+t} \geq \sup_{\substack{v \in H_{s-t} \\ v \neq 0}} \frac{|\langle u,v\rangle_s|}{\|v\|_{s-t}}.$$

To prove the equality in (*f*), let $v = K^t u$. Then by (d),

$$\|u\|_{s+t} = \frac{\langle u,u\rangle_{s+t}}{\|u\|_{s+t}} = \frac{\langle u,v\rangle_s}{\|v\|_{s-t}}.$$

To prove the Peter-Paul inequality (g), first observe that for any positive number y,

$$1 \leq y^{t''-t} + \left(\frac{1}{y}\right)^{t-t'} \tag{14}$$

since either y or $1/y$ is greater than or equal to 1. If in (14) we let

$$y = \varepsilon^{1/(t''-t)}(1 + |\xi|^2),$$

we obtain

$$(1 + |\xi|^2)^t \leq \varepsilon(1 + |\xi|^2)^{t''} + \varepsilon^{(t'-t)/(t''-t)}(1 + |\xi|^2)^{t'},$$

from which the Peter-Paul inequality follows with

$$c(\varepsilon) = \varepsilon^{(t'-t)/(t''-t)}.$$

The inequality in (h) follows from the inequality (11).

In part (i), the inequality (7) follows immediately from part (b) and inequality (6). To prove (6), we first consider the case in which $s \geq 0$. Let $\varphi \in \mathscr{P}$. Then by applying (1), we obtain

$$\begin{aligned}
\|\omega\varphi\|_s &\leq \text{const} \sum_{[\alpha]=0}^{s} \|D^\alpha \omega\varphi\| \\
&\leq \text{const} \sum_{[\alpha]=0}^{s} \|\omega D^\alpha \varphi\| + \text{const} \sum_{[\alpha]=0}^{s} \|D^\alpha \omega\varphi - \omega D^\alpha \varphi\| \\
&\leq c\, \|\omega\|_\infty\, \|\varphi\|_s + \text{const} \sum_{[\alpha]=0}^{s-1} \|D^\alpha \varphi\| \\
&\leq c\, \|\omega\|_\infty\, \|\varphi\|_s + c'\, \|\varphi\|_{s-1}.
\end{aligned}$$

At the third stage we have used the fact that $D^\alpha\omega\varphi - \omega D^\alpha\varphi$ only contains derivatives of φ up to order $[\alpha] - 1$. For $s < 0$ we have

$$\begin{aligned}
\|\omega\varphi\|_s^2 &= \langle \omega\varphi,\omega\varphi\rangle_s = \langle \omega K^{-s}K^s\varphi, K^s\omega\varphi\rangle_0 \\
&= \langle K^{-s}\omega K^s\varphi, K^s\omega\varphi\rangle_0 + \langle (\omega K^{-s} - K^{-s}\omega)K^s\varphi, K^s\omega\varphi\rangle_0 \\
&\leq |\langle K^{-s}\omega K^s\varphi, K^s\omega\varphi\rangle_0| + \left|\left\langle \sum_{[\alpha]=0}^{-2s-1} a_\alpha D^\alpha K^s\varphi, K^s\omega\varphi \right\rangle_0\right|,
\end{aligned} \tag{15}$$

where the a_α are combinations of derivatives of ω. Now, by the case $s \geq 0$ of (i) we obtain

$$(16)\quad \begin{aligned} |\langle K^{-s}\omega K^s\varphi, K^s\omega\varphi\rangle_0| &= |\langle \omega K^s\varphi, K^s\omega\varphi\rangle_{-s}| \leq \|\omega K^s\varphi\|_{-s}\,\|K^s\omega\varphi\|_{-s} \\ &\leq (c\,\|\omega\|_\infty\,\|K^s\varphi\|_{-s} + k'\,\|K^s\varphi\|_{-s-1})(\|K^s\omega\varphi\|_{-s}) \\ &= (c\,\|\omega\|_\infty\,\|\varphi\|_s + k'\,\|\varphi\|_{s-1})\,\|\omega\varphi\|_s. \end{aligned}$$

As for the last term in (15),

$$(17)\quad \begin{aligned} \left|\left\langle \sum_{[\alpha]=0}^{-2s-1} a_\alpha D^\alpha K^s\varphi, K^s\omega\varphi \right\rangle_0\right| &\leq \text{const} \sum_{[\alpha]=0}^{-2s-1} |\langle D^\alpha K^s\varphi, K^s\omega\varphi\rangle_0| \\ &\leq \text{const} \sum_{[\alpha]=0}^{-2s-1} \|D^\alpha K^s\varphi\|_s\,\|K^s\omega\varphi\|_{-s} \\ &\leq \text{const} \sum_{[\alpha]=0}^{-2s-1} \|K^s\varphi\|_{s+[\alpha]}\,\|K^s\omega\varphi\|_{-s} \\ &\leq \text{const}\,\|K^s\varphi\|_{-s-1}\,\|K^s\omega\varphi\|_{-s} \\ &\leq \text{const}\,\|\varphi\|_{s-1}\,\|\omega\varphi\|_s. \end{aligned}$$

Thus for $s < 0$, (6) follows from (15), (16), and (17).

It is sufficient to prove (8) and (9) in part (j) on the dense subspace $\mathscr{P}$ of H_s. Observe that (9) follows immediately from the fact (2) that the L_2-norm is identical with the Sobolev norm $\|\ \|_0$ on $\mathscr{P}$. Let $\varphi, \psi \in \mathscr{P}$. We consider the case of (8) in which s is negative. Using part (d) and equation (9), we obtain

$$\begin{aligned} \langle \omega\varphi,\psi\rangle_s &= \langle \omega K^{-s}K^s\varphi, K^s\psi\rangle_0 \\ &= \langle K^{-s}K^s\varphi, \bar{\omega}K^s\psi\rangle_0 = \langle K^s\varphi, K^{-s}\bar{\omega}K^s\psi\rangle_0 \\ &= \langle \varphi, \bar{\omega}\psi\rangle_s + \langle K^s\varphi, (K^{-s}\bar{\omega} - \bar{\omega}K^{-s})K^s\psi\rangle_0. \end{aligned}$$

It follows as in (17) that

$$|\langle \omega\varphi,\psi\rangle_s - \langle \varphi,\bar{\omega}\psi\rangle_s| \leq \text{const}\,\|\varphi\|_s\,\|\psi\|_{s-1}.$$

By symmetry we also have

$$|\langle \omega\varphi,\psi\rangle_s - \langle \varphi,\bar{\omega}\psi\rangle_s| \leq \text{const}\,\|\psi\|_s\,\|\varphi\|_{s-1}.$$

Thus (8) is proved for s negative. The proof is similar for s positive. The proof of Theorem 6.18 is complete.

6.19 Difference Quotients If $\varphi \in \mathscr{P}$, then the ξth Fourier coefficient of the translate $\varphi(x+h)$ of φ by an element $h \in \mathbb{R}^n$ is $e^{ih\cdot\xi}\varphi_\xi$. Thus if $u \in \mathcal{S}$ and $h \in \mathbb{R}^n$, we define the *translate of u by h* to be the element

$$(1)\qquad T_h(u) = \{e^{ih\cdot\xi}u_\xi\} \in \mathcal{S}.$$

The *difference quotient of u determined by a non-zero h* is the element

$$(2)\qquad u^h = \frac{T_h(u) - u}{|h|} = \left\{\left(\frac{e^{ih\cdot\xi} - 1}{|h|}\right)u_\xi\right\} \in \mathcal{S}.$$

So if $\varphi \in \mathscr{P}$, then $T_h(\varphi)(x) = \varphi(x + h)$ and

$$\varphi^h(x) = \frac{\varphi(x + h) - \varphi(x)}{|h|}.$$

Observe that if $u \in H_s$, then

$$(3)\qquad \|T_h(u)\|_s = \|u\|_s,$$

so T_h is an isometry on H_s. Thus, in particular, if $u \in H_s$, then also $u^h \in H_s$ for each h. It follows from the inequality

$$(4)\qquad \left|\frac{e^{ih\cdot\xi} - 1}{|h|}\right|^2 = \left|\frac{(\cos h\cdot\xi) - 1}{|h|}\right|^2 + \left|\frac{\sin h\cdot\xi}{|h|}\right|^2 = \frac{2(1 - \cos h\cdot\xi)}{|h|^2}$$

$$= \frac{4\sin^2(\frac{1}{2}h\cdot\xi)}{|h|^2} \le \frac{(h\cdot\xi)^2}{|h|^2} \le (1 + |\xi|^2)$$

that if $u \in H_{s+1}$, then the u^h are uniformly bounded in the s-norm. In fact,

$$(5)\qquad \|u^h\|_s \le \|u\|_{s+1}.$$

We shall need the converse of this, as follows.

6.20 Lemma *Let $u \in H_s$, and assume that there is a constant k such that $\|u^h\|_s \le k$ for all non-zero $h \in \mathbb{R}^n$. Then $u \in H_{s+1}$.*

PROOF For each positive integer N we let u_N be the element of H_s obtained by truncating u at N; that is,

$$(1)\qquad (u_N)_\xi = \begin{cases} u_\xi & \text{if } |\xi| < N \\ 0 & \text{otherwise.} \end{cases}$$

We need only prove that the $\|u_N\|_{s+1}$ are uniformly bounded. Let $(e_1, \ldots, e_n)$ be the standard orthonormal basis of $\mathbb{R}^n$, and let $h = te_i$. Then

$$(2)\qquad \left|\frac{e^{ih\cdot\xi} - 1}{|h|}\right|^2 = \left|\frac{e^{it\xi_i} - 1}{t}\right|^2 \to |\xi_i|^2 \quad \text{as} \quad t \to 0.$$

Since there are only finitely many ξ with $|\xi| < N$, and since by hypothesis

$$\sum_{|\xi| < N} (1 + |\xi|^2)^s |u_\xi|^2 \left|\frac{e^{ih\cdot\xi} - 1}{|h|}\right|^2 \le k^2,$$

it follows from (2) that

$$\sum_{|\xi| < N} (1 + |\xi|^2)^s |u_\xi|^2 |\xi_i|^2 \le k^2.$$

Thus $\|u_N\|_{s+1} = \sum_{|\xi|<N} (1 + |\xi|^2)^{s+1}|u_\xi|^2 \le nk^2 + \|u\|_s^2$, so the $\|u_N\|_{s+1}$ are uniformly bounded, and consequently $u \in H_{s+1}$.

6.21 Associate with each $u \in H_{-\infty}$ the series $\sum_\xi u_\xi e^{ix\cdot\xi}$. It will be of fundamental importance in what follows to know when this series converges, and (if it converges) how differentiable its limit is. The answer is supplied by the following lemma due to Sobolev.

6.22 Sobolev Lemma *If $t \ge [n/2] + 1$ and $u \in H_t$, then the series $\sum_\xi u_\xi e^{ix\cdot\xi}$ converges uniformly. Thus each $u \in H_t$ for $t \ge [n/2] + 1$ corresponds to a continuous function.*

PROOF It is sufficient to demonstrate that the series converges absolutely, $\sum |u_\xi| < \infty$. Now,

$$\sum_{|\xi|<N} |u_\xi| = \sum_{|\xi|<N} (1+ |\xi|^2)^{-t/2}(1 + |\xi|^2)^{t/2}\,|u_\xi|$$

$$\le \left(\sum_{|\xi|<N} (1 + |\xi|^2)^{-t}\right)^{1/2} \left(\sum_{|\xi|<N} (1 + |\xi|^2)^t\,|u_\xi|^2\right)^{1/2}$$

$$\le \left(\sum_{|\xi|<N} (1 + |\xi|^2)^{-t}\right)^{1/2} \|u\|_t.$$

Thus the result follows from 6.16(7).

Corollary (a) *If $u \in H_t$ where $t \ge [n/2] + 1 + m$, then $D^\alpha u = \sum_\xi \xi^\alpha u_\xi e^{ix\cdot\xi}$ converges uniformly for $[\alpha] \le m$. Thus each $u \in H_t$ for this range of t corresponds to a function $\sum_\xi u_\xi e^{ix\cdot\xi}$ of class C^m.*

PROOF With $u \in H_t$ for $t \ge [n/2] + 1 + m$ and with $[\alpha] \le m$, it follows from 6.18(h) that $D^\alpha u \in H_{t-[\alpha]}$, where $t - [\alpha] \ge [n/2] + 1$. Thus it follows from the Sobolev lemma that $\sum_\xi \xi^\alpha u_\xi e^{ix\cdot\xi}$ converges uniformly. Now this series is the αth "formal" derivative of the series $\sum_\xi u_\xi e^{ix\cdot\xi}$. Thus $\sum_\xi u_\xi e^{ix\cdot\xi}$ is of class C^m.

Corollary (b) *From the proof of 6.22 it follows that if $t \ge [n/2] + 1$, then there is a constant $c > 0$ such that if $\varphi \in \mathscr{P}$, then*

$$\|\varphi\|_\infty \le c\,\|\varphi\|_t. \tag{1}$$

Applying (1) to $D^\alpha\varphi$ and using 6.18(h), we obtain

$$\|D^\alpha\varphi\|_\infty \le c\,\|\varphi\|_{t+[\alpha]}. \tag{2}$$

6.23 Rellich Lemma *Let $\{u^i\}$ be a sequence of elements of H_t with $\|u^i\|_t \le 1$. If $s < t$, then there is a subsequence of $\{u^i\}$ which converges in H_s.*

PROOF By assumption,

$$\sum_\xi (1 + |\xi|^2)^t\,|u_\xi^i|^2 \le 1. \tag{1}$$

For each fixed ξ, elements of the sequence $\{|(1 + |\xi|^2)^{t/2}u^i_\xi|\}$ are all bounded by 1, so the sequence $\{(1 + |\xi|^2)^{t/2}u^i_\xi\}$ has a convergent subsequence in $\mathbb{C}^m$. By the usual diagonal process, one can select a subsequence $\{u^{j_i}\}$ such that the sequence $(1 + |\xi|^2)^{t/2}u^{j_i}_\xi$ converges in $\mathbb{C}^m$ for each fixed ξ. We claim that $\{u^{j_i}\}$ is Cauchy, and therefore convergent, in H_s if $s < t$. Let $\varepsilon > 0$ be given. Now,

$$\text{(2)} \qquad \|u^{j_i} - u^{j_k}\|_s^2 = \sum_{|\xi| < N} (1 + |\xi|^2)^{s-t}(1 + |\xi|^2)^t |u^{j_i}_\xi - u^{j_k}_\xi|^2$$
$$+ \sum_{|\xi| \geq N} (1 + |\xi|^2)^{s-t}(1 + |\xi|^2)^t |u^{j_i}_\xi - u^{j_k}_\xi|^2.$$

The second sum in (2) is bounded by

$$N^{2(s-t)} \sum_{|\xi| \geq N} (1 + |\xi|^2)^t(|u^{j_i}_\xi|^2 + 2\,|u^{j_i}_\xi|\,|u^{j_k}_\xi| + |u^{j_k}_\xi|^2),$$

which is $\leq 4N^{2(s-t)}$ in view of (1). Since $s - t < 0$, $4N^{2(s-t)}$ can be made less than $\varepsilon/2$ by taking N large enough, say $N = N_0$. The first sum in (2) is then bounded by

$$\text{(3)} \qquad \sum_{|\xi| < N_0} (1 + |\xi|^2)^t |u^{j_i}_\xi - u^{j_k}_\xi|^2,$$

and since there are only a finite number of terms in this sum and since the sequences $(1 + |\xi|^2)^{t/2}u^{j_i}_\xi$ converge for each fixed ξ, there is a constant $J > 0$ such that if j_i and j_k are greater than J, then (3) is less than $\varepsilon/2$. Thus for $j_i, j_k > J$, we have $\|u^{j_i} - u^{j_k}\|_s^2 < \varepsilon$, and the proof is complete.

6.24 Definitions A (*linear*) *differential operator L of order l* on the $\mathbb{C}^m$-valued C^∞ functions on $\mathbb{R}^n$ consists of an $m \times m$ matrix (L_{ij}) in which

$$\text{(1)} \qquad L_{ij} = \sum_{[\alpha]=0}^{l} a^\alpha_{ij} D^\alpha,$$

and the a^α_{ij} are C^∞ complex-valued functions on $\mathbb{R}^n$, with at least one $a^\alpha_{ij} \not\equiv 0$ for some i, j and for some α for which $[\alpha] = l$. A differential operator L is a *periodic differential operator*, or an *operator on $\mathscr{P}$*, if, in addition, the a^α_{ij} are periodic functions.

Let L be a periodic differential operator, and let $\varphi \in \mathscr{P}$ with component functions $\varphi_1, \ldots, \varphi_m$. Then

$$\text{(2)} \qquad L\varphi = \left(\sum_j L_{1j}\varphi_j, \ldots, \sum_j L_{mj}\varphi_j\right).$$

It follows from integration by parts that if we define the operator L^* on $\mathscr{P}$ by

$$\text{(3)} \qquad L^*_{ij} = \sum_{[\alpha]=0}^{l} D^\alpha \overline{a^\alpha_{ji}}$$

so that the ith component of $L^*\varphi$ is given by

(4)
$$(L^*\varphi)_i = \sum_{j=0}^{m} \sum_{[\alpha]=0}^{l} D^\alpha(\overline{a_{ji}^\alpha \varphi_j}),$$

then in the L_2 inner product on $\mathscr{P}$ we have

(5)
$$\langle L\varphi,\psi\rangle = \langle \varphi, L^*\psi\rangle$$

for $\varphi, \psi \in \mathscr{P}$. L^* is called the *formal adjoint of L*. The word "formal" is used here to emphasize that L^* is not the adjoint of L on a Hilbert space. It is simply the adjoint relative to the L_2 inner product on $\mathscr{P}$.

6.25 Proposition *Let L be a partial differential operator on $\mathscr{P}$ of order l, and let s be an integer. Then there are positive constants c, k, and c′, where c depends only on n, m, l, and s, where k is a bound on the absolute values of the coefficients of the highest order terms in L, and where c′ depends on n, m, l, s and on all the coefficients of L and their derivatives up to order l, such that*

(1)
$$\|L\varphi\|_s \le ck \, \|\varphi\|_{s+l} + c' \, \|\varphi\|_{s+l-1}$$

for all $\varphi \in \mathscr{P}$. In particular, there is a constant c″ such that

(2)
$$\|L\varphi\|_s \le c'' \, \|\varphi\|_{s+l}$$

for all $\varphi \in \mathscr{P}$, so L extends by continuity to a bounded operator from H_{s+l} to H_s for each s.

PROOF The inequality (2) is an immediate consequence of (1) and 6.18(b). As for (1), the case $m = 1$ (in which elements of $\mathscr{P}$ are $\mathbb{C}^1$ valued functions and the operator L is a single partial differential operator $\sum_\alpha a^\alpha D^\alpha$ rather than a matrix) follows immediately from 6.18(h) and (i). The inequality (1) for general m follows from the case $m = 1$ and from the inequality $\|L\varphi\|_s \le \text{const} \sum_{i,j} \|L_{ij}\varphi_j\|_s$, where $\varphi = (\varphi_1, \ldots, \varphi_m) \in \mathscr{P}$, and the constant depends only on m.

6.26 Remark If L is an operator on $\mathscr{P}$ of order l, and if ω is a C^∞ complex-valued function on $\mathbb{R}^n$, then the operator $M = \omega L - L\omega$, where $M\varphi = \omega(L\varphi) - L(\omega\varphi)$, is of order at most $l - 1$. Consequently, given s, there is a positive constant such that

(1)
$$\|M\varphi\|_s \le \text{const} \, \|\varphi\|_{s+l-1}$$

for all $\varphi \in \mathscr{P}$.

6.27 Lemma *If ω is a real-valued periodic C^∞ function, and L is a differential operator of order l on $\mathscr{P}$, then there is a positive constant such that*

(1)
$$|\langle L(\omega^2 u), Lu\rangle_s - \|L(\omega u)\|_s^2| \le \text{const} \, (\|u\|_{s+l} \, \|u\|_{s+l-1})$$

for all $u \in H_{s+l}$.

PROOF $|\langle L(\omega^2 u), Lu\rangle_s - \langle L(\omega u), L(\omega u)\rangle_s|$

$$\leq |\langle \omega L(\omega u), Lu\rangle_s - \langle L(\omega u), \omega Lu\rangle_s| + |\langle L(\omega u), (\omega L - L\omega)u\rangle_s| + |\langle (L\omega - \omega L)(\omega u), Lu\rangle_s|,$$

and (1) follows by applying 6.18(j), 6.25(2), and 6.18(7) to the first term and applying the Schwartz inequality, 6.26(1), 6.25(2), and 6.18(7) to the last two terms.

ELLIPTIC OPERATORS

6.28 Definition Let L be a partial differential operator of order l. We write L as

(1) $$L = P_l(D) + \cdots + P_0(D),$$

where $P_j(D)$ is an $m \times m$ matrix each entry of which is a differential operator $\sum_{[\alpha]=j} a_\alpha D^\alpha$, homogeneous of order j, and where the a_α are C^∞ complex-valued functions on $\mathbb{R}^n$. We let $P_j(\xi)$ denote the matrix obtained by substituting ξ^α for D^α in $P_j(D)$, where $\xi = (\xi_1, \ldots, \xi_n)$ is a point in $\mathbb{R}^n$. L is said to be *elliptic at the point* $x \in \mathbb{R}^n$ if the matrix $P_l(\xi)$ is non-singular at x for each non-zero ξ. L is *elliptic* if it is elliptic at each x. Observe that ellipticity is a condition on the highest-order part of L only. Observe also that L is elliptic at x if and only if

(2) $$L(\varphi^l u)(x) \neq 0$$

for each $\mathbb{C}^m$-valued C^∞ function u such that $u(x) \neq 0$ and each smooth real-valued function φ such that $\varphi(x) = 0$ but $d\varphi(x) \neq 0$, since for each such φ and u,

(3) $$L(\varphi^l u)(x) = P_l(D)(\varphi^l u)(x) = P_l(d\varphi|_x)(u(x)).$$

The advantage of this criterion (2) for ellipticity is that it generalizes to a coordinate-free definition of ellipticity on manifolds, as we shall see later.

The basic analytic property of an elliptic operator that we shall need is the following.

6.29 Fundamental Inequality *Let L be an elliptic operator on $\mathscr{P}$ of order l, and let s be an integer. Then there is a constant $c > 0$ such that*

(1) $$\|u\|_{s+l} \leq c(\|Lu\|_s + \|u\|_s)$$

for all $u \in H_{s+l}$.

PROOF It is sufficient to prove (1) for all $\varphi \in \mathscr{P}$. The proof consists of several parts. We first consider the case of an elliptic operator L_0 on $\mathscr{P}$ with constant coefficients, which consists of the leading term

$P_l(D)$ only. If $u \in \mathbb{R}^n$ with $u \neq 0$, and if $\xi \neq 0$, then since $P_l(\xi)$ is non-singular, we have $|P_l(\xi)u|^2 > 0$. It follows from the compactness of the unit sphere in $\mathbb{R}^n$ that there is a constant $c > 0$ such that

$$|P_l(\xi)u|^2 \geq c$$

for all u and ξ such that $|u| = |\xi| = 1$. From this it follows that

$$|P_l(\xi)u|^2 \geq c|\xi|^{2l}\,|u|^2 \tag{2}$$

for all u and ξ in $\mathbb{R}^n$. Thus for $\varphi \in \mathscr{P}$, it follows from (2) and the fact that L_0 has constant coefficients that

$$\begin{aligned} \|L_0\varphi\|_s^{\,2} &= \sum_\xi |P_l(\xi)\varphi_\xi|^2\,(1 + |\xi|^2)^s \\ &\geq \text{const} \sum_\xi |\xi|^{2l}\,|\varphi_\xi|^2(1 + |\xi|^2)^s. \end{aligned} \tag{3}$$

Hence

$$\begin{aligned} (\|L_0\varphi\|_s + \|\varphi\|_s)^2 &\geq \|L_0\varphi\|_s^{\,2} + \|\varphi\|_s^{\,2} \\ &\geq \sum_\xi |\varphi_\xi|^2(1 + |\xi|^2)^s(1 + \text{const}\,|\xi|^{2l}) \\ &\geq \text{const} \sum_\xi |\varphi_\xi|^2\,(1 + |\xi|^2)^{s+l} \\ &= \text{const}\,\|\varphi\|_{s+l}^2. \end{aligned} \tag{4}$$

Secondly, consider a general periodic elliptic operator L of order l, and let $p \in \mathbb{R}^n$. We shall prove that there is a neighborhood U of p such that (1) holds for all $\varphi \in \mathscr{P}$ with support in U. (By a slight abuse of terminology we say that the support of a periodic function φ lies in U if the support of φ lies in the union of U with all of the periodic translates of U.) Let L_0 denote the constant coefficient elliptic operator, homogeneous of order l, determined by the highest-order part of L at the point p. Then it follows from (4) that for each $\varphi \in \mathscr{P}$,

$$\begin{aligned} \|\varphi\|_{s+l} &\leq \text{const}\,(\|L_0\varphi\|_s + \|\varphi\|_s) \\ &\leq \text{const}\,(\|L\varphi\|_s + \|(L_0 - L)\varphi\|_s + \|\varphi\|_s). \end{aligned} \tag{5}$$

Let k denote the constant in (5). Then choose a positive ε smaller than $1/(2ck)$, where c denotes the constant c of 6.25. On a small enough neighborhood of p the coefficients of the highest-order part of $L_0 - L$ are less than ε in absolute value. Let $\tilde{L}$ be a periodic operator agreeing with $L_0 - L$ on a possibly smaller neighborhood U of p and with coefficients in the highest-order part everywhere less than ε in absolute value. Then it follows from (5), 6.25(1), and the choice of ε, that for an element $\varphi \in \mathscr{P}$ whose support lies in U,

$$\begin{aligned} \|\varphi\|_{s+l} &\leq \text{const}\,(\|L\varphi\|_s + \|\tilde{L}\varphi\|_s + \|\varphi\|_s) \\ &\leq \text{const}\,\|L\varphi\|_s + \tfrac{1}{2}\,\|\varphi\|_{s+l} + \text{const}\,\|\varphi\|_{s+l-1} + \text{const}\,\|\varphi\|_s. \end{aligned}$$

Applying the Peter-Paul inequality to the term $\|\varphi\|_{s+l-1}$, we obtain

$$\|\varphi\|_{s+l} \leq \text{const } \|L\varphi\|_s + \tfrac{3}{4}\|\varphi\|_{s+l} + \text{const } \|\varphi\|_s,$$

which proves (1) for these φ.

Let T^n be the torus obtained as the quotient of $\mathbb{R}^n$ by the lattice consisting of all points $2\pi\xi$, where ξ is an n-tuple of integers. The open sets U obtained above for each $p \in \mathbb{R}^n$ project to an open cover of T^n. Let $U_1, \ldots, U_k$ be a finite subcover, and let $\omega_1, \ldots, \omega_k$ be a partition of unity fitting this cover, of the special form

$$\sum_{i=1}^{k} \omega_i^2 = 1. \tag{6}$$

This is easily arranged—see the proofs of 1.11 and 1.10. Now consider the ω_i as periodic C^∞ (real-valued) functions on $\mathbb{R}^n$. Let $\varphi \in \mathscr{P}$. Then by (6), and by 6.18(i) and (j),

$$\|\varphi\|_{s+l}^2 = \langle\varphi,\varphi\rangle_{s+l} = \left\langle\sum_i \omega_i^2\varphi,\varphi\right\rangle_{s+l}$$
$$\leq \sum_i \langle\omega_i\varphi,\omega_i\varphi\rangle_{s+l} + \text{const } \|\varphi\|_{s+l}\,\|\varphi\|_{s+l-1}.$$

Now since $\omega_i\varphi$ has support in one of the small open sets U obtained above, and since there are only finitely many of the ω_i, there are constants such that the last displayed line above is

$$\leq \text{const } \sum_i \|L\omega_i\varphi\|_s^2 + \text{const } \|\varphi\|_s^2 + \text{const } \|\varphi\|_{s+l}\,\|\varphi\|_{s+l-1}$$
$$\leq \text{const } \sum_i \langle L(\omega_i^2\varphi),L\varphi\rangle_s + \text{const } \|\varphi\|_s^2 + \text{const } \|\varphi\|_{s+l}\,\|\varphi\|_{s+l-1}$$

(by 6.27)

$$= \text{const } \|L\varphi\|_s^2 + \text{const } \|\varphi\|_s^2 + \text{const } \|\varphi\|_{s+l}\,\|\varphi\|_{s+l-1}$$
$$\leq \text{const } \|L\varphi\|_s^2 + \text{const } \|\varphi\|_s^2 + \tfrac{1}{2}\|\varphi\|_{s+l}^2 + \text{const } \|\varphi\|_{s+l-1}^2$$
$$\leq \text{const } \|L\varphi\|_s^2 + \text{const } \|\varphi\|_s^2 + \tfrac{3}{4}\|\varphi\|_{s+l}^2 + \text{const } \|\varphi\|_s^2,$$

(by the Peter-Paul inequality).

It follows that (1) holds for all $\varphi \in \mathscr{P}$ and hence for all $u \in H_{s+l}$.

6.30 Theorem (Regularity for Periodic Elliptic Operators) *Let L be a periodic elliptic operator of order l. Assume that $u \in H_{-\infty}$, $v \in H_t$, and*

$$Lu = v. \tag{1}$$

Then $u \in H_{t+l}$.

PROOF It is sufficient to prove that if $u \in H_s$ and $v = Lu \in H_{s-l+1}$, then $u \in H_{s+1}$. Let $h \in \mathbb{R}^n$ with $h \neq 0$, and let L^h represent the operator obtained from L by replacing each coefficient α by its difference quotient

$$\frac{\alpha(x+h)-\alpha(x)}{|h|}.$$

Then it follows for $\varphi \in \mathscr{P}$, and hence by continuity for all $u \in H_{-\infty}$, that

$$(2) \qquad L(u^h) = (Lu)^h - L^h(T_h u) \qquad \text{(cf. 6.19)}.$$

It follows from (2) and the Fundamental Inequality 6.29(1), that

$$\begin{aligned}(3) \quad \|u^h\|_s &\leq \text{const}\,\|L(u^h)\|_{s-l} + \text{const}\,\|u^h\|_{s-l} \\ &\leq \text{const}\,\|(Lu)^h\|_{s-l} + \text{const}\,\|L^h(T_h u)\|_{s-l} + \text{const}\,\|u^h\|_{s-l}.\end{aligned}$$

Now since the coefficients of the operator L are periodic C^∞ functions, their difference quotients are uniformly bounded, so that

$$(4) \qquad \|L^h(T_h u)\|_{s-l} \leq \text{const}\,\|T_h(u)\|_s$$

where the constant does not depend on h. It follows from (3), (4), and 6.19(3) and (5) that

$$\|u^h\|_s \leq \text{const}\,\|Lu\|_{s-l+1} + \text{const}\,\|u\|_s,$$

where the right-hand side is independent of h. Thus, by 6.20, $u \in H_{s+1}$, and the theorem is proved.

REDUCTION TO THE PERIODIC CASE

6.31 Remarks Before beginning the proof of Theorem 6.5 we need to establish some convenient notation and to make a few observations.

We are going to let C^∞ denote the set of all complex m-space valued C^∞ functions on $\mathbb{R}^n$. C_0^∞ will denote those of compact support, and $C_0^\infty(V)$ those whose compact support lies in V. By the L_2 inner product on C_0^∞ we shall mean

$$(1) \qquad \langle u,v\rangle = \frac{1}{(2\pi)^n}\int_{R^n} u \cdot v,$$

where $u \cdot v$ as before denotes the Hermitian product $u_1\overline{v_1} + \cdots + u_m\overline{v_m}$.

Let V be an open set in $\mathbb{R}^n$ with $\overline{V}$ contained in some 2π cube. Then by extending periodically we can (and do) identify $C_0^\infty(V)$ with a subspace of $\mathscr{P}$. Observe that the L_2 inner product on $C_0^\infty(V)$ agrees with the L_2 inner product on $C_0^\infty(V)$ considered as a subset of $\mathscr{P}$, which in turn agrees with the Sobolev inner product $\| \quad \|_0$.

Now suppose that L is an elliptic partial differential operator of order l on C^∞ (no assumption of periodic coefficients). L has a formal adjoint L^* on C_0^∞, with respect to the ordinary L_2 inner product on C_0^∞, obtained by integration by parts. Just as in 6.24, if $L = (L_{ij})$ where $L_{ij} = \sum a_{ij}^\alpha D^\alpha$, then $L^* = (L_{ij}^*)$ where $L_{ij}^* = \sum D^\alpha \overline{a_{ji}^\alpha}$. L^* is also a differential operator of order l.

Now let $p \in \mathbb{R}^n$. Then there is a sufficiently small neighborhood V of p and a periodic elliptic operator $\tilde{L}$ such that $\tilde{L}$ agrees with L on V. For let L_0 denote the constant coefficient operator determined by L at p. Then since L is elliptic at p, there is some $\varepsilon > 0$ such that any operator whose coefficients are everywhere within ε in absolute value of the corresponding coefficients of L_0 is elliptic. So let U be a neighborhood of p, small enough to be contained in some 2π cube Q, on which the coefficients of L differ from the corresponding coefficients of L_0 by at most ε; and let $V \subset \overline{V} \subset U$. Choose a C^∞ function φ with $0 \leq \varphi \leq 1$ such that φ is 1 on V and has support in U. Then the operator

$$\varphi \cdot L + (1 - \varphi)L_0$$

is elliptic on all of $\mathbb{R}^n$ and clearly can be extended from Q to be a periodic elliptic operator $\tilde{L}$ which agrees with L on V.

We shall need the following slight extension of the formal adjoint property of L^*. Let $u \in H_s$, and let $\varphi \in C_0^\infty(V)$. Then

$$\langle \tilde{L}u,\varphi\rangle_0 = \langle u,L^*\varphi\rangle_0. \tag{2}$$

For let $\psi_j \to u$ in $\| \ \|_s$ with $\psi_j \in \mathscr{P}$. Then

$$\begin{aligned}\langle \tilde{L}\psi_j,\varphi\rangle_0 &= \langle \tilde{L}\psi_j,\varphi\rangle = \langle L\psi_j,\varphi\rangle\\ &= \langle \psi_j,L^*\varphi\rangle = \langle \psi_j,L^*\varphi\rangle_0,\end{aligned}$$

so that

$$\begin{aligned}|\langle \tilde{L}u,\varphi\rangle_0 - \langle u,L^*\varphi\rangle_0| &= |\langle \tilde{L}(u - \psi_j), \varphi\rangle_0 - \langle u - \psi_j, L^*\varphi\rangle_0|\\ &\leq \|\tilde{L}(u - \psi_j)\|_{s-l}\, \|\varphi\|_{-s+l} + \|u - \psi_j\|_s\, \|L^*\varphi\|_{-s}\\ &\leq \text{const}\, \|u - \psi_j\|_s\, \|\varphi\|_{-s+l} + \|u - \psi_j\|_s\, \|L^*\varphi\|_{-s},\end{aligned}$$

which converges to zero as $j \to \infty$.

Let V, as above, be an open set whose closure lies in some 2π cube. Let u and v belong to H_s. Then we shall say that u and v are *equal on* V if

$$\langle u - v, \varphi\rangle_0 = 0 \tag{3}$$

for all $\varphi \in C_0^\infty(V)$. We shall say that a *periodic operator* L *has support in* V if the coefficients of L belong to $C_0^\infty(V) \subset \mathscr{P}$. Now, if L has support in V, and if the elements u and v of H_s are equal on V, then

$$Lu = Lv. \tag{4}$$

For by 6.18(f), it is sufficient to prove that

(5) $$\langle L(u-v), \varphi \rangle_0 = 0$$

for all $\varphi \in \mathscr{P}$. But applying (2) (with $\tilde{L} = L$), we obtain

$$\langle L(u-v), \varphi \rangle_0 = \langle (u-v), L^* \varphi \rangle_0$$

which is zero by (3) since $L^* \varphi \in C_0^\infty(V) \subset \mathscr{P}$.

6.32 Proof of the Regularity Theorem 6.5 We are going to restate Theorem 6.5 in slightly different notation, adapted to the local problem in $\mathbb{R}^n$ to which the theorem will immediately be reduced. We shall use $\langle\ ,\ \rangle'$ to denote the inner product 6.1(5) on $E^p(M)$. We shall prove:

(1) *Given a C^∞ p-form f on M and a bounded linear functional $l' \colon E^p(M) \to \mathbb{R}$ such that $l'(\Delta^* \varphi) = \langle f, \varphi \rangle'$ for every $\varphi \in E^p(M)$, then there exists a C^∞ p-form u on M such that $l'(t) = \langle u, t \rangle'$ for every $t \in E^p(M)$.*

We note that the last statement of Theorem 6.5, which says that $\Delta u = f$, is an immediate consequence of (1), since, as we have already observed in 6.4(6), $\langle \Delta u, \varphi \rangle' = \langle u, \Delta^* \varphi \rangle' = l'(\Delta^* \varphi) = \langle f, \varphi \rangle'$ for all $\varphi \in E^p(M)$.

We now reduce Theorem 6.5 to a local problem. Let U be a coordinate patch on M with coordinate map γ such that $\gamma(U) = \mathbb{R}^n$. Via this coordinate system, differentiable p-forms become vector-valued functions from $\mathbb{R}^n$ to $\mathbb{R}^m \subset \mathbb{C}^m$ where $m = \binom{n}{p}$. We adopt the notation of 6.31. So via the coordinate system (U, γ), p-forms on M yield elements of C^∞; and in the reverse direction, each element of C_0^∞ extends by zero to a complex-valued p-form on all of M. The Laplacian Δ induces a partial differential operator L of order 2 on C^∞. The basic fact that we shall need concerning L is that L is an elliptic operator. This we assume for the moment and shall establish in 6.35. We let L^* denote the formal adjoint of L with respect to the L_2 inner product on C_0^∞.

We extend the inner product $\langle\ ,\ \rangle'$ to complex-valued p-forms in the obvious way, so that if u_1, u_2, v_1, and v_2 are real-valued p-forms, then

$$\langle u_1 + iu_2, v_1 + iv_2 \rangle' = \langle u_1, v_1 \rangle' + \langle u_2, v_2 \rangle' + i(\langle u_2, v_1 \rangle' - \langle u_1, v_2 \rangle').$$

Then by transferring this to Euclidean space we obtain another inner product $\langle\ ,\ \rangle'$ on C_0^∞ induced from the inner product on complex-valued p-forms on M. It follows from the observation that both the L_2 inner product $\langle\ ,\ \rangle$ and the inner product $\langle\ ,\ \rangle'$ on C_0^∞ are integrals of pointwise inner products, that there exists a matrix A of smooth functions on $\mathbb{R}^n$, Hermitian and positive definite at each point, such that

(2) $$\langle \varphi, \psi \rangle' = \langle \varphi, A\psi \rangle$$

for all $\varphi, \psi \in C_0^\infty$.

The adjoint of L on C_0^∞ with respect to $\langle\ ,\ \rangle'$ is simply Δ^* restricted to C_0^∞. We claim that for $\varphi \in C_0^\infty$,

(3) $$L^*\varphi = A\Delta^* A^{-1}\varphi.$$

Indeed, for arbitrary $\psi \in C_0^\infty$,

$$\begin{aligned}\langle L^*\varphi,\psi\rangle &= \langle \varphi, L\psi\rangle = \langle A^{-1}\varphi, L\psi\rangle' \\ &= \langle \Delta^* A^{-1}\varphi, \psi\rangle' = \langle A\Delta^* A^{-1}\varphi, \psi\rangle.\end{aligned}$$

We extend the linear functional $l'\colon E^p(M) \to \mathbb{R}$ complex linearily to complex-valued differential forms, and we define a complex-valued linear functional l on C_0^∞ by setting

(4) $$l(\varphi) = l'(A^{-1}\varphi).$$

We claim that l is locally represented by a C^∞ function. Precisely, we shall prove:

(5) *If $p \in \mathbb{R}^n$, then there is a neighborhood W_p of p and an element $u_p \in \mathscr{P}$ such that $l(t) = \langle u_p, t\rangle$ for every $t \in C_0^\infty(W_p)$.*

First we show that (5) implies (1). It follows from (5) that for each $p, q \in \mathbb{R}^n$, $u_p \,|\, W_p \cap W_q = u_q |\, W_p \cap W_q$ since both u_p and u_q have the same L_2 inner product with all elements of $C_0^\infty(W_p \cap W_q)$. Thus the u_p piece together to give $u \in C^\infty$ such that $u \,|\, W_p = u_p |\, W_p$ for each $p \in \mathbb{R}^n$. Now let $\{\varphi_i\}$ be a partition of unity on $\mathbb{R}^n$ subordinate to the $\{W_p\}$. Then if $t \in C_0^\infty$, $l(t) = \sum_i l(\varphi_i t) = \sum_i \langle u, \varphi_i t\rangle = \langle u, t\rangle$. Now if φ is a smooth p-form on M with support in U, then $l'(\varphi) = l(A\varphi) = \langle u, A\varphi\rangle = \langle u, \varphi\rangle'$. By the same argument as above, the u's for various coordinate systems on M piece together to form a C^∞ p-form u (necessarily real-valued) on M such that $l'(\varphi) = \langle u, \varphi\rangle'$ for every $\varphi \in E^p(M)$. Thus we have reduced the proof of (1) to the proof of (5), which follows.

Let $p \in \mathbb{R}^n$ be fixed, and let Q' be some open 2π cube containing p. Choose an open set V such that $p \in V \subset \bar{V} \subset Q'$, and let

(6) $$\tilde{l} = l \,|\, C_0^\infty(V).$$

First of all, we observe that $\tilde{l}$ is a bounded linear functional on $C_0^\infty(V)$. For since $\bar{V}$ is compact, the matrix norms $\|A_x{}^{-1}\|$ have a maximum as x ranges over $\bar{V}$, and thus, using the fact that l' is bounded, we obtain

$$\begin{aligned}|\tilde{l}(\varphi)| &= |l(\varphi)| = |l'(A^{-1}\varphi)| \le \text{const } \|A^{-1}\varphi\|' \\ &= \text{const } (\langle A^{-1}\varphi, A^{-1}\varphi\rangle')^{1/2} = \text{const } \langle \varphi, A^{-1}\varphi\rangle^{1/2} \\ &\le \text{const } (\|\varphi\|\ \|A^{-1}\varphi\|)^{1/2} \le \text{const } \|\varphi\|\end{aligned}$$

for all $\varphi \in C_0^\infty(V)$. Second, we observe that it follows from (6), (4), (3), (2), and (1) that

(7) $$\tilde{l}(L^*\varphi) = l'(A^{-1}L^*\varphi) = l'(\Delta^* A^{-1}\varphi) = \langle f, A^{-1}\varphi\rangle' = \langle f, \varphi\rangle$$

for all $\varphi \in C_0^\infty(V)$. Thus $\tilde{l}$ is a weak solution of $Lu = f$.

Since $\tilde{l}$ is bounded, $\tilde{l}$ extends to a bounded linear functional on H_0. It follows that there is an element $\tilde{u} \in H_0$ such that

$$\tilde{l}(t) = \langle \tilde{u}, t \rangle_0 \quad \text{for all } t \in H_0. \tag{8}$$

Our task is to show that on a small enough neighborhood of p, the element $\tilde{u}$ agrees with an element of $\mathscr{P}$.

Choose a neighborhood O_0 of p with $\overline{O_0} \subset V$ small enough so that there exists a periodic elliptic operator $\tilde{L}$ which agrees with L on O_0 (cf. 6.31). Choose a neighborhood O of p such that $\overline{O} \subset O_0$, and then choose a sequence

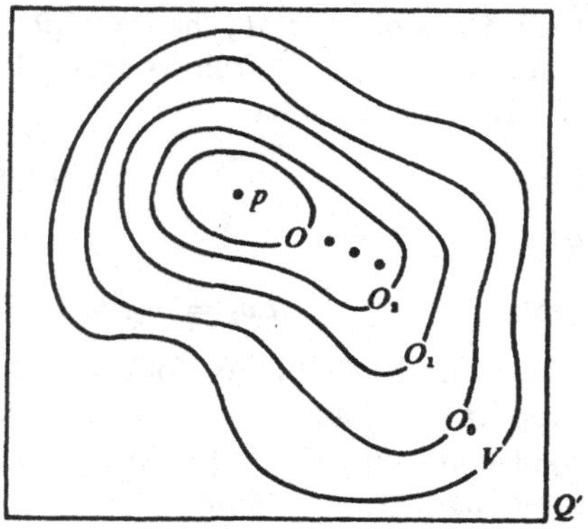

of neighborhoods O_n of p such that $\overline{O} \subset O_n$ and $\overline{O_n} \subset O_{n-1}$ for each $n = 1, 2, \ldots$. For each integer $n \geq 1$, choose a C^∞ function ω_n which is identically equal to 1 on O_n, has values between 0 and 1, and has support in O_{n-1}. Let

$$v_1 = \omega_1 \tilde{u} \in H_0. \tag{9}$$

Then

$$\tilde{L}v_1 = \tilde{L}\omega_1\tilde{u} = \omega_1\tilde{L}\tilde{u} + M_1\tilde{u}, \tag{10}$$

where

$$M_1 = \tilde{L}\omega_1 - \omega_1\tilde{L}. \tag{11}$$

In order to apply 6.30, we must first determine to which Sobolev space the right-hand side of (10) belongs.

First, we claim that

$$\omega_1\tilde{L}\tilde{u} = \omega_1 f, \tag{12}$$

from which it follows that $\omega_1 \tilde{L}\tilde{u} \in C_0^\infty(O_0)$ and thus belongs to H_s for each s. Now both sides of (12) belong to some H_s, and it follows from 6.18(f) that for (12) to hold, it is sufficient to prove that

$$\langle \omega_1 \tilde{L}\tilde{u} - \omega_1 f, \varphi \rangle_0 = 0 \tag{13}$$

for all $\varphi \in \mathscr{P}$. We compute, using (7), (8), 6.18(9), and 6.31(2):

$$\begin{aligned} \langle \omega_1 \tilde{L}\tilde{u} - \omega_1 f, \varphi \rangle_0 &= \langle \omega_1 \tilde{L}\tilde{u}, \varphi \rangle_0 - \langle \omega_1 f, \varphi \rangle_0 \\ &= \langle \tilde{L}\tilde{u}, \omega_1 \varphi \rangle_0 - \langle f, \omega_1 \varphi \rangle_0 \\ &= \langle \tilde{u}, L^* \omega_1 \varphi \rangle_0 - \tilde{l}(L^* \omega_1 \varphi) \\ &= \tilde{l}(L^* \omega_1 \varphi) - \tilde{l}(L^* \omega_1 \varphi) = 0. \end{aligned}$$

Thus (12) holds. Now since L (and hence $\tilde{L}$) is an operator of order 2, then M_1 is of order 1; hence $M_1 \tilde{u} \in H_{-1}$. Thus the right-hand side of (10) belongs to H_{-1}. It follows from the periodic regularity theorem 6.30 that $v_1 \in H_1$. Now we let

$$v_2 = \omega_2 \tilde{u}. \tag{14}$$

Then

$$\tilde{L} v_2 = \omega_2 \tilde{L}\tilde{u} + M_2 \tilde{u} = \omega_2 \tilde{L}\tilde{u} + M_2 v_1, \tag{15}$$

where the last equality follows from 6.31(4) since $M_2 = \tilde{L}\omega_2 - \omega_2 \tilde{L}$ has support in O_1 and $\tilde{u} = v_1$ on O_1. Now, arguing as before, we see that the right-hand side of (15) lies in H_0. Thus by 6.30, $v_2 \in H_2$. Continuing in the same manner, we obtain

$$v_n = \omega_n \tilde{u} \quad \text{with } v_n \in H_n. \tag{16}$$

Finally, let W_p be an open neighborhood of p with $\overline{W_p} \subset O$, and let ω be identically 1 on W_p with values between 0 and 1, and with support in O. Then $\omega\tilde{u} = \omega\omega_n \tilde{u}$ for each n; and so by (16), $\omega\tilde{u} \in H_n$ for every n. By the corollary to the Sobolev lemma 6.22, $\omega\tilde{u}$ represents a C^∞ function $u \in \mathscr{P}$. Now if $t \in C_0^\infty(W_p)$, then

$$\begin{aligned} l(t) = \tilde{l}(t) &= \langle \tilde{u}, t \rangle_0 = \langle \tilde{u}, \omega t \rangle_0 \\ &= \langle \omega\tilde{u}, t \rangle_0 = \langle u, t \rangle, \end{aligned}$$

and (5) is proved. Except for a proof of the fact that L is elliptic, this completes the proof of Theorem 6.5.

6.33 Proof of Theorem 6.6 It suffices to show that if $m \in M$, then there is some neighborhood of m such that if φ is any C^∞ function on M with support in that neighborhood, then $\{\varphi\alpha_n\}$ has a Cauchy subsequence. For then we simply cover M by a finite number of such neighborhoods and take a partition of unity $\varphi_1, \ldots, \varphi_N$ subordinate to this cover. One can select a subsequence α_{n_k} such that $\varphi_j\alpha_{n_k}$ is Cauchy for each j. Then $\{\alpha_{n_k}\}$ is Cauchy for

$$\|\alpha_{n_k} - \alpha_{n_l}\| = \left\| \sum_{j=1}^{N} \varphi_j(\alpha_{n_k} - \alpha_{n_l}) \right\| \leq \sum_{j=1}^{N} \|\varphi_j\alpha_{n_k} - \varphi_j\alpha_{n_l}\|.$$

We reduce the problem to Euclidean space by choosing a coordinate neighborhood of m with m going to $p \in \mathbb{R}^n$. We continue with the notation and setup as developed in 6.31 and 6.32; in particular, the norm on M and the corresponding induced norm on C_0^∞ will now be denoted by $\| \quad \|'$. Let φ be a real-valued C^∞ function with support in O_0. It suffices to show that the sequence $\{\varphi\alpha_n\}$ of elements of $\mathscr{P}$ has a Cauchy subsequence $\{\varphi\alpha_{n_k}\}$ in the 0-norm, since the 0-norm and the L_2-norm agree on $C_0^\infty(O_0)$, and the L_2-norm and the norm $\| \quad \|'$ are equivalent on $C_0^\infty(O_0)$. According to the Rellich lemma 6.23, in order to prove that $\{\varphi\alpha_n\}$ has a Cauchy subsequence in the 0-norm, it is sufficient to show that the sequence is bounded in H_1. It follows from the Fundamental Inequality that

$$\begin{aligned}(1) \quad \|\varphi\alpha_n\|_1 &\leq \text{const}\,(\|\tilde{L}\varphi\alpha_n\|_{-1} + \|\varphi\alpha_n\|_{-1}) \\ &= \text{const}\,(\|L\varphi\alpha_n\|_{-1} + \|\varphi\alpha_n\|_{-1}) \\ &\leq \text{const}\,\|\varphi L\alpha_n\|_{-1} + \text{const}\,\|(L\varphi - \varphi L)\alpha_n\|_{-1} + \text{const}\,\|\varphi\alpha_n\|_{-1}.\end{aligned}$$

Now

$$\begin{aligned}(2) \qquad \|\varphi L\alpha_n\|_{-1} &\leq \|\varphi L\alpha_n\|_0 = \|\varphi L\alpha_n\| \\ &\leq \text{const}\,\|\varphi L\alpha_n\|' \leq \text{const}\,\|L\alpha_n\|' \leq \text{const}\,\|\Delta\alpha_n\|'.\end{aligned}$$

Let τ be a C^∞ function with values between 0 and 1 which equals 1 on O_0 and has support in V. Then

$$(L\varphi - \varphi L)\alpha_n = (L\varphi - \varphi L)(\tau\alpha_n),$$

so that

$$\begin{aligned}(3) \qquad \|(L\varphi - \varphi L)\alpha_n\|_{-1} &= \|(L\varphi - \varphi L)(\tau\alpha_n)\|_{-1} \\ &\leq \text{const}\,\|\tau\alpha_n\|_0 = \text{const}\,\|\tau\alpha_n\| \\ &\leq \text{const}\,\|\tau\alpha_n\|' \leq \text{const}\,\|\alpha_n\|'.\end{aligned}$$

Finally,

$$(4) \qquad \|\varphi\alpha_n\|_{-1} \leq \|\varphi\alpha_n\|_0 = \|\varphi\alpha_n\| \leq \text{const}\,\|\varphi\alpha_n\|' \leq \text{const}\,\|\alpha_n\|'.$$

So from (1), (2), (3), and (4) we obtain

$$(5) \qquad \|\varphi\alpha_n\|_1 \leq \text{const}\,(\|\Delta\alpha_n\|' + \|\alpha_n\|').$$

But by assumption $\|\Delta\alpha_n\|'$ and $\|\alpha_n\|'$ are bounded. Therefore, the sequence $\{\varphi\alpha_n\}$ is bounded in H_1, and the proof of Theorem 6.6 is complete (assuming the ellipticity of L).

ELLIPTICITY OF THE LAPLACE-BELTRAMI OPERATOR

6.34 Remark We shall make use of the following observation. Let U, V, and W be finite dimensional inner product spaces, and suppose that

$$U \xrightarrow{A} V \xrightarrow{B} W$$

is exact. Let $A^*: V \to U$ and $B^*: W \to V$ be the adjoints of A and B respectively. Then $B^*B + AA^*$ is an isomorphism on V. For let v be a non-zero element of V. We need only show that $(B^*B + AA^*)v \neq 0$. Now

$$\langle (B^*B + AA^*)v, v\rangle = \langle Bv, Bv\rangle + \langle A^*v, A^*v\rangle.$$

If $Bv \neq 0$, then $(B^*B + AA^*)v \neq 0$. If $Bv = 0$, then by the exactness, v lies in the image of A. But A^* is injective on the image of A. Thus $A^*v \neq 0$, which implies that $(B^*B + AA^*)v \neq 0$.

We shall apply this to the special case in which U, V, and W are $\Lambda_{p-1}(M_m^*)$, $\Lambda_p(M_m^*)$, and $\Lambda_{p+1}(M_m^*)$ respectively, with the inner products

$$\langle \omega, \tau\rangle = *(\omega \wedge *\tau),$$

and with A and B both left exterior multiplication by $\xi \in M_m^*$:

$$\Lambda_{p-1}(M_m^*) \xrightarrow{\xi} \Lambda_p(M_m^*) \xrightarrow{\xi} \Lambda_{p+1}(M_m^*). \tag{1}$$

According to Exercise 15 of Chapter 2, the sequence (1) is exact; and according to Exercise 14 of Chapter 2, the adjoint of $\xi: \Lambda_p(M_m^*) \to \Lambda_{p+1}(M_m^*)$ is

$$(-1)^{np} * \xi *: \Lambda_{p+1}(M_m^*) \to \Lambda_p(M_m^*). \tag{2}$$

Thus it follows from the above remarks that

$$(-1)^{np} * \xi * \xi + (-1)^{n(p-1)} \xi * \xi * \tag{3}$$

is an isomorphism on $\Lambda_p(M_m^*)$.

6.35 The Laplacian is Elliptic In order to complete the proofs of Theorems 6.5 and 6.6, we need to prove that the operator L of 6.32 induced on Euclidean space by the Laplace-Beltrami operator Δ via a coordinate system is elliptic. Proving this, according to 6.28(2), is equivalent to showing that for each $m \in M$,

$$\Delta(\varphi^2\alpha)(m) \neq 0 \tag{1}$$

for each smooth form α such that $\alpha(m) \neq 0$ and for each C^∞ function φ on M such that $\varphi(m) = 0$ but $d\varphi(m) \neq 0$. Assume that α is a p-form, and let $0 \neq d\varphi = \xi \in M_m^*$. Recall that

$$\Delta = (-1)^{n(p+1)+1} d * d * + (-1)^{np+1} * d * d.$$

We compute the left-hand side of (1), keeping in mind at each stage that $\varphi(m) = 0$. Thus

$$d{*}d{*}(\varphi^2\alpha)(m) = (d{*}d(\varphi^2){*}\alpha)(m) = (2d{*}\varphi(d\varphi){*}\alpha)(m)$$
$$= (2(d\varphi){*}(d\varphi){*}\alpha)(m) = 2\xi{*}\xi{*}(\alpha(m)).$$

Similarly,

$${*}d{*}d(\varphi^2\alpha)(m) = 2{*}\xi{*}\xi(\alpha(m)).$$

Thus

$$\Delta(\varphi^2\alpha)(m) = -2[(-1)^{np}{*}\xi{*}\xi + (-1)^{n(p-1)}\xi{*}\xi{*}](\alpha(m)),$$

which is not zero, according to 6.34(3). Hence Δ is elliptic, and the proof of the Hodge theorem is at last complete.

6.36 Remark We have seen that for each ξ in M_m^* there is a well-defined linear transformation $\sigma_\Delta(\xi)$ on $\Lambda_p(M_m^*)$ defined by

(1) $$\sigma_\Delta(\xi)(v) = \Delta(\varphi^2\alpha)(m),$$

where $v \in \Lambda_p(M_m^*)$, where α is any p-form such that $\alpha(m) = v$, and where φ is any C^∞ function such that $\varphi(m) = 0$ and $d\varphi(m) = \xi$. This linear transformation $\sigma_\Delta(\xi)$ is known as the *symbol of the operator* Δ. The ellipticity of all the operators L on $\mathbb{R}^n$ obtained from Δ via coordinate systems is equivalent to the property that the symbol $\sigma_\Delta(\xi)$ is an isomorphism at each point m for each non-zero $\xi \in M_m^*$. Our computations in 6.35 contain a proof of the fact that the symbol of the exterior derivative operator d, namely $\sigma_d(\xi)\colon \Lambda(M_m^*) \to \Lambda(M_m^*)$, is simply left exterior multiplication by ξ; whereas the symbol of the adjoint δ of d is the adjoint of left exterior multiplication by ξ. In the theory of general partial differential operators on vector bundles over M, ellipticity is defined, as above, in terms of the symbol.

EXERCISES

1 Prove that ${*}\Delta = \Delta{*}$.

2 (a) Prove that the Green's operator G is a bounded linear operator
(b) Prove that G is self-adjoint on $(H^p)^\perp$.
(c) Prove that G takes bounded sequences into sequences with Cauchy subsequences.

3 By using an integral test, prove that the series $\sum (1 + |\xi|^2)^{-k}$, where ξ ranges over all n-tuples of integers $(\xi_1, \ldots, \xi_n)$, converges for $k \geq [n/2] + 1$. (*Hint:* Induct on n, and evaluate the appropriate integrals in terms of spherical coordinates.)

4 Prove in detail the inequality 6.16(15).

5 Establish the existence of the matrix A of 6.32(2) together with its stated properties.

6 Derive explicit formulas for d, $*$, δ, and Δ in Euclidean space. In particular, show that if

$$\alpha = \sum_{i_1 < \cdots < i_p} \alpha_I \, dx_{i_1} \wedge \cdots \wedge dx_{i_p},$$

then

$$\Delta\alpha = (-1) \sum_{i_1 < \cdots < i_p} \left(\sum_{i=1}^{n} \frac{\partial^2 \alpha_I}{\partial x_i^2} \right) dx_{i_1} \wedge \cdots \wedge dx_{i_p}.$$

7 Let φ belong to the C^∞ periodic functions $\mathscr{P}$ on the plane. Prove that

$$\left\| \frac{\partial^2 \varphi}{\partial x \, \partial y} \right\| \leq \tfrac{1}{2} \|\Delta\varphi\|.$$

8 The Rellich lemma 6.23 says that the natural injection $i: H_t \to H_s$ for $s < t$ is a *compact operator;* that is, it takes bounded sequences into sequences with convergent subsequences. An analogous example of this phenomenon is the following. Let C denote the Banach space of periodic continuous functions on the real line, say with period 2π, and with norm the sup norm $\| \quad \|_\infty$. Let C^1 be the subset of C consisting of functions with continuous first derivative. As a norm for C^1 we take

$$\|f\| = \|f\|_\infty + \left\| \frac{df}{dx} \right\|_\infty.$$

Use the Arzelà-Ascoli theorem [27, p. 126] to prove that the natural injection $i: C^1 \to C$ is a compact operator.

9 We shall consider a number of elliptic equations of the form $Lu = f$ on the real line. In each case, f will be smooth and periodic of period 1, and we look for solutions u also periodic of period 1. This restriction to periodic functions makes this in essence a problem on a *compact* space, the unit circle. We let $u' = du/dx$, etc.

(a) $u' = f$. This is the simplest example of an elliptic operator which exhibits all of the essential ingredients of the theory. What is the formal adjoint of this differential operator? Show that there is a solution u (periodic) if and only if f is orthogonal to the kernel of this adjoint.

(b) $u' - u = f$. What is the kernel (in the periodic functions) in this case? What are the necessary and sufficient conditions on f for there to exist a periodic solution?

(c) $u'' = f$. Show that this operator is formally self-adjoint. Show that there is a periodic solution if and only if f is orthogonal to the kernel; and using the fact that

$$\int_0^x \left(\int_0^t f(s) \, ds \right) dt = \int_0^x f(s) \left(\int_s^x dt \right) ds,$$

show that the unique solution orthogonal to the kernel is

$$u(x) = \int_0^x t(x-1)f(t)\,dt \quad + \quad \int_x^1 x(t-1)f(t)\,dt$$
$$- \frac{1}{2}\int_0^1 t(t-1)f(t)\,dt.$$

This explicitly exhibits the Green's operator for this case.

(d) $u'' + 4\pi^2 u = f$. Show that this operator is formally self-adjoint. What is its kernel? Derive an explicit formula for the solution u, and show that u is periodic if and only if f is orthogonal to the kernel.

10 In Theorem 6.11 we used the fact that the Green's operator commutes with d in proving that each closed form on a compact orientable Riemannian manifold differs from a harmonic form by an exact form. Give a proof of this result directly from the Hodge decomposition theorem without using the Green's operator.

11 A *periodic distribution* l is a linear functional on $\mathscr{P}$ for which there exists some integer $k \geq 0$ and a positive constant c such that

(1) $$|l(\varphi)| \leq c \sum_{|\alpha| \leq k} \|D^\alpha \varphi\|_\infty$$

for all $\varphi \in \mathscr{P}$. Prove that if l is a periodic distribution, then there is an element $u \in H_{-\infty}$ such that

(2) $$l(\varphi) = \langle u, \varphi \rangle_0$$

for all $\varphi \in \mathscr{P}$. Conversely show that every element of $H_{-\infty}$ determines a periodic distribution via (2).

12 Let α and β be n-forms on a compact oriented manifold M^n such that $\int_M \alpha = \int_M \beta$. Prove that α and β differ by an exact form.

13 Show that Theorem 6.6 cannot be strengthened to the assertion of the existence of a subsequence which is convergent in $E^p(M)$.

14 One should observe that in the course of proving Theorem 6.5 we have proved the following regularity theorem. *Let L be an elliptic operator on C^∞ (the complex m-space valued smooth functions on $\mathbb{R}^n$). Suppose that u is sufficiently differentiable for Lu to make sense, and suppose that $Lu = f$ where $f \in C^\infty$. Then $u \in C^\infty$.*

In fact, we have proved more—*every weak solution l is smooth.* Precisely: *Let l be a linear functional on C_0^∞ which is bounded on $C_0^\infty(V)$ whenever $\overline{V}$ is compact, and which satisfies $l(L^*\varphi) = \langle f, \varphi \rangle$ for every $\varphi \in C_0^\infty$. Then l is smooth in the sense that there exists $u \in C^\infty$ which represents l; that is, $l(t) = \langle u, t \rangle$ for every $t \in C_0^\infty$. Such a smooth representative u of a weak solution l is an actual solution, $Lu = f$.*

15 Observe that the Cauchy-Riemann operator $(\partial/\partial x) + i(\partial/\partial y)$ is elliptic. Conclude from the general theory of elliptic operators that every holomorphic function is C^∞. Prove that every holomorphic function is a complex-valued harmonic function on the plane, and use Green's 1st identity (Exercise 5, Chapter 4) to prove that a holomorphic function with compact support must be identically zero.

16 **The Eigenvalues of the Laplacian** This is an extended exercise in which the fundamental properties of the eigenfunctions and eigenvalues of the Laplacian are developed. Proofs for the more difficult parts are outlined, and in some cases are given nearly in full.

Consider the Laplace-Beltrami operator Δ acting on the p-forms $E^p(M)$ for some fixed p. A real number λ corresponding to which there exists a not identically zero p-form u such that $\Delta u = \lambda u$ is called an *eigenvalue* of Δ. If λ is an eigenvalue, then any p-form u such that $\Delta u = \lambda u$ is called an *eigenfunction* of Δ corresponding to the eigenvalue λ. The eigenfunctions corresponding to a fixed λ form a subspace of $E^p(M)$ called the *eigenspace* of the eigenvalue λ.

(a) Prove that the eigenvalues of Δ are non-negative.

(b) Prove that the eigenspaces of Δ are finite dimensional.

(c) Prove that the eigenvalues have no finite accumulation point.

(d) Prove that eigenfunctions corresponding to distinct eigenvalues are orthogonal.

(e) *Existence.* In order for the above statements to have substance we must prove that there exist eigenvalues of Δ. First of all, zero is an eigenvalue if and only if there are non-trivial harmonic p-forms on M, and the corresponding eigenspace is precisely the space H^p of harmonic forms. We shall now establish that Δ has a positive eigenvalue—in fact, a whole sequence of eigenvalues diverging to $+\infty$. Consider Δ to be restricted to $(H^p)^\perp$. Then we have $\Delta\colon (H^p)^\perp \to (H^p)^\perp$, and also we have the Green's operator $G\colon (H^p)^\perp \to (H^p)^\perp$, and $\Delta G\alpha = \alpha$, $G\Delta\alpha = \alpha$ for all $\alpha \in (H^p)^\perp$. Observe that the eigenvalues of $G \mid (H^p)^\perp$ are the reciprocals of the eigenvalues of $\Delta \mid (H^p)^\perp$. Let

$$\eta = \sup_{\substack{\|\varphi\|=1 \\ \varphi\in(H^p)^\perp}} \|G\varphi\| .$$

Then $\eta > 0$ and $\|G\varphi\| \leq \eta \,\|\varphi\|$ for every $\varphi \in (H^p)^\perp$. We shall prove that $1/\eta$ is an eigenvalue of Δ. Let $\{\varphi_j\} \in (H^p)^\perp$ be a maximizing sequence for η; that is, $\|\varphi_j\| = 1$ and $\|G\varphi_j\| \to \eta$.

First, we claim that $\|G^2\varphi_j - \eta^2\varphi_j\| \to 0$, for

$$\|G^2\varphi_j - \eta^2\varphi_j\|^2 = \|G^2\varphi_j\|^2 - 2\eta^2\langle G^2\varphi_j,\varphi_j\rangle + \eta^4$$
$$\leq \eta^2\,\|G\varphi_j\|^2 - 2\eta^2\,\|G\varphi_j\|^2 + \eta^4 \to 0.$$

Second, we claim that $\|G\varphi_j - \eta\varphi_j\| \to 0$. For if we let $\psi_j = G\varphi_j - \eta\varphi_j$, then

$$0 \leftarrow \langle\psi_j, G^2\varphi_j - \eta^2\varphi_j\rangle$$
$$= \langle\psi_j, G\psi_j + \eta\psi_j\rangle$$
$$= \langle\psi_j,G\psi_j\rangle + \eta\,\|\psi_j\|^2 \geq \eta\,\|\psi_j\|^2,$$

where we have used the fact that $\langle\psi_j,G\psi_j\rangle \geq 0$. (Why?) Now there is a subsequence of the φ_j, call it $\{\varphi_j\}$, such that $\{G\varphi_j\}$ is Cauchy. Define a linear functional l on $E^p(M)$ by setting

$$l(\beta) = \lim_{j\to\infty} \eta\langle G\varphi_j,\beta\rangle \quad \text{for} \quad \beta \in E^p(M).$$

Prove that l is a non-trivial weak solution of

$$(\Delta - (1/\eta))u = 0.$$

From this and the fact that $\Delta - 1/\eta$ is elliptic, conclude that $\lambda = 1/\eta$ is an eigenvalue of Δ.

(f) *Existence of Other Eigenvalues.* Suppose that we have eigenvalues $\lambda_1 \leq \lambda_2 \leq \cdots \leq \lambda_n$ and corresponding eigenfunctions $u_1, u_2, \ldots, u_n$ (orthonormalized) for $\Delta \mid (H^p)^\perp$. Let R_n be the subspace of $(H^p)^\perp$ spanned by $\{u_1, \ldots, u_n\}$. Observe that G and Δ map $(H^p \oplus R_n)^\perp$ into itself, then define

$$\eta_{n+1} = \sup_{\substack{\|\varphi\|=1 \\ \varphi\in(H^p\oplus R_n)^\perp}} \|G\varphi\|$$

and proceed as in part (e) to establish that $\lambda_{n+1} = 1/\eta_{n+1}$ is an eigenvalue of Δ. Clearly, $\lambda_{n+1} \geq \lambda_n$.

(g) *L_2 Completeness.* Let $\lambda_1 \leq \lambda_2 \leq \cdots$ be the eigenvalues of Δ on $E^p(M)$, where each eigenvalue is included as many times as the dimension of its eigenspace, with a corresponding orthonormalized sequence of eigenfunctions $\{u_i\}$. Let $\alpha \in E^p(M)$. Then

$$\lim_{n\to\infty} \left\| \alpha - \sum_{i=1}^{n} \langle\alpha,u_i\rangle u_i \right\| = 0.$$

For the proof, let k be the dimension of H^p. Then there exists $\beta \in (H^p)^\perp$ such that $G\beta = \alpha - \sum_{i=1}^{k} \langle\alpha,u_i\rangle u_i$. It follows that

$$\left\| \alpha - \sum_{i=1}^{n} \langle \alpha, u_i \rangle u_i \right\| = \left\| G\left(\beta - \sum_{i=k+1}^{n} \langle \beta, u_i \rangle u_i\right) \right\|$$

for $n > k$. But, by the definition of λ_{n+1},

$$\left\| G\left(\beta - \sum_{i=k+1}^{n} \langle \beta, u_i \rangle u_i\right) \right\| \leq \frac{1}{\lambda_{n+1}} \left\| \beta - \sum_{i=k+1}^{n} \langle \beta, u_i \rangle u_i \right\|$$
$$\leq \frac{1}{\lambda_{n+1}} \|\beta\| \to 0 \quad \text{as} \quad n \to \infty.$$

(h) *Uniform Completeness.* The uniform norm $\|\alpha\|_\infty$ is defined on $E^p(M)$ by

$$\|\alpha\|_\infty = \sup_{m \in M} (*(\alpha \wedge *\alpha)(m))^{1/2}.$$

From the Sobolev inequality 6.22(1) and the compactness of M, one can conclude that there exists a large enough integer k and a constant $c > 0$ such that

$$\|\alpha\|_\infty \leq c \, \|(1 + \Delta)^k \alpha\|$$

for every $\alpha \in E^p(M)$. Let $\alpha \in E^p(M)$, and let $P_n(\alpha) = \sum_{i=1}^{n} \langle \alpha, u_i \rangle u_i$, where we are continuing with the notation of part (g). Now $\Delta P_n = P_n \Delta$, so that

$$\|\alpha - P_n(\alpha)\|_\infty \leq c \, \|(1 + \Delta)^k [\alpha - P_n(\alpha)]\|$$
$$= \|\varphi - P_n \varphi\| \to 0,$$

where $\varphi = (1 + \Delta)^k \alpha$.

17 We define the operator $\Delta^2 \colon E^p(M) \to E^p(M)$ by $\Delta^2(\alpha) = \Delta(\Delta\alpha)$. Discuss the solvability of $\Delta^2 \alpha = \beta$.

18 Consider the operator

$$L = \sum_{i,j=1}^{n} a_{ij} \frac{\partial^2}{\partial x_i \, \partial x_j} + \sum_{i=1}^{n} b_i \frac{\partial}{\partial x_i} + c$$

acting on $C^2(\mathbb{R}^n)$. Show that there is no loss of generality in assuming that $a_{ij} = a_{ji}$, and prove that L is elliptic at a point x if and only if the matrix $(a_{ij}(x))$ is (positive or negative) definite. In particular, show that the *wave equation*

$$\Box u = \frac{\partial^2 u}{\partial x^2} - \frac{\partial^2 u}{\partial y^2} = f$$

is not elliptic, and give an example where

$$\Box u = f \in C^\infty \quad \text{but} \quad u \notin C^\infty.$$

19 Consider $\Delta\colon E^p(M) \to E^p(M)$. Prove that if λ is the minimum eigenvalue of Δ, and if $c > -\lambda$, then $(\Delta + c)\alpha = \beta$ can be solved for every $\beta \in E^p(M)$.

20 **The Peter-Weyl Theorem** The *representative ring* of a compact Lie group G is the ring generated over the complex numbers by the set of all continuous functions f for which there is a continuous homomorphism $\rho\colon G \to Gl(n,\mathbb{C})$ for some n such that $f = \rho_{ij}$ for some choice of i and j. The Peter-Weyl theorem states that *the representative ring is dense in the space of complex-valued continuous functions on G in the uniform norm.* That is, if g is a complex-valued continuous function on G, and if $\varepsilon > 0$ is given, then there is a function f in the representative ring such that $|f(\sigma) - g(\sigma)| < \varepsilon$ for all $\sigma \in G$. We outline a proof of this theorem which is based on the uniform completeness of the eigenfunctions of the Laplacian. One can choose a Riemannian structure on G such that each of the diffeomorphisms l_σ for $\sigma \in G$ (left translation by σ) is an isometry (that is, $\langle v,w\rangle_\tau = \langle dl_\sigma v, dl_\sigma w\rangle_{\sigma\tau}$ for all $\tau \in G$ and all $v, w \in G_\tau$). Since the C^∞ functions are dense in the space of continuous functions in the uniform norm, and since by Exercise 16(h) the direct sum of the eigenspaces of the Laplacian is dense in the space of C^∞ functions in the uniform norm, it suffices for the Peter-Weyl theorem to prove that each eigenfunction of the Laplacian $\Delta\colon C^\infty(G) \to C^\infty(G)$ belongs to the representative ring.

Now, G acts on the C^∞ functions on G by

$$\sigma(f) = f \circ l_\sigma \quad \text{for } \sigma \in G.$$

Prove that since the l_σ are isometries, this action commutes with the Laplacian:

$$\Delta(f \circ l_\sigma) = (\Delta f) \circ l_\sigma \qquad (\sigma \in G).$$

Let V_λ be the (finite dimensional) eigenspace associated with the eigenvalue λ of $\Delta\colon C^\infty(G) \to C^\infty(G)$. Prove that the action of G leaves V_λ invariant. Then let $\varphi_1, \ldots, \varphi_n$ be a basis of V_λ, and let

$$\sigma(\varphi_i) = \sum_j g_{ji}(\sigma)\varphi_j.$$

Then $\sigma \mapsto \{g_{ij}(\sigma)\}$ is a homomorphism of $G \to Gl(n,\mathbb{R})$. Prove that this homomorphism is continuous. Then observe that

$$\varphi_i(\sigma) = \varphi_i \circ l_\sigma(e) = \sum_j g_{ji}(\sigma)\varphi_j(e),$$

so that φ_i belongs to the representative ring.

21 The reader who is familiar with the theory of vector bundles should observe that the results in this chapter on the Laplace-Beltrami operator are valid for general elliptic operators on vector bundles. We outline the statements of the results in this situation:

Let E and F be (real or complex) vector bundles over a compact orientable manifold M. Let $C^\infty(E)$ and $C^\infty(F)$ denote the vector spaces of smooth sections of E and F respectively. A (*linear*) *differential operator L of order l* from E to F is a linear map from $C^\infty(E)$ to $C^\infty(F)$ which when expressed in terms of local trivializations of E and F yields an ordinary linear partial differential operator of order l. The operator L is *elliptic* if it is locally elliptic. Equivalently, ellipticity of L can be defined in terms of the symbol as in 6.36. Observe that for L to be elliptic, the fibre dimensions of E and F must be equal. Choose inner products in the fibres E_m and F_m which vary smoothly with m, and choose a Riemannian structure and an orientation on M with respect to which smooth functions can be integrated over M. By integrating the fibre inner products over M, we obtain inner products on $C^\infty(E)$ and $C^\infty(F)$. Let $L\colon C^\infty(E) \to C^\infty(F)$ be a differential operator. Prove that L has a formal adjoint $L^*\colon C^\infty(F) \to C^\infty(E)$. Assume that L is elliptic.

(a) Prove that $\ker L$ and $\ker L^*$ are finite dimensional.

(b) Prove that $C^\infty(E)$ and $C^\infty(F)$ have the following orthogonal direct sum decompositions:

$$C^\infty(F) = L\big(C^\infty(E)\big) \oplus \ker L^*,$$
$$C^\infty(E) = L^*\big(C^\infty(F)\big) \oplus \ker L.$$

22 Let φ_n, for $n = 1, 2, 3, \ldots$, be a periodic C^∞ function on the plane which agrees with $\log\log(1/(r + (1/n)))$ for $0 \le r \le \frac{1}{2}$, where $r = \sqrt{x^2 + y^2}$. Show that there is no constant $c > 0$ such that

$$\|\varphi_n\|_\infty \le c\,\|\varphi_n\|_1 \qquad \text{for all } n.$$

This shows, in the case $n = 2$, that the restriction $t \ge [n/2] + 1$ in Corollary (b) of the Sobolev lemma 6.22 is essential.

23 Let L be a (linear) elliptic operator on the C^∞ functions on a compact oriented Riemannian manifold M. Let γ be a diffeomorphism of M which preserves the volume form on M. We say that a C^∞ function f on M is *invariant* under γ if $f \circ \gamma = f$, and we say that the operator L is *invariant* under γ if $Lu \circ \gamma = L(u \circ \gamma)$ for all C^∞ functions u on M. Suppose that L and f are invariant under γ and that f is orthogonal to the kernel of L^*. Prove that there is an invariant solution u of $Lu = f$.

Bibliography

Index of Notation

Index

Bibliography

[1] Bers, L., F. John, and M. Schechter. *Partial Differential Equations.* New York: John Wiley & Sons, Inc., 1964.

[2] Bishop, R. L., and R. J. Crittenden. *Geometry of Manifolds.* New York: Academic Press, 1964.

[3] Bredon, G. E. *Sheaf Theory.* New York: McGraw-Hill, 1967.

[4] Cartan, H. *Séminaire 1950/1951.* Paris: Ecole Normale Supérieure, 1955.

[5] Chevalley, C. *Theory of Lie Groups* I. Princeton, N.J.: Princeton University Press, 1946.

[6] Fleming, W. H. *Functions of Several Variables.* Reading, Mass.: Addison-Wesley, 1965. (2nd ed. Undergraduate Texts in Mathematics, Springer-Verlag, New York, 1977.)

[7] Godement, R. *Topologie Algébrique et Théorie des Faisceaux.* Paris: Hermann, 1958.

[8] Gunning, R. C. *Lectures on Riemann Surfaces.* Princeton, N.J.: Princeton University Press, 1966.

[9] Helgason, S. *Differential Geometry Lie Groups and Symmetric Spaces.* New York: Academic Press, 1978.

[10] Hodge, W. V. D. *The Theory and Applications of Harmonic Integrals.* 2d ed. Cambridge: Cambridge University Press, 1952.

[11] Hurewicz, W. *Lectures on Ordinary Differential Equations.* New York and Cambridge, Mass.: John Wiley & Sons, Inc., and MIT Press, 1958.

[12] Jacobson, N. *Lie Algebras.* New York: John Wiley & Sons, Inc., 1962. (Reprinted by Dover, 1979.)

[13] Kelley, J. L. *General Topology.* Princeton, N.J.: Van Nostrand Company, Inc., 1955. (Graduate Texts in Mathematics, vol. 27, Springer-Verlag, New York, 1975.)

[14] Kervaire, M. A Manifold which does not admit any differentiable structure. *Comment. Math. Helv.*, **35**(1961), 1–14.

[15] Kobayashi, S., and K. Nomizu. *Foundations of Differential Geometry*, 2 vols. New York: John Wiley & Sons, Inc., 1963 and 1969.

[16] Kohn, J. J. *Introduccion a la teoria de integrales harmonicas.* Lecture notes issued by the Centro de Investigacion del IPN, Mexico, 1963.

[17] Lang, S. *Introduction to Differentiable Manifolds.* New York: John Wiley & Sons, Inc., 1962.

[18] Loomis, L. H., and S. Sternberg. *Advanced Calculus.* Reading, Mass.: Addison-Wesley, 1968.

[19] Milnor, J. On manifolds homeomorphic to the 7-sphere. *Ann. of Math.*, **64**(1956), 399–405.

[20] Montgomery, D., and L. Zippin. *Topological Transformation Groups.* New York: Interscience, 1955. (Reproduction by Krieger, Melbourne, Florida, 1974.)

[21] Newns, N., and A. Walker. Tangent planes to a differentiable manifold. *J. London Math. Soc.*, **31**(1956), 400–407.

[22] Nirenberg, L. On elliptic partial differential equations. *Ann. Scuola Norm. Sup. Pisa*, **13**(1959), 115–162.

[23] Pontrjagin, L. S. *Topological Groups.* Princeton, N.J.: Princeton University Press, 1939.

[24] de Rham, G. *Variétés Différentiables.* Paris: Hermann, 1960. (3rd Ed., 1973.)

[25] Samelson, H. Uber die Sphären die als Gruppenräume auftreten. *Comment Math. Helv.*, **13**(1940), 144–155.

[26] Simmons, G. F. *Introduction to Topology and Modern Analysis.* New York: McGraw-Hill, 1963.

[27] Singer, I. M., and J. Thorpe. *Lecture Notes on Elementary Topology and Geometry.* Glenview, Ill.: Scott, Foresman and Company, 1967. (Undergraduate Texts in Mathematics, Springer-Verlag, New York, 1976.)

[28] Spanier, E. H. *Algebraic Topology.* New York: McGraw-Hill 1966.

[29] Spivak, M. *Calculus on Manifolds.* New York: W. A. Benjamin, Inc., 1965.

[30] Sternberg, S. *Lectures on Differential Geometry.* Englewood Cliffs, N.J.: Prentice-Hall, Inc., 1964.

[31] Woll, J. W., Jr. *Functions of Several Variables.* New York: Harcourt, Brace & World, Inc., 1966.

Supplement to the Bibliography

No attempt at completeness was made for either the original bibliography or the following supplement which is being added for the Springer edition. My goal is simply to provide the reader with a few basic references and sources for alternate treatments or additional readings. Quite extensive bibliographies on differentiable manifolds and their role in the many aspects of modern analysis and geometry can be found in volume II of Kobayashi-Nomizu, listed above, and volume 5 of Spivak, listed below.

For a comprehensive, but leisurely and very readable treatment of differential and Riemannian geometry, I recommend

Spivak, M. *Differential Geometry*, 5 vols. Boston, Mass.: Publish or Perish, Inc., 1970–1975.

For a beautiful introduction to manifolds and a geometric treatment of many central topics of differential topology, see

Guillemin, V., and A. Pollack. *Differential Topology.* Englewood Cliffs, N.J.: Prentice-Hall, Inc., 1974.

A rapid introduction to manifolds and a very beautiful development of basic Riemannian geometry is given in

do Carmo, M. *Geometria Riemanniana.* Rio de Janeiro: IMPA, 1979.

For a very recent treatment of smooth manifolds furnished with metric tensors of arbitrary signature (pseudo-Riemannian manifolds), with applications to the theory of relativity, see

O'Neill, B. *Semi-Riemannian Geometry.* New York: Academic Press, 1983.

The reader who wishes to study the theory of characteristic classes should consult the excellent exposition in

Milnor, J., and J. Stasheff. *Characteristic Classes.* Annals of Mathematics Studies, no. 76. Princeton, N.J.: Princeton University Press, 1974.

The following three texts are additional sources of relatively self-contained treatments of basic elliptic theory. The Griffiths/Harris and Wells texts develop Hodge theory for compact complex manifolds. The

Lang text has an appendix on elliptic partial differential equations including the regularity theory on a torus and in Euclidean space.

Griffiths, P., and J. Harris. *Principles of Algebraic Geometry.* New York: John Wiley & Sons, Inc., 1978.

Lang, S. $SL_2(R)$. Reading, Mass.: Addison-Wesley, 1975.

Wells, R.O., Jr. *Differential Geometry on Complex Manifolds.* Englewood Cliffs, N.J.: Prentice-Hall, Inc., 1973. (Graduate Texts in Mathematics, vol. 65, Springer-Verlag, New York, 1979.)

For an application of elliptic operator theory to very remarkable and far-reaching connections between analysis and topology, see

Palais, R. S. *Seminar on the Atiyah-Singer Index Theorem.* Annals of Mathematics Studies, no. 57. Princeton, N.J.: Princeton University Press, 1965.

A systematic development of the general theory of second order quasilinear elliptic partial differential equations and the required linear elliptic theory is given in

Gilbarg, D., and N. Trudinger. *Elliptic Partial Differential Equations of Second Order.* 2nd ed. Berlin Heidelberg New York Tokyo: Springer-Verlag, 1983, in prep.

For very recent and striking applications of analysis to problems in geometry, see

Aubin, T. *Nonlinear Analysis on Manifolds. Monge–Ampère Equations.* Grundlehren der mathematischen Wissenschaften, vol. 252. New York Heidelberg Berlin: Springer-Verlag, 1982.

Yau, S-T, ed. *Seminar on Differential Geometry.* Annals of Mathematics Studies, no. 102. Princeton, N.J.: Princeton University Press, 1982.

Lie groups are central to the theory of fibre bundles and connections. For some recent and very significant applications of these theories to the study of gauge theories in physics, one should consult the following two references:

Drechsler, W., and M. E. Mayer. *Fiber Bundle Techniques in Gauge Theories.* Lecture Notes in Physics, no. 67. Berlin Heidelberg New York: Springer-Verlag, 1977.

Bleecker, D. *Gauge Theory and Variational Principles.* Reading, Mass.: Addison-Wesley, 1981.

Index of Notation

Index

Graduate Texts in Mathematics

continued from page ii

119 ROTMAN. An Introduction to Algebraic Topology.
120 ZIEMER. Weakly Differentiable Functions: Sobolev Spaces and Functions of Bounded Variation.
121 LANG. Cyclotomic Fields I and II. Combined 2nd ed.
122 REMMERT. Theory of Complex Functions. *Readings in Mathematics*
123 EBBINGHAUS/HERMES et al. Numbers. *Readings in Mathematics*
124 DUBROVIN/FOMENKO/NOVIKOV. Modern Geometry—Methods and Applications. Part III.
125 BERENSTEIN/GAY. Complex Variables: An Introduction.
126 BOREL. Linear Algebraic Groups. 2nd ed.
127 MASSEY. A Basic Course in Algebraic Topology.
128 RAUCH. Partial Differential Equations.
129 FULTON/HARRIS. Representation Theory: A First Course. *Readings in Mathematics*
130 DODSON/POSTON. Tensor Geometry.
131 LAM. A First Course in Noncommutative Rings.
132 BEARDON. Iteration of Rational Functions.
133 HARRIS. Algebraic Geometry: A First Course.
134 ROMAN. Coding and Information Theory.
135 ROMAN. Advanced Linear Algebra.
136 ADKINS/WEINTRAUB. Algebra: An Approach via Module Theory.
137 AXLER/BOURDON/RAMEY. Harmonic Function Theory.
138 COHEN. A Course in Computational Algebraic Number Theory.
139 BREDON. Topology and Geometry.
140 AUBIN. Optima and Equilibria. An Introduction to Nonlinear Analysis.
141 BECKER/WEISPFENNING/KREDEL. Gröbner Bases. A Computational Approach to Commutative Algebra.
142 LANG. Real and Functional Analysis. 3rd ed.
143 DOOB. Measure Theory.
144 DENNIS/FARB. Noncommutative Algebra.
145 VICK. Homology Theory. An Introduction to Algebraic Topology. 2nd ed.
146 BRIDGES. Computability: A Mathematical Sketchbook.
147 ROSENBERG. Algebraic *K*-Theory and Its Applications.
148 ROTMAN. An Introduction to the Theory of Groups. 4th ed.
149 RATCLIFFE. Foundations of Hyperbolic Manifolds.
150 EISENBUD. Commutative Algebra with a View Toward Algebraic Geometry.
151 SILVERMAN. Advanced Topics in the Arithmetic of Elliptic Curves.
152 ZIEGLER. Lectures on Polytopes.
153 FULTON. Algebraic Topology: A First Course.
154 BROWN/PEARCY. An Introduction to Analysis.
155 KASSEL. Quantum Groups.
156 KECHRIS. Classical Descriptive Set Theory.
157 MALLIAVIN. Integration and Probability.
158 ROMAN. Field Theory.
159 CONWAY. Functions of One Complex Variable II.
160 LANG. Differential and Riemannian Manifolds.
161 BORWEIN/ERDÉLYI. Polynomials and Polynomial Inequalities.
162 ALPERIN/BELL. Groups and Representations.
163 DIXON/MORTIMER. Permutation Groups.
164 NATHANSON. Additive Number Theory: The Classical Bases.
165 NATHANSON. Additive Number Theory: Inverse Problems and the Geometry of Sumsets.
166 SHARPE. Differential Geometry: Cartan's Generalization of Klein's Erlangen Program.
167 MORANDI. Field and Galois Theory.
168 EWALD. Combinatorial Convexity and Algebraic Geometry.
169 BHATIA. Matrix Analysis.
170 BREDON. Sheaf Theory. 2nd ed.
171 PETERSEN. Riemannian Geometry.
172 REMMERT. Classical Topics in Complex Function Theory.
173 DIESTEL. Graph Theory. 2nd ed.
174 BRIDGES. Foundations of Real and Abstract Analysis.
175 LICKORISH. An Introduction to Knot Theory.
176 LEE. Riemannian Manifolds.
177 NEWMAN. Analytic Number Theory.
178 CLARKE/LEDYAEV/STERN/WOLENSKI. Nonsmooth Analysis and Control Theory.
179 DOUGLAS. Banach Algebra Techniques in Operator Theory. 2nd ed.
180 SRIVASTAVA. A Course on Borel Sets.

181 KRESS. Numerical Analysis.
182 WALTER. Ordinary Differential Equations.
183 MEGGINSON. An Introduction to Banach Space Theory.
184 BOLLOBAS. Modern Graph Theory.
185 COX/LITTLE/O'SHEA. Using Algebraic Geometry.
186 RAMAKRISHNAN/VALENZA. Fourier Analysis on Number Fields.
187 HARRIS/MORRISON. Moduli of Curves.
188 GOLDBLATT. Lectures on the Hyperreals: An Introduction to Nonstandard Analysis.
189 LAM. Lectures on Modules and Rings.
190 ESMONDE/MURTY. Problems in Algebraic Number Theory.
191 LANG. Fundamentals of Differential Geometry.
192 HIRSCH/LACOMBE. Elements of Functional Analysis.
193 COHEN. Advanced Topics in Computational Number Theory.
194 ENGEL/NAGEL. One-Parameter Semigroups for Linear Evolution Equations.
195 NATHANSON. Elementary Methods in Number Theory.
196 OSBORNE. Basic Homological Algebra.
197 EISENBUD/ HARRIS. The Geometry of Schemes.
198 ROBERT. A Course in *p*-adic Analysis.
199 HEDENMALM/KORENBLUM/ZHU. Theory of Bergman Spaces.
200 BAO/CHERN/SHEN. An Introduction to Riemann–Finsler Geometry.
201 HINDRY/SILVERMAN. Diophantine Geometry: An Introduction.
202 LEE. Introduction to Topological Manifolds.

9 780387 90894